The Bridge

Natural Gas in a Redivided Europe

THANE GUSTAFSON

Harvard University Press
Cambridge, Massachusetts, and London, England 2020

Printed in the United States of America

First printing

Library of Congress Cataloging-in-Publication Data

Names: Gustafson, Thane, author.
Title: The bridge : natural gas in a redivided Europe / Thane Gustafson.
Description: Cambridge, Massachusetts : Harvard University Press, 2020. |
Includes bibliographical references and index.
Identifiers: LCCN 2019014637 | ISBN 9780674987951
Subjects: LCSH: Natural gas—Political aspects—Europe—History. |
Natural gas—Political aspects—Soviet Union—History. | Natural gas—Political aspects—Russia (Federation)—History. | Natural gas—Economic aspects—Europe—History. | Natural gas—Economic aspects—Soviet Union—History. |
Natural gas—Economic aspects—Russia (Federation)—History. |
Natural gas pipelines—Europe. | Geopolitics—Europe.
Classification: LCC HD9581.E82 G87 2020 | DDC 338.2/7285094—dc23
LC record available at https://lccn.loc.gov/2019014637

This book is dedicated to the gazoviki, *both East and West,*

And to Simon Blakey,

Who sparked the idea for this book and guided it to the end

Contents

A Note on Transliteration

Transliteration from Slavic to Latin script is always a challenge, particularly where proper nouns are concerned. Throughout this book I have attempted to follow the guidelines of the US Library of Congress wherever possible. However, for place names the American Association of Geographers has its own conventions, and in most places I have followed those. (Thus: Ob River instead of Ob'.) Finally, if a place or person has been mentioned frequently in the Western media, then I follow the spelling used in the media. (Thus: Mikhelson instead of Mikhel'son.) Unfortunately, the result is a running series of compromises, which will make no one entirely satisfied.

The Bridge

Introduction

Three decades after the fall of the Berlin Wall and the end of the Soviet empire, the West faces what it thought it would never see again—a new era of East-West tensions in a redivided Europe. The vision that inspired the 1990s and the early 2000s, of a modern Russia integrated into the world economy and aligned in peaceful partnership with a reunited Europe, has abruptly vanished. Instead, a new line of conflict is being drawn across the map. But a stable and peaceful Europe will not be rebuilt with tanks and barbed wire, or sanctions. It requires a renewed search for shared interests, and above all economic interest. That is the starting point for this book.

The search for common economic ground began well before the fall of the Iron Curtain, and it played an important part in the rise of détente and the beginnings of change in Russia. As Russia emerged from its isolation in the 1990s, Russian-European business relations flourished. It was not beyond reasonable belief that normal economic ties would foster normal political relations, and vice versa—and for a time that appeared to be happening.

At the center of this story was natural gas. But natural gas has now become part of the problem, for reasons that could never have been anticipated during the Cold War.

Natural Gas: From Bridge to Divide?

From the 1960s, when a coalition of Soviet and Austrian players negotiated the first Soviet-European pipeline deal in the middle of the Cold War, Russian gas exports to Europe grew over five decades to become (along with gas from the North Sea) one of the foundations of Europe's energy economy. Today, a vast web of gas fields, pipelines, and compressors serves thousands of factories and millions of consumers, binding them together in a dense network. By any standard the construction of the Russian and European gas systems ranks as one of the most important engineering and commercial achievements of the past half century.

By the 1980s, East-West relations had developed into a ritual: freezes and thaws came and went, but business carried on. In Western Europe, natural gas became a fuel of choice, and demand boomed, met increasingly with the seemingly limitless West Siberian gas reserves. On both sides of the Iron Curtain, the East-West gas trade was governed by stable monopolies that were closely connected to their respective governments. The contracts between them were long-term affairs. The norm was twenty to twenty-five years, but some were thirty years plus—negotiated by career professionals who came to know and respect one another as fellow members of a "Gas Club" that put business first and ideology second.

The 1990s and early 2000s brought four fundamental changes in succession. The first was the fall of the Soviet Union and the Soviet empire. The second was the strengthening and expansion of the European Union and, under its active sponsorship, the spread of market-oriented liberalization and regulation within a unified Europe. The third was the restoration of a strong Russian state under Vladimir Putin, with ambitions to rebuild a Russian sphere of influence in the former Soviet space. The fourth was the emergence of environmentalism as a powerful political force in Europe, especially in Germany. Together these four events have changed, and are continuing to change, the role of gas in Russian-European relations. The familiar world of state-backed monopolies and long-term contracts is disappearing. The rules by which gas is bought and sold have been turned upside down.[1] And the place of energy itself in politics and public opinion—the very future of gas in a Western Europe increasingly focused on decarbonization—is being transformed.

Along the way, what was a shared interest and a factor of stability between Europe and Russia became a subject of strife, both as a symbol and a contributing cause.

How did this happen? The initial effect of the fall of the Iron Curtain and the creation of the Single Internal Market in Europe was to open the European energy sector to penetration by players from both sides. Western European companies moved into Eastern Europe and bought up local utilities and transmission systems, largely supplied by Russian gas. Russian companies—especially the Russian gas monopoly Gazprom—moved west and south into Europe, forming partnerships and new ventures, as they sought to build an integrated gas value chain in Europe, from the European borders to the burner tip. At first this was hailed as a positive development. In the early 1990s there was a flowering of optimism and entrepreneurship across the previous divide.

But it did not last. In Russia, the initial commitment to a market-led economy in the 1990s gave way by the mid-2000s to an increasingly nationalistic version of state-led capitalism, and energy became for the Kremlin an instrument for the reestablishment of a Russian sphere of influence in the Former Soviet Union and Eastern Europe. In Western Europe, the drive to create a single European space evolved by the early 2000s into a supranational legal and regulatory structure based in Brussels, driven by a militant market-oriented doctrine and armed with formidable enforcement powers. Even as Russian state capitalism moved west, the European Union (EU) and its *acquis* moved east, reaching the borders of the Former Soviet Union by the mid-2000s.

The two soon clashed—especially in the area of gas. By the early 2000s they were colliding along a broad front, from Germany to the Baltic Republics. Economic relations that had initially supported cooperation and partnership became causes of discord and conflict, which intensified as Europe's indigenous sources of gas began to decline and its dependence on imports grew. In Eastern Europe, dependence on Russian gas came to seem increasingly threatening. Growing conflicts over gas in Ukraine poisoned the atmosphere. In parallel, natural gas itself came to be viewed with hostility by the rising environmental movement, especially in Germany, adding yet another source of conflict over Russian gas.

Much of the commentary about the causes of conflict focuses on the geopolitics. There are two competing narratives about gas, one Western and the other Russian. The Western narrative focuses on Europe's perceived unhealthy dependence on Russian gas and Russia's manipulation of the "gas weapon" for political purposes, especially in Eastern Europe, the Baltics, and Ukraine. But in the Russian narrative, the source of conflict is the aggressive eastward spread of EU law and regulation motivated by anti-Russian hostility, in disregard of commercial precedent and economic rationality—especially in Brussels. Each side denounces the other's arguments as specious and self-serving.

Yet to boil down the subject of Russian-European gas relations to geopolitics is to miss a large part of the story. In parallel with the competing geopolitical narratives, there has been an upheaval in the commercial structures and business models by which gas is bought and sold in Europe. The gas revolution in Europe has deep roots, which originated quite independently of Russia, and are only distantly related to geopolitics. And finally, there is the emerging environmental issue in Europe and the growing boom in renewable power, to which the Russians until now have hardly been a party at all. The aim of this book is to disentangle these separate strands of the story and show how they interact in shaping the future of Russian gas in Europe.

The economic fundamentals remain compelling: Europe needs gas and will continue to do so for decades, and Russia has gas in abundance. Despite the heated geopolitical rhetoric, the gas business carries on. Indeed, the gas bridge is being rebuilt around new principles, and at this writing it is prospering. Recent gas negotiations have shown flexibility and adaptation between Russian sellers and European buyers, and commercial logic has driven significant compromises—particularly on the Russian side, as Gazprom has responded to commercial and regulatory pressures. In Eastern Europe, governments are taking steps to diversify their energy sources and manage risk.

In addition, developments outside Europe—the rise of shale gas in the United States, the expanded use of liquefied natural gas (LNG) as an alternative way of shipping gas, China's booming demand for gas, and Russia's "pivot to the east"—hold the long-range promise of lessening the centrality and sensitivity of gas politics in Europe.

But geopolitical risk is real. If Russian intervention in eastern Ukraine continues and East-West relations become increasingly toxic, the space for constructive commercial dialogue will narrow. The danger ahead is that the community of economic interest that sustains the gas bridge, and indeed much of the web of Western-Russian economic relations, will be weakened by the escalation of tensions, sanctions and countersanctions, proxy warfare, and worse. Alternatively, the rise of environmentalism in Western Europe may ultimately weaken Europe's call for Russian gas. What then?

The Plan of the Book

The Bridge is a history of the Russian and European gas industry and the Russian-European gas trade. It begins by telling the story of the rise of the gas industry in the Soviet Union and Europe and the evolution of the Russian-European gas trade over the last half century, followed by an analysis of current trends and possible futures. The book weaves together three stories:

The first is the story of the development and evolution of Soviet and Russian gas policy, showing how the gas industry originated and evolved in the Soviet Union. The birth of the Soviet gas sector—in contrast to the Norwegian case, which the book also describes—was due to a uniquely Soviet combination of high politics and state entrepreneurship, from which foreign companies were entirely absent but which was acutely dependent on foreign technology and capital. The creation of the gas bridge in an economy that was largely closed to the outside world was motivated by the need to develop the vast discoveries of gas in West Siberia and to offset the threat of declining oil production, as well as developing a new source of export revenue. Despite the fears of many in the West, the gas bridge was not viewed in the Soviet Union as an instrument of geopolitical leverage over the West, although it definitely played that role in the Soviet Union's relations with its Eastern European satellites.

After the fall of the Soviet Union, gas rents inherited from the Soviet legacy played a vital role in maintaining the Russian economy, through both exports and subsidized domestic distribution. Inside Russia, this process was highly political: Russian energy policy from the beginning consisted of the capture and allocation of legacy rents and their translation into political benefits. In Western Europe, in contrast, Gazprom showed an impressive degree of entrepreneurship and innovativeness as it sought to capture a place in gas transmission and distribution. Finally, in Ukraine and the Baltic Republics, Russian gas policy was a complex mix of political and commercial motives, but the central aim was to maintain Russia's dominant position.

The second story is about the evolution and expansion of the European Union since the early 1980s, showing how today's European energy policy and competition doctrine arose out of the drive for a single European market, and how these were subsequently applied to individual member-states—particularly Germany, where resistance to the new thinking by established companies was especially strong. The market-oriented ideas underlying the single European market largely originated in Great Britain and were translated into EU legislation by a succession of mostly British and German commissioners and officials working together. As a result, for the past two decades market-oriented competition law has been by far the most vigorous component of European energy policy. It is only recently that the two other key objectives of EU energy policy—security and environmental protection—have gained equal status with competition. But even now the three objectives coexist uneasily, defended by separate EU institutions. The consequence is frequently unstable and contradictory policy.

The third story is about the axis of crisis: Germany and Ukraine. Today, although the bulk of Russian gas exports comes from West Siberia, nearly half of it is still shipped through Ukraine, the historic cradle of the Soviet gas industry (although just how large the share shall be in the future is one of the most difficult points in Russian-Ukrainian relations). The chronic difficulties of managing the Ukrainian connection after the Soviet breakup and the combination of conflict and collusion on both sides have contributed to destabilizing the Russian-European gas trade in recent decades.

Why Gas Is Special

Natural gas is fundamentally different from oil. Oil is an arm's-length commodity: once produced, it is sold into a global market, and there need be no further relationship between producer and consumer. In contrast, gas is traditionally a relationship commodity: producer and consumer are linked by a pipeline, which automatically creates a mutual dependence—and if there is a transit country in between, the result is a complex triangle. It is only with the advent of new technologies (chiefly LNG, but also computerized trading floors for gas) that gas has begun to take on some of the commercial characteristics of oil, and is rapidly becoming a global commodity.

Traditionally, gas relationships played out over long periods of time. Fields and pipelines took up to two decades to plan, finance, and build, and once built they operated for a half century or more. In contrast to oil, where most of the complexity is concentrated in the field, in gas the complexity is in the network. Terms have to be renegotiated periodically, to adjust to changing conditions of weather and the shifting prices of competing fuels. As a result, gas people grew to know each other intimately. Friendships formed, and partnerships were long-lasting. However, if the two sides were unequal in means and political power, asymmetric in motives, or divided by mistrust, the resulting relationships were ones of hostility or mutual corruption. This was particularly the case in Russia's gas relations with Ukraine.

But technology today is changing the nature of relationships in the gas business. Gas shipped by pipeline is limited to a regional market. However, LNG, shipped by tanker, is no longer tied to a single destination. In these and other ways, the traditional gas trade by pipeline is being challenged, along with the commercial and political—and personal—relationships that go with it. But technological change takes many forms: for example, the advent of online gas trading would not be possible without advanced computers and algorithms. These developments are changing the nature of the business. One can already foresee a day when much of the gas trade will be conducted by buyers and sellers who do not know one another, linked by intermediaries who have no personal or emotional stake in the long-term stability of the business.

Gas is central to both the European and the Russian economies and their energy policies, but it plays very different roles in the two regions:

1. *In Russia, gas is the other face of Russia's hydrocarbon dependence.* Gas was the last major achievement of the Soviet planned economy. Faced with an impending crisis in the oil sector, the Soviet leadership invested heavily in gas. In remarkably short order, gas became the fastest-growing sector of the Soviet economy and its most important primary fuel. European companies played a key role, supplying pipe and funds to help develop what became the world's largest gas industry within two decades. When the Soviet system collapsed, it was gas, even more than oil, that kept the Russian economy going through the 1990s.

Russia today produces about as much gas as it does oil, but it uses the two differently: oil is mostly exported, while gas is mainly consumed at home, and gas exports yield only one-fifth as much export revenue as oil. Yet in a broader sense the rents from both are equally vital to the Russian state. After 2000 the gas industry was quickly recentralized under Putin.

Inside Russia, Gazprom's essential role remained the same: to supply gas at subsidized prices to Russian's inefficient domestic economy. However, Gazprom faces growing competition from so-called independent producers, leading to the ironic result that there is an increasingly competitive domestic gas market, while Gazprom plays the thankless role of swing producer, serving the least profitable end of the value chain. Outside Russia, the picture is more complex: Gazprom quickly evolved into one of the most entrepreneurial of Russian companies, taking equity positions in transmission and distribution throughout Western Europe, forming partnerships with major European companies, and developing new lines of business. Yet this commercial dynamism was combined with a highly conservative business model, and Gazprom has fought a steady rearguard action against the changes sweeping through the European gas sector. Only recently has Gazprom begun adapting its business strategy in Europe to changing times. These two faces of Russian gas policy, and of Gazprom itself, account for much of the initial success of Russian gas in Europe through the mid-2000s and for much of the growing conflict since then. The tension between the domestic and foreign faces of Gazprom is one of the major themes of this book.

2. *In Europe, gas is the most obvious bridge fuel to the low-carbon future:* the rise of natural gas in Europe since World War II has been an essential part of Europe's search for energy security, efficiency, and environmental quality. As Europe moves toward a lower-carbon economy based on renewable sources of energy (chiefly wind and solar), natural gas is the logical bridge fuel—the cleanest and most economical means currently available to get there. This would suggest that Russian gas will continue to play a strong role in Europe, especially as Europe's own sources of natural gas, particularly from the Netherlands, continue to decline.

Yet natural gas is under challenge in Europe. It is unpopular with environmentalists, who favor renewables instead. This calls into question the very future of Russian gas in Europe.

The original gas bridge of the 1960s and 1970s was the product of a different world. It was born at a time when central planning and state-driven technocracy in both Europe and the Soviet Union were at the height of their prestige and success. That world is long gone. The world that has replaced it has the virtues of openness, flexibility, and transparency but also the flaws of instability and unpredictability. The long-term role of gas in Europe is uncertain, and thus so is that of the gas bridge. Depending on which narrative one listens to, there are two possible futures.

The geopolitical scenario holds that the cross-border gas trade is ultimately hostage to the overall state of political relations between Russia and Europe. The new divide across Central Europe is hardening. So long as Russia persists in its revival of nineteenth-century power politics, the borders of Europe will be insecure. According to this telling, the gas bridge represents a worrisome source of dependence that Russia will exploit, and the result will be further conflict.

In contrast, the economic scenario holds that gas is ultimately a business, and like all businesses its future will be determined by economics and technology, as well the regulatory regime under which they are deployed. The revolutionary changes in the structure of the gas business will bring disruption, because that is the result of change. But the outcome will ultimately be determined by economic interest, as it has been with remarkable reliability for a half century.

Yet how strong will that shared economic interest remain in the decades ahead? Might it fade away, not because of geopolitics but because of deep-seated changes in the place of gas and the structure of the gas business? Thus there are two opposing narratives of the future of Russian gas in Europe. The gas bridge is at the center of each. Which one will prevail in the coming decades—or will it be another narrative altogether, the environmental one? That is the subject of this book.

CHAPTER 1

Two Worlds of Gas

In the background of this story is a simple molecule, one atom of carbon and four atoms of hydrogen, arranged in space in a tetrahedron. There is something appealing in its purity, its elegance of understatement. When it mates with oxygen it burns cleanly, with a bright blue flame, and then vanishes, leaving only water and carbon dioxide, the stuff of soda water. When cooled, it condenses into a clear, shimmering liquid, only one–six hundredth of its original volume, ready to spring back as a gas again when needed. Oil, by comparison, is a sludge of chains and hexagons that is largely useless until refined and sorted out, disciplined, so to speak. Coal is even worse. Natural gas is a princess.

That kind of talk is not much heard these days. In Europe, natural gas is increasingly blamed today as a culprit in climate change, in the same dock as its fossil fuel cousins. One has to travel back in one's mind to the 1960s to recapture a time when natural gas was the next new thing, the virtuous fuel, the welcome liberator of kitchens and happy housewives, the magic source of warm baths and hot showers at the mere touch of a button, not to mention its benefits to industry. The advent of natural gas summed up the optimism of the postwar boom. The three decades from the mid-1940s to the mid-1970s—which have come to be known by their French name, *Les Trente Glorieuses*—could just as well have been called *les trente gazeuses,* a time of bubbly optimism, when natural gas became synonymous with progress. Enthusiasm over natural gas was no less effervescent in the Soviet Union.

The era of natural gas had begun much earlier in the United States, where all of the needed ingredients had come together in the late nineteenth century: an abundant resource typically found next door to oil, advanced metallurgy and engineering, vigorous market-driven companies, and (initially at least) a regulatory void. In the earliest days of the US oil industry, natural gas had been regarded a waste product and was flared off. But beginning in the 1880s a Pennsylvania oil family, the Pews, began collecting natural gas and selling it as a fuel for oil field operations. By 1883 they had built a pipeline to Pittsburgh, the first to supply natural gas to a major city (it would be a half century before anything similar was constructed in Western Europe). Thanks in part to their early efforts, Standard Oil became interested and, true to form, created the Natural Gas Trust, which captured the growing volumes of natural gas thrown off by the oil industry and began piping it throughout the eastern half of the United States.[1] In 1940, on the eve of World War II, the natural gas industry in the United States produced 3.162 quadrillion Btu, or 12.1 percent of the country's total fossil fuel production.[2]

In Europe and the Soviet Union, by contrast, the era of natural gas did not really begin until after World War II. From the beginning there were two worlds of gas, on either side of the Iron Curtain. The underlying physics, chemistry, and technology of natural gas are much the same everywhere. Yet the ways in which natural gas was discovered and developed, produced, priced, and sold in Europe and the Soviet Union arose out of two completely different universes of thought and politics and, correspondingly, two contrasting sorts of structures and people. The result was two very different industries. These initial differences between East and West continue to echo down through the decades and explain much of the mix of cooperation and conflict in the Russian-European gas trade that we observe today—which in turn has played no small part in a new time of troubles in Europe.

Two Worlds of Gas

Even now, a generation after the end of the Soviet Union, the differences between the gas systems of Western Europe and the former Soviet empire can be seen from space at night, in the contrasting infrared glow—the heat signature—of East and West. Russia looks like a system of large arteries, connecting the bright patches of gas flares in Siberia with large cities and industrial centers in western Russia. Western Europe looks more like a web, a dense network of capillaries feeding closely spaced cities and towns, while the arteries come in at the periphery. Eastern Europe and the western states of the Former Soviet Union (FSU) are still lands of coal and form a darker no-man's-land in between, crossed by the Russian gas pipes.

Much of the contrast is due to differences in distances and population densities. About 90 percent of Russia's gas is located east of the Urals—technically in Asia—while 80 percent of its consumption is in the western third of the country. The average molecule of Russian gas travels nearly 2,500 kilometers from field to burner tip. Like traces of ancient settlements that can be seen only from the air, the pattern bears witness to the different political and economic environments in which natural gas grew up in the two halves of Europe. Just as the Soviet Union was a world apart from Europe, so was its gas industry.[3]

The differences were fundamental from the outset. In Western Europe natural gas was preceded by so-called town gas (produced by burning coal in large gas plants) and coke gas (a by-product of the steel industry). Thus when natural gas first appeared in the 1950s, a century-old infrastructure already existed, with a multitude of public and private players—municipalities, labor unions, small family-owned enterprises, steel and coke producers, and so on. It was local or at most regional, and it grew organically from the bottom up, building on the existing base as markets arose. Only a decade or more after World War II, as the first major natural gas resources were discovered in Western Europe and at its edges, did governments become involved as major players, and then only at the regional or national levels. The European Union, or European Economic Community as it was then, as yet played virtually no role.

In the Soviet Union natural gas started practically from a blank slate. For obvious reasons, there were no private players, and the growth of the industry was driven entirely by state action, specifically by a single specialized bureaucracy with strong political sponsorship. From the earliest days it was conceived as a vertically integrated network on a national scale (though limited at first to the western third of the country). The Soviet gas industry was all about central planning, but in the Russian style, with its signature blend of improvisation and haphazard coordination and the fierce competition of state agencies and personal ambitions.[4]

The result was two very different approaches to the fundamental issues of the industry—the allocation of resources, the definition of price and value, and above all the management of risk. The two systems also generated two characteristically different sorts of leaders. In the Soviet Union these were larger-than-life bureaucratic entrepreneurs who were tough enough to manage the bureaucracy and ruled their sectors for decades at a time, subject only to the fortunes of their political bosses. In Europe the early heroes of the gas industry were mostly unsung business figures and civil servants who came and went as their careers peaked and whose critical roles, often behind the scenes, are remembered mainly in corporate histories and obscure monographs. The handful of key exceptions—Willy Brandt, Jacques Delors, and Margaret Thatcher—made their mark as political leaders whose impact on the European gas industry, though profound, was incidental to other things. The only example of a European oil and gas entrepreneur who came at all close to the Soviet model in the 1950s and 1960s was Enrico Mattei, the swashbuckling head of the Italian state hydrocarbon monopoly ENI, whose career was cut short by his untimely death in a plane crash in 1962.

Thus Europe was a mosaic of unique situations: each country was different, and natural gas developed differently in each, with initially very little connection among them, either physically or at the level of policy. In contrast, the Soviet Union was a monolith (or at least so it appeared from the outside). That remained the case right through the 1990s, and it shaped the relations between them.

Europe's Gas Industry on the Eve of the Great Discoveries

In the late 1950s, on the eve of the great natural gas discoveries in the Netherlands and the North Sea, the European gas industry was still essentially a derivative of the coal and steel industry. For over a century, European cities had been lighted by town gas. Jelle Zijlstra, who as Dutch minister of finance in the 1950s had witnessed the beginning of the transition from town gas to natural gas, recalled: "Countless districts, each with its own little gas plant! Even one with less than eight thousand inhabitants. Imagine! It was dirty, it was expensive, it was inefficient."[5] In France there were over seven hundred such small plants, most of them primitive and wheezingly inefficient. Even today, the French expression for a system that is overly complicated and inefficient is *une épouvantable usine à gaz* (a frightful town-gas plant).

The gas plants and the coal that fueled them were commonly located near city centers, in unsightly clusters that dominated the urban landscape and added to the heavy pollution of city air. As late as the 1950s, Paris—which owed its fame as *La Ville Lumière* (the City of Light) to its early adoption of town gas for public lighting in the nineteenth century—still depended on the same antiquated arrangement, which not only heated but also lighted the city with coal. Massive quantities of the stuff were hauled into the city by rail from the east and stored in depots near the railroad stations. A plan to bring coke gas by pipeline from eastern France to Paris had first been proposed as early as 1886. But owing to stout resistance from the railroads and labor unions—not to mention the disruptions caused by two world wars—it was not built until seven decades later, finally bringing gas by pipeline to the capital in 1954. Even then, it was still coke gas, not natural gas.[6]

But Paris by then was an extreme case. Already before World War II, the ancient system of local gas plants had begun to fade, as gas derived from coking plants, shipped through pipeline systems that sometimes stretched for over a hundred kilometers, began to replace town gas from the old gasworks. The unsightly gasworks streets of the nineteenth century gradually disappeared. The key was the steel industry and the coking plants that supported the smelting of steel, which generated large quantities of coke gas as a by-product. The rapid growth of steel production stimulated the development of regional pipeline networks. In Belgium and the Netherlands by this time there were countrywide networks of gas pipelines, and here and there, as in Switzerland, other regional networks had begun to appear. Some famous company names appear at this time, such as Belgium's Distrigaz, which was founded in 1929 by a group of British investors to supply coke gas by pipeline to the city of Antwerp.[7] In Germany, also in the 1920s, the ancestor of the postwar giant Ruhrgas was formed by the steel industry to supply surplus coke gas to the surrounding towns.[8] In 1929 the new company contracted to supply the city of Cologne; the document was signed by the mayor of the city, Konrad Adenauer.[9] On the eve of World War II, some twelve billion cubic meters of gas—one-third of the German consumption—was delivered annually by pipeline, mostly to municipalities in the coal and steel region of the Ruhr.

In much of Europe gas distribution was controlled by numerous small companies, typically owned completely or in part by local municipalities. In France and Italy this structure was swept away after World War II by nationalization and consolidation into state-owned monopolies. But in Germany the municipally owned utilities (known as *Stadtwerke,* or city utilities) remained a permanent feature of the landscape, reflecting the highly decentralized nature of postwar German politics. As we shall see, the role of the municipalities in German energy policy remains crucial today, particularly with the rise of the environmental movement in the 1970s and later the drive for renewables in Germany.

The transition to natural gas happened slowly at first, the main reason being the lack of large discoveries on European soil, as well as the entrenched position of the coal industry. Where hydrocarbon companies were active, they typically preferred to focus on oil instead of gas. Through the end of the 1950s, only in a handful of places such as the Netherlands, where convenient local sources of natural gas existed, were they developed and connected to nearby towns.[10] An early exception was Italy, where major finds in the Po Valley after World War II touched off a rapid expansion that reached 3.6 billion cubic meters by 1956. The pipelines to support the new Italian production were built and operated by two state-owned companies, Agip and Snam.[11] Revenues from natural gas played a major role in financing the ambitions of Enrico Mattei in his drive to make his company—today's Eni—one of the oil giants of the world. Another local exception was France, where the first natural gas was developed in the middle of World War II, with a small discovery that brought natural gas to Toulouse and Bordeaux. But the era of natural gas really began in France only in 1957, with the discovery of a major field at Lacq in the southwest of the country. In all of these cases, however, the discoveries were not large enough to supply more than regional-level systems and did not stimulate cross-border exports.

Gas in Britain

In the aftermath of World War II, gas in Britain and the men who ran it were a backward vestige of the nineteenth century. Production was in headlong decline, owing to growing competition from imported oil—which, as Kenneth Hutchinson of the British Gas Council put it in his entertaining autobiography, was "equally smokeless, requiring no stokers and becoming relatively cheaper all the time."[12] All efforts to find significant sources of natural gas in Britain had failed, and by the mid-1950s the British gas industry was losing the fight.[13] As Dieter Helm described the British gas industry at the end of the town gas era, "It was an industry with a very limited market, and an even more limited future."[14]

In 1954, a Chicago businessman named William Wood Prince appeared on the doorstep of the British Gas Council with the revolutionary idea of bringing natural gas to Great Britain in liquid form in refrigerated tankers. Wood Prince, who owned the largest stockyards in Chicago, had already experimented with bringing liquefied methane by barge from gas fields on the US Gulf Coast, both to chill meat and to provide fuel for his meatpacking plants. Up to this point there had been only limited experience with liquefying methane. Indeed, the only functioning methane liquefaction and storage plant in the world at that time was located near Moscow. It had been built for the Russians in 1947 by an American company, Dresser Industries of Dallas, Texas.

Although the British had never shipped liquid methane, they were not entirely without experience with liquid gases. Kenneth Hutchinson at the Gas Council had worked at the Air Ministry during the war to adapt road tankers to take liquid oxygen—essential for pilots of fighters and bombers—across the Channel to France for the military advance through Europe into Germany. As he wrote in his memoirs, when the idea of importing liquefied natural gas (LNG) from the United States was first proposed to the Gas Council, he recalled that the US Air Force, in preparation for an invasion of Japan, had developed a technique to tow liquid oxygen in large barges across the Pacific Ocean. As Hutchinson wrote in his memoirs, "I could see no insuperable difficulty in the plan to bring liquid methane across the Atlantic."[15]

The Gas Council was convinced, and a company was formed in partnership with Wood Prince to turn the idea into reality. The next five years were taken up in designing and testing the tankers and overcoming many problems along the way—steel, for example, became brittle at the low temperature of liquid methane (–161.6° F), and aluminum was used for the tanks instead.[16] Finally, a converted Liberty ship, now reequipped as a tanker and rechristened the *Methane Pioneer*, left Lake Charles in Louisiana in January 1959 with the first load of LNG. Along the way, it encountered Force 10 winds, with twenty degrees of roll and seven degrees of pitch. But "as it came out of the mist on a bitterly cold morning to dock at Canvey Island on 20 February 1959," it opened the era of natural gas in the British gas industry.[17]

The experimental shipment of US LNG to Great Britain had been followed with growing interest by Shell. Major discoveries of natural gas had been made by the French in Algeria at Hassi R'Mel, but there appeared to be no practical means of bringing it to market in Europe. Various schemes had been discussed, including a long pipeline across the Sahara and under the Mediterranean to Spain or France, but these were neither economic nor safe—the Algerian war for independence was raging at the time, and the Algerian coastline was not secure. Transport of liquid gas by tanker offered the ideal way to bring Algerian gas to France and Britain. A joint company called Conch was quickly formed, which allied Shell and Conoco; two new methane tankers were built; and storage and regasification facilities were prepared in Britain and France. The first cargo of Algerian gas reached Great Britain in 1964, and another shipment reached France the following year. For a brief time in the mid-1960s, Algerian LNG accounted for one-tenth of the total gas supply (natural gas and town gas combined) of Great Britain.

But even as the first cargo of Algerian LNG reached Europe, another revolution was brewing in northern Europe. In 1959, the year the *Methane Pioneer* reached Britain, the first discovery of natural gas was made at Slochteren, in the northern Netherlands province of Groningen, and 1965 marked the first major offshore discovery in the North Sea—by BP at West Sole, in the British sector. Thus for Britain, LNG in the 1960s and early 1970s was a brief interlude; the era of pipeline gas from the North Sea was about to dawn. Only in France, with its historic if troubled ties to Algeria, did LNG imports remain a major source of supply throughout the subsequent decades.

Slochteren Changes Everything

As usual, the upstream explorers had been looking for oil. Back at the home office, any news of gas was distinctly unwelcome. Nevertheless, gas is what they found. Seismic studies of the area around the village of Slochteren had suggested that there was something beneath a thick salt layer. But it could have been oil, gas, or water—or nothing. In 1955, a first exploratory well, drilled ten kilometers from Slochteren, provided the first clue that the answer might be gas. But the well nearly went out of control, and further work stopped while the Nederlandse Aardolie Maatschappij (NAM)—the joint venture of Exxon and Shell that led exploration in the Netherlands after the war[18]—turned its focus to maximizing oil production during the Suez crisis. NAM did not return to Slochteren until May 1959. Two months later gas began to flow, and at such a prodigious rate that this time there could be no doubt that a major field lay beneath. The flame from the first well could be seen twenty kilometers away in the city of Groningen.

For more than a year after the first discovery, NAM imposed a complete blackout on any news from Slochteren. NAM did not own exclusive drilling rights, and it feared a gas rush by other oil companies, or possibly a government takeover or nationalization. But during that time it carried out careful studies, and even its cautious early estimates made it clear that life in the Netherlands was about to undergo a spectacular change. And indeed it was. By the time Groningen reached its peak in the early 1970s, it was generating 20 percent of the government's revenues per year.

Why the Dutch Model Was Crucial

Natural gas was not entirely new to the Dutch in the late 1950s. Small deposits had already been developed, mainly in the eastern Netherlands. In 1951 the first Dutch town was supplied with natural gas. By 1960 over one-quarter of Dutch gas consumers were already burning natural gas. But the discovery at Groningen changed the landscape practically overnight. Within just a few years, the known reserves of natural gas in Europe grew from a few billion cubic meters to trillions. The Netherlands was not alone: in 1965 the British made the first large discoveries in the North Sea, and they were soon followed by the Norwegians. The great Soviet discoveries in West Siberia occurred at about the same time, as did those in the southern Saharan desert regions of Algeria.

But in continental Europe, the Dutch discoveries were crucial. Dutch gas basically created today's continental European gas industry. For the first time, Europe had a world-class gas field, large enough to support exports and cheap enough to compete with coal. This raised challenges that the energy industry in the Netherlands and the Dutch government had never had to face before. The successful way they dealt with them, as we shall see in a moment, not only opened the way for natural gas in Europe but also created the entire conceptual framework and the very vocabulary used to price and sell natural gas throughout Europe over the following forty years—including Russian imports, when their time came. It is only in the past decade that the pragmatic arrangements developed in the Netherlands in the early 1960s have gradually yielded to new commercial and regulatory models based on very different premises, with far-reaching economic and political consequences. But we are getting ahead of our story.

In the wake of the discovery of Groningen, one basic question had to be settled first: How would the state and the private sector relate to each other? Gas, as we have said, grew up as a business of relationships. Explorers, producers, distributors, investors, and consumers were soon connected by a single chain of value. But who would own and control it?

From this flowed a host of questions that had to be decided: Who would invest, and who would bear the risks and receive the rewards? What was the market? How would gas be priced to different users? The answers first developed in the Netherlands provided the commercial and regulatory model—what the Russians still call the "Groningen model"—around which the entire West European gas industry was built. It is only in the last two decades, with the advent of computerized trading technologies and new EU regulations governing competition and company structures, that the traditional model has been swept away.

In the United States, these questions had been debated for half a century; but the American experience was all but irrelevant in Europe, where the role and prerogatives of the state were traditionally far stronger.[19] For example, in Europe, since the Napoleonic mining law of 1810, private-sector producers have not owned the rights to mineral resources in the ground; unlike in the United States, companies operate under concessions granted by the state. After World War II the already strong position of governments was further reinforced by widespread enthusiasm for state planning. This was not due to the Soviet example, which for most Europeans remained remote and alien. Rather, the renewed faith in the powers of the state drew on earlier reform traditions native to Europe, combined with the prevailing view that prewar capitalism and politics had failed. In the words of the historian Tony Judt, "The disasters of the inter-war decades . . . all seemed to be connected by the utter failure to organize society better. If democracy was to work, if it was to recover its appeal, it would have to be *planned*."[20] As natural gas made its appearance in a Europe perennially short of energy, it seemed like the prime candidate for a planned approach.

Just what planning meant in practice varied from country to country. In Britain, it consisted mainly of nationalization.[21] In Germany, the state constructed elaborate corporatist structures between companies and labor unions. But it was in France that planning took its most elaborate form. The "commanding heights" (in Lenin's famous phrase) were nationalized across a wide range of sectors, ranging from utilities to automobiles. Public investment was guided by the state. In 1946, President Charles de Gaulle approved a proposal by Jean Monnet—subsequently one of the founding fathers of the European Community—to create a Commissariat Général du Plan, and the first national plan was adopted—unanimously—by the French cabinet in January 1947. But where natural gas was concerned, the results proved to be contentious, which held its development back by several years.

Thus it was a fortunate accident that the one truly major gas discovery in continental Europe happened to occur in the one country where the prevailing political culture and political alignments were uniquely favorable for the rapid development of a new resource and its early move into exports. That country was the Netherlands.

"Jan Compromise"

In the Netherlands conflict was avoided. Instead, an ingenious and pragmatic partnership was negotiated between the private oil companies that had discovered Slochteren and the state, which owned the resource. That everything came together so quickly and smoothly was due most of all to one man, Jan Willem de Pous, the Dutch economics minister from 1959 to 1963. In just fifteen months, between August 1961 and October 1962, de Pous masterminded the negotiations and assembled the coalition of players that would own and run the new gas business in the Netherlands. His role was fundamental: he crystallized the business plan and the pricing principles, fended off skeptics from his own party who wanted to nationalize natural gas, and shepherded the whole package through the Dutch parliament, where it was approved unanimously. After this feat, de Pous vanishes from history.

De Pous was a classic product of the Dutch system of consensus-based politics that prevailed from the end of World War II to the mid-1960s.[22] The son of a Dutch flower grower, he came from the Christian Historical Union (CHU), one of two Calvinist parties that along with other confessional parties represented the main traditional pillars of Dutch society and had occupied the political center since the end of the war. Since no party could command an absolute majority, they had to work together in a succession of pragmatic-minded center-left and center-right coalition governments. The CHU, with its small handful of seats, cheerfully joined both. It was a system that put a premium on negotiation and compromise, skills that de Pous had mastered to a high degree. (Such was his skill as a negotiator that he was nicknamed "Jan Compromise.")

It is little wonder that De Pous's contribution has become a legend in the gas industry. Half a century later—in 2013, on the fiftieth anniversary of the creation of the Dutch gas monopoly GasUnie—a distant successor as economics minister, two generations younger than de Pous, paid tribute to his unique role as "the architect of gas cooperation in the Netherlands": "De Pous set it all up with a single ten-page policy document, which he got through parliament after a debate lasting only half a day." The minister added feelingly, "And to think of all the hours and days I've spent in parliament discussing energy policy!"[23]

In reality, though, de Pous was never quite the commanding figure portrayed by subsequent mythology. The smooth passage of the Dutch gas law of 1962, on closer examination, was due less to his individual efforts than to those of the party managers in parliament. The Dutch political system, as was the case through the late 1960s, was based on strict proportional representation. The real power in the system belonged not to cabinet ministers—not even the prime minister—but to the group leaders in parliament, who controlled the party agendas and got together to name the prime ministers and their cabinets. Ministers were frequently second-rank party members or even outsiders, viewed by the parliamentary managers more as delegates than as policymakers. Thus they were effectively disposable. This proved to be the case with de Pous: after the center-right coalition of which he was a part fell in 1963, he was never minister again, and he effectively left politics.[24]

In 1966, the entire system blew apart. In the parliamentary elections of that year, both Labour and the confessional parties went down to unprecedented defeat in what has been described as "the most dramatic event in modern Dutch politics."[25] The result was the end of the traditional machinery of parliamentary balancing of which de Pous had been a part. From that point on, Dutch politics became much more polarized, reflecting the growing affluence of Dutch society and the coming of age of a new generation of voters. New small parties appeared, with antiestablishment agendas. In the new atmosphere of the late 1960s, the entire model of consensus-based politics came under attack: Democracy-66, one of the new parties, denounced the old system as "unprincipled, unaccountable, and arthritic."[26] Compromise had become a dirty word.

Thus the discovery at Slochteren came at a fortuitous time. The consensus embodied in the "nota de Pous," and the smooth creation of GasUnie, codified by a unanimous vote in parliament after only symbolic debate, might not have been possible after 1967. Indeed, just three years later, when the first offshore discoveries were made and the Dutch government debated a new gas bill, consensus was becoming harder to reach. There was sentiment in the Labour Party for outright nationalization of the oil and gas industry. In 1965–1966 a coalition led by Labour came to power, which included a handful of ministers who were prepared to fight for nationalization. But again electoral politics intervened: in 1966 the Labour Party was defeated, the radical finance minister was swept away, and there was no further talk of nationalization.

The Challenges of the Dutch Gas Market

The first challenge faced by the Dutch was deciding what to do with the gas. Up to that time the oil companies had not thought much about gas, except to conclude that it was a costly bother. A former Shell director remembered the advice given to him by a senior colleague in The Hague in the late 1950s, on the eve of the great discoveries: "Stay away from gas. There's no profit in it. The State regards it as a public utility." At the time, natural gas was indeed unprofitable, both for the sellers and for the buyers.[27]

Upon learning the first news from Slochteren, Standard of New Jersey's corporate headquarters sent a pair of junior economists to The Hague, where they teamed up with two Dutch colleagues to study the situation.[28] Together they came up with a revolutionary plan. Whereas higher management, particularly at Shell, was inclined to target industry and power as the main markets, the "gang of four" after exhaustive calculations argued against that. By far the most profitable market for the gas in Europe, their numbers showed, was households, especially for home heating. Their advice, accepted only after much initial skepticism, proved crucial for the development of gas in continental Europe. It provided a strong incentive for rapid development, and it catalyzed a vast investment in gas pipelines that in short order connected millions of consumers to gas.

Gas for households changed an entire way of life in the Netherlands. On the eve of Slochteren, only 10 percent of Dutch households had central heating (the typical Dutch home was heated in only one room, by a coal stove), but within a decade natural gas had driven coal out of the house. The strategy of focusing on home use was followed in other countries as well, giving the gas map of Western Europe its signature pattern—a few large arteries and a vast system of capillaries—that sets it dramatically apart from that of Russia and the Former Soviet Union. This basic difference goes back to the earliest days of the Dutch model and to the four junior economists, barely remembered today, who first realized where the market lay.

Who Takes the Risk?

The second major challenge had to do with the management of risk. The consumer's need for gas is as changeable as the weather—literally, since a cold winter increases demand—and varies also with economic conditions. But this creates a risk for both the producer and the consumer. The producer needs protection against fluctuations in demand, and so does the consumer, who needs to know that on a cold day the gas will be there. The classic solution, embodied in gas contracts since the 1960s, is known as "take or pay." If the buyer does not actually "take" the volumes specified in the contract, he must pay anyway. The unused gas may be taken at a later date, but the basic principle is clear: the buyer traditionally bears the volume risk.

The first example of take or pay appeared as early as 1954, in the founding agreement between NAM and the Dutch state, and in that arrangement the state plainly bore the risk. NAM was bound by the terms of its concession to sell whatever modest amounts of gas it produced to the Dutch government, but at a price so low it was almost symbolic. The government had little use for it and therefore ended up paying for gas that remained in the ground. That was awkward enough, but with the discovery of gas at Slochteren, what had been costing the state millions of guilders annually would soon become billions. One veteran Shell director later recalled: "We could have bankrupted the State of the Netherlands in one fell swoop. We used to joke about it together."[29] The result, after some discussion, was the modern take or pay agreement, under which limits are set on the buyer's obligation to take, so as not to cause undue losses.

What's the Price?

Part of the mythology surrounding the "nota de Pous" is that it created overnight the doctrine for pricing gas that has been used in every European gas contract practically down to the present day. The basic idea is that gas should be priced according to its value to the user, not its cost to the producer.[30] In practical terms, this means that the price of gas must not exceed that of competing fuels at the point of consumption, a principle that came to be known in the gas business by a German name, *Anlegbarkeit.*[31] When applied to gas exports, it logically leads to the idea that the price at the wellhead should be the "netback"—that is, the price to the consumer minus the costs of transportation and distribution. In practical terms, this means that the price of gas at the Dutch border depends on whom it is going to.

These pricing principles did not spring full-blown from the founding Dutch legislation but evolved over time by trial and error, as the market developed and experience accumulated. Domestic prices for Dutch households were a particularly sensitive issue, and it took several years of debate before compromises were worked out between the producers and the local municipally owned distributors. It was eventually decided to price the gas according to a sliding scale: the more you used, the lower the price. This turned into a bargain for household heating, which accordingly boomed, and for small businesses—most spectacularly for greenhouses, which soon dotted the Dutch landscape. Once the prices to consumers had been set, they proved difficult to change. As a result, Dutch household prices remained unchanged well into the 1970s.

Export pricing also evolved gradually. The first export prices were fixed, based largely on cost. Thus the first Dutch exports, beginning in 1965 to Germany and Belgium and in 1966 to France, charged identical fixed prices at the Dutch border.[32] It was not until 1971, as it became apparent that energy prices were on a rising trend, that the Dutch began pressing foreign buyers for contract revisions to tie the price of gas exports to those of oil products. Buyers understandably pushed back, and it took until 1974—after the first oil price shock—for oil-product benchmarks to be incorporated into the export contracts, and even then that happened only gradually.[33]

Italy was a special problem. Concerned that the Italians were about to sign a large gas contract with the Soviet Union, the United States, via NATO and the Dutch Ministry of Foreign Affairs, pressed the Dutch to agree to a lower export price to Italy than to other importing countries. Raising the price later on proved delicate, since the initial contract had contained no revision clause. It was not until 1975 that a new contract was concluded with the Italians, containing both a revision clause and prices tied to oil products, as well as the concept that the longer distances from the Netherlands to Italy justified a lower wellhead price—in short, the netback principle. Thus it took a decade for the core concepts of export pricing to be worked out.

The Significance of Contracts

What made the gas relationship work was the body of understandings built into contracts. The multiple risks and uncertainties faced by buyer and seller—the unknowns generated by weather, markets, and geology—were managed through standard arrangements that were first developed in the wake of the first major gas discoveries in the Netherlands. The significance of the contract as the foundation of the gas business cannot be overestimated. The contract created a degree of unity that offset the early decentralization of the industry. Long before the European gas industry was interconnected physically by pipelines, it was interconnected by the common intellectual structure of contracts, enabling very different players to manage risks and reach agreements. It is no wonder that these long and unwieldy texts, written in a technical language unintelligible to outsiders, have over time become invested with deep emotional significance. As the energy expert Simon Blakey puts it, "hard-bitten, commercially driven men can grow misty-eyed at the recollection of what they see as the almost magical ways in which the contracts succeeded in bridging impossible gaps and managing unmanageable risks."[34]

Thanks to the favorable arrangements concluded in the Netherlands, natural gas boomed. There followed an explosion of pipeline construction. This reached a peak in the second half of the 1960s with the construction of major arteries into France and Italy—but above all in northern Europe, as gas from Groningen reached cities throughout the Netherlands and then in Belgium and Germany. From less than 500 million cubic meters of gas in 1963, the year of its creation, GasUnie was shipping nearly 95 billion cubic meters a year by the mid-1970s, more than half of which was exported.[35]

The effort stretched to the limit the ability of the pipe producers to keep up. To support the growing network, the gas industry needed not only vastly greater amounts of pipe, but also larger and larger diameters, capable of operating at ever-higher pressures. The result was a burst of innovation in the European steel industry, which made possible the subsequent gas-for-pipe deals of the 1970s.

The decision to favor the international market, contained in the "nota de Pous," caused Dutch exports to grow even faster than domestic sales. The first Dutch gas reached Germany in 1964, Belgium in 1965, and France in 1966. Within a decade, Dutch exports had expanded to a peak of nearly 50 billion cubic meters a year, of which two-thirds went to Germany and Belgium. Yet the path was not entirely smooth. Dutch gas initially faced strong resistance in Germany. There were mutterings about "foreign infiltration" *(Überfremdung)*; portions of the German media campaigned against the invader; and the German Ministry of the Economy worried about the threat to the coal industry. In those days, fear of dependence meant dependence on Dutch gas.

The boom that followed was driven largely by market demand. Natural gas proved immensely popular, and the industry could barely keep up. Gas followed the market; the growth was bottom-up, not top-down. National and local governments acted as facilitators (especially by keeping household prices low and, in the Dutch case, favoring exports), but they were not the drivers. The action was all at the national or regional levels. Nowhere in Western Europe was there an overarching vision for the continent-wide integration of the gas network, or any power structure in existence that could have implemented one. Nevertheless, European gas as it developed was knit together by the invisible conceptual structure that guided its growth—the system of contracts between sellers and buyers, underpinned by an emerging legal structure according to which gas was sold and traded.

The Soviet Union in all these respects was as different as night from day. Beginning in the 1950s, the Soviet natural gas industry began to develop in parallel with that of Europe, but it was driven by different forces and guided by different principles. When these two worlds came into contact, by the mid-1960s, it was like the meeting of two alien civilizations.

The Soviet Case

It took a palace revolt against the tsar to bring city lighting to Russia's capital, Saint Petersburg. Nicholas I, "the Gendarme," alarmed by an attempted coup against him by young army officers in December 1825, complained of the security problems posed by dark streets. A British firm was hired to build the capital's first coal-gas plant and pipe system—all imported from Britain, as was the coal burned in the plant.[36] But progress was slow. On the eve of World War I, only about 3,000 apartments had gas. As a Soviet-era source drily notes, gas was "only for privileged families."[37]

The picture was much the same in Moscow. By 1867, foreign investors had installed 6,000 streetlamps, but over the following twenty-five years only 3,000 more appeared. There was no interest in proceeding further. The tsarist government was indifferent, and building owners didn't want to invest in piping gas to their properties. On the eve of World War I, Moscow lagged far behind the major cities of Europe: London used 176 cubic meters per capita, Brussels 116, and Paris 108, but Moscow only 10.[38] The situation did not change much under the Bolsheviks, even after they moved the capital to Moscow in 1918. In 1945, at the end of World War II, Moscow still had only one gas plant. In a sense, this was an advantage. Unlike the Europeans, the Soviets did not have to adapt an existing gas system; they went more or less straight to natural gas.

It is an ironic fact, in light of later events, that the modern Soviet natural gas industry originated in Ukraine—or, to be more precise, in what was at the time Polish Galicia. Early in the twentieth century, in the last remaining years of the Austro-Hungarian Empire, gas had been discovered near the Galician village of Dashava. But the find remained undeveloped until after World War I, by which time both the Russian and Austro-Hungarian empires had disappeared. By 1921, when the first gas well was drilled at Dashava, it was part of a newly reconstituted Poland, a gift of the mapmakers at Versailles. In 1929, Polish and German engineers built one of the first natural gas pipelines in Eastern Europe, sixty-eight kilometers long, to supply gas from Dashava to the capital of what was then Polish Galicia—which until 1939 and the Molotov-Ribbentrop Pact would be known by its Polish spelling of Lwów. Dashava changed hands twice more during World War II. During the Nazi occupation, German engineers built another pipeline two hundred kilometers westward to a metallurgical plant at Stalowa Wola, which the Soviets then continued after the war. Thus in a sense, as Per Högselius writes in his history of the origins of the gas bridge, *Red Gas,* the Soviet Union was a "born gas exporter"—thanks to Ukraine.[39]

Up to that time, the Soviet economy was still largely based on coal.[40] Joseph Stalin, despite his origins as a young revolutionary in the oil fields of Baku, to the end of his life favored coal over hydrocarbons, as did a significant part of the Soviet energy establishment.[41] But coal faced growing problems: available supplies were of poor quality, declining, or too remote.[42] Consequently, gas from coal was scarce as well. Coal-based gas supplied only a handful of the major cities. Even gas for city lighting in Leningrad (the once and future Saint Petersburg) came from Estonian oil shale, a poor-quality source, and town gas in Moscow relied on brown coal from the surrounding province. Only in the coal-and-steel complex of Ukraine's Donbas was there an adequate source of coke gas as a by-product of the steel industry.

But the Soviet occupation of Galicia had brought with it a welcome trophy—the Dashava gas field, which returned to Soviet control in 1944. Even before the Red Army reached Berlin, construction began on a 500-kilometer pipeline to Kiev, where gas arrived in 1948, followed three years later by an 800-kilometer pipeline to Moscow. Within a few more years another pipeline from Dashava took Ukrainian gas to neighboring Belorussia (Belarus) and the Baltic Republics. As production from Galicia neared a peak, new gas fields in discovered eastern Ukraine (notably, the gas giant Shebelinka) took over. Thus, although supplies soon diversified as new sources were developed elsewhere, in southern Russia, the Caucasus, and Central Asia, the Ukrainian gas fields remained the workhorse of the Soviet gas industry for its first quarter-century.[43] If Ukraine depends on imported gas today, it is because its gas resources were used to supply the Soviet economy—and, as we shall see in Chapter 2, the first Soviet gas exports.

Working for Beria

The early gas industry grew out of the oil industry, and the oil industry, like every other strategic sector of Soviet industry from the late 1930s to the end of the Stalin era, was under the KGB (known in those days as the NKVD) and its fearsome chief, Lavrentii Beria. When captured Soviet archives were examined by American analysts after the war, it came to light that nearly half of all Soviet industrial output on the eve of the war was controlled by the NKVD and overseen by Beria.[44] In the Great Purge of 1936–1938, the better part of the Soviet military, political, and administrative elite had been swept away, as millions fell victim to Stalin's paranoia and the scythe of the secret police. In their place came a new generation of young—sometimes astonishingly young—engineers and professionals, who rocketed up to positions of power. It was said of them that they stepped over the dead bodies. They came to be known as the Generation of 1939, and in many cases they remained in charge until the mid-1980s and the advent of Mikhail Gorbachev.

One of those was Nikolai Baibakov, the perennial head of state planning from 1965 to 1985. In 1938, as the Great Purge reached its bloodiest phase, the twenty-seven-year-old petroleum engineer from Baku got his first big break, when he was named to lead the effort to develop the next generation of Soviet oil, in the Volga-Urals Basin. By 1940, he was deputy minister of oil (or "people's commissar," as they called it then), and in 1944, at the age of thirty-three, he was made head of the entire industry. (In such meteoric careers, a year or two made all the difference: Baibakov's friend and contemporary at the Azerbaijan oil institute, Sabit Orudzhev, who happened to be slightly younger than Baibakov and graduated two years later, remained behind in Baku for another decade.)

A key protégé of Baibakov was Aleksey Kortunov, a remarkably talented engineer and organizer who became the first head of the Soviet gas industry and was in many ways the architect of the gas bridge, as we shall see in Chapter 2. Kortunov was that rarity in the Soviet system, a person trained in a wide variety of engineering fields (as opposed to the narrow curricular background more characteristic of most Soviet, and indeed today's Russian, engineers), which served him well when he was sent to work in Bashkiriia—then the heart of the Russian oil industry. His skill as a builder of oil projects and pipelines brought him to the attention of Baibakov, who promoted him to Moscow in 1950 as deputy minister for construction. When a few years later Baibakov won political support to create a separate organization for gas, he put Kortunov at its head. Kortunov remained there for the next two decades as the chief driving force behind the development of the new industry.[45]

Baibakov, as minister of oil, reported directly to Beria. It was a terrifying job, but it was also a guarantee of results. The story goes that when Stalin learned from the mayor of Moscow that the first gas pipeline to the city, from a Volga gas field near Saratov (Yelshanka-Kurdyum), was experiencing interruptions after it was first commissioned in 1946, he turned the problem over to Beria. In very short order the pipeline was working properly. There were many such stories. If Stalin, in Alexander Solzhenitsyn's famous phrase, "gave the entire country insomnia," it was because of ruthless henchmen like Beria. When Stalin died in March 1953, Beria appeared about to take over—until one of his rivals, Nikita Khrushchev, had him arrested and shot, to the tremendous relief of his Politburo colleagues. From that point on, until his overthrow in 1964, Khrushchev ruled over the Soviet Union as first secretary of the Communist Party.

Nikita Khrushchev and the Difficult Birth of the Gas Industry

Above Khrushchev's grave in Moscow's Novodevichii Monastery stands a bust of the late Soviet leader by the modernist sculptor Ernst Neizvestnyi. It is a remarkable work, made of two blocks of marble, one white and the other black, joined vertically at the midline of Khrushchev's face. For Russians it captures the man's dual personality and mixed legacy.[46] As one of Stalin's chief henchmen, Khrushchev was the Party boss in Ukraine at the time of the Great Purge, yet he was also the man who attempted to conjure away Stalin's ghost by denouncing Stalin's crimes.

A pudgy, bald little man who wore suits that always seemed two sizes too large for him, Khrushchev looked like a collective farm chairman or a factory manager of the 1930s, a throwback to the era of first five-year plans—which in many respects he was. Many of the traits for which we remember him came from his penchant for Bolshevik-style *shturmovshchina*:[47] his colorful harangues and pungent scoldings, his love of the spectacular gamble, his tub-thumping campaigns and whirlwind inspection tours, and his endless and frantic reorganizations. Yet he was a highly intelligent man who, though he had served Stalin loyally, realized that the system had to change. But he knew no other style than the one he had practiced all his life, and the results were frequently disastrous. A system in which so much depended on the good will and support of the top leaders had its advantages but also its drawbacks. Khrushchev illustrated both in extreme form, and his role in the early years of the gas industry was a typical example.

Khrushchev, unlike Stalin, was a great believer in the future of natural gas. The landmark year was 1956. At the 20th Congress of the Communist Party, the same historic gathering at which Khrushchev denounced Stalin's crimes (including the mass famine of the 1930s), Khrushchev unveiled ambitious targets for the gas industry. In the same year the giant gas field in eastern Ukraine, Shebelinka, produced its first gas. A long-distance pipeline from the Volga region soon brought it to Moscow. Other new long-distance gas pipelines followed in the late 1950s, including the first of what would be a four-string system carrying North Caucasus gas to Moscow and Leningrad. Yet more pipelines were built from the Volga region, including to Nizhniy Novgorod (then Gorkii) and Cherepovets. A decision was made to build a pipeline from Dashava through Belorussia (now Belarus) to the Baltic Republics. Soviet gas production soared from 9 billion cubic meters in 1955 to 198 billion in 1970.[48]

For Khrushchev, the rapid rise of natural gas was one of the clearest signs that the Soviet Union was catching up with the United States and indeed would shortly surpass it. On a trip to America in 1959 he used natural gas as a prime example of coming Soviet superiority: "Our geologists have discovered gas resources so immense that they will suffice for decades. This gives us the possibility to increase even more the production and consumption of gas and to overtake you also in this respect."[49]

But Khrushchev's enthusiasm was a mixed blessing. He was impulsive, changeable, and utterly unpredictable. Once an idea had entered his head, it took over completely, and he was ready to gamble the resources of the entire system to realize it. In 1957, he became convinced that the Stalinist economy was overly rigid and centralized—which it certainly was—and unbalanced in favor of heavy industry. His solution was to refashion the whole economy. He abruptly dissolved the central ministries and replaced them with regional economic councils *(sovnarkhozy)* that would coordinate the economy locally and give more attention to light industry, consumer goods, and housing.

In the resulting chaos, the nascent gas industry was nearly broken up. This happened just at the moment when proven gas reserves were ballooning, thanks to increased spending on exploration ordered by Baibakov in the first half of the decade. Baibakov's vision, and that of the gas enthusiasts in the oil industry led by Kortunov, was to build an integrated pipeline system to bring gas to the industrial heartland from the newly discovered fields of Ukraine, southern Russia, and Central Asia—although not yet West Siberia, where the great discoveries still lay ahead. Gas production was expanding rapidly, and it now seemed certain that the Soviet Union would shortly become a gas superpower.

But this vision for gas was the very essence of classic Soviet-style centralization, and under Khrushchev in 1957 centralization was suddenly out of fashion. Baibakov fiercely opposed the whole *sovnarkhoz* idea: he believed that strong central ministries were of crucial importance to keep the economy on track. A gruff, plainspoken man, he did not hesitate to criticize Khrushchev's plan to his face. In 1957, as part of the meltdown of the central ministries, Glavgaz (the specialized department responsible for gas within the Oil Ministry) was abolished, and it was only through desperate appeals to Khrushchev himself that Baibakov was able to have it restored and to place Kortunov, his protégé, at its head. Yet the more

Baibakov resisted the Party line, the more he lost favor and the more his influence weakened. Khrushchev increasingly perceived him as disloyal, and in 1958 he exiled Baibakov to the southern provinces—ironically, to head two of the *sovnarkhozy* that Baibakov so despised. Kortunov remained in Moscow (since Glavgaz was not officially a ministry it was not broken up) and was able to return Baibakov's support in the most classic Soviet way, by preserving Baibakov's Moscow apartment for him until his return from exile.[50] As it turned out, Baibakov was able to reclaim his apartment sooner than expected. Behind the scenes Khrushchev's closest associates, led by his closest protégé, Leonid Brezhnev, were plotting a coup against him. In October 1964 the trap closed, and Khrushchev was overthrown.[51]

Khrushchev's disastrous experiment with *sovnarkhozy* had a damaging effect on the oil and gas construction sector—but especially on oil, which was far more important than gas at that time. "The oil-construction organizations attached to the *sovnarkhozy*," wrote Baibakov in retirement years later, "were constantly missing their targets for completing oil infrastructure. As a result, oil-production targets were being missed as well." "I felt that it was not enough to move additional resources to West Siberia," Baibakov goes on, "but also to seriously strengthen their capacity. But first they would have to be grouped together under a single powerful hand *(dlan')*, or to put it more simply, to transfer them to an agency on which one could count seriously in such an important matter."

But on whom could one "count seriously"? Baibakov had an answer ready to hand—Kortunov and his Glavgaz organization, which had already built a reputation under Kortunov's energetic leadership as an outfit that could get jobs done. In fact, Glavgaz had been doing oil projects right along, in parallel with gas projects, all over the Soviet Union. For example, the Druzhba oil export pipeline to Eastern Europe was largely a Glavgaz project, which it built at the same time as the Bukhara-Ural pipeline for Central Asian gas.

As soon as Baibakov became chairman of Gosplan in October 1965, he writes, "I considered that the time had come for a conversation with Kortunov. To give him proper tribute, he didn't require much persuading." Baibakov, after all, had the rank of a deputy prime minister; and Kortunov knew that Baibakov's support was crucial for his plans for the West

Siberian gas industry. Thus the two men arrived at what Baibakov describes as a "gentlemen's agreement." Baibakov endorsed the creation of a dedicated construction agency for West Siberia (dubbed Glavtiumenneftegazstroi), placed under Kortunov's control. In exchange, Kortunov took on responsibility for building all of the hydrocarbon infrastructure in the region, oil as well as gas. This arrangement lasted for the following eight years, until 1973. By that time Kortunov's construction empire within what was by that time the Ministry of the Gas Industry (Mingazprom) employed over 30,000 people and had built the infrastructure for over 200 million tons a year (4 million barrels a day) of West Siberian oil. This was the price of Baibakov's support for West Siberian gas,[52] But he did not keep his part of the deal.

Meanwhile, Khrushchev was not done with his "harebrained schemes" (as his former colleagues later dubbed them). In 1963, following a massive crop failure, Khrushchev concluded that the key to improving the Soviet food supply lay in boosting the production of fertilizers. As always with Khrushchev, his latest fixation became a single-minded campaign. Suddenly it was all fertilizers all the time—and the gas industry was expected to play the lead role. In addition, Khrushchev had convinced himself that the Soviet Union's gas reserves needed to be preserved for future generations. To the horror of Kortunov and those who supported the development of the gas industry, Khrushchev issued orders to cut investment in gas exploration and pipeline construction in the coming five-year plan for the second half of the 1960s. Existing gas production was to be rerouted to the chemical industry. Kortunov, never one to back down from a fight, scaled the barricades to defend his industry. Remarkably, he was able to prevail in an open confrontation with the general secretary. But revealingly, he had help. By this time, many of Khrushchev's colleagues had already concluded that his schemes were severely damaging the economy, and Kortunov found sympathetic support behind the scenes. In the end, the investment plan for the gas industry was rescued.[53]

Thus the early history of the Soviet gas industry was completely different from that of Europe. It was political rather than commercial, driven from the top, and from the first it was led by technocratic entrepreneurs who envisioned natural gas as a single integrated system that would bind the entire Soviet economy together. In addition, it benefited at its beginnings from a strong platform in the oil industry. The great strength of the Soviet economy at its best was that when these elements were present, it could produce impressive results.

What the Soviet gas technocrats had in common with the Europeans was enthusiasm over the superiority of natural gas as the fuel of the future. Iulii Bokserman, Kortunov's right-hand man for technology policy in Glavgaz, waxed lyrical about the virtues of natural gas: Natural gas is a "smoke-free fuel, which burns up completely and does not emit any polluting gases into the atmosphere."[54] Both the Soviets and the Dutch from the earliest days had a vision of the gas network of the future. Natural gas was a cause on both sides. The difference was that the centrally planned system was able to implement the vision on a countrywide scale, while the Dutch model spread into Europe organically—from the bottom up, so to speak—and the integration of the system into a Europe-wide web was not completed for another generation or even two.

The great problem for the Soviet energy sector was that the system's *vertikal'* (its vertically integrated system of power) existed primarily to serve one goal, that of defense. The Soviet command economy was less an economy than a political pump, which channeled resources and talent preferentially toward the military-industrial complex. The civilian sector was in constant competition with the military and military industry. As the growth rate of the Soviet economy began to slow, which it did from the mid-1960s, the competition for scarce resources and talents became more and more acute, and the military were consistent winners.[55] From the 1950s through the 1980s, the share of GDP devoted to the military-industrial sector increased steadily. The contest was particularly acute in the area of machine-building and basic industrial goods such as steel pipe. In these sectors, investment in machinery for civilian purposes chronically lagged behind that in machinery for military purposes.

This is the fundamental reason why, as the needs of the gas sector grew from the 1950s on, the industrial support system was unable to deliver. New gas discoveries, mainly in Central Asia and West Siberia, called for longer pipelines with larger-diameter pipe and higher pressures. These required better steel and compressors—but in a system in which the best metallurgists and engineers worked for the military-industrial complex, they were not available in sufficient quantity or quality to support the gas industry's rapidly growing requirements. In the Soviet system of allocation, missiles and submarines took precedence over pipelines and compressors.

No one knew this better than Nikolai Baibakov. As minister of oil, he had been the chief patriot of the gas industry. But as head of Gosplan, he had to reckon with the *oboronka* (the nine powerful military-industrial ministries that commanded the top priority in the Soviet system) as well as the formidable Military-Industrial Commission that oversaw them, all under the watchful eye of a special department of the Communist Party Central Committee headed by a senior member of the ruling Politburo. Even within Gosplan, military-industrial planning was managed by a special department headed by a senior figure from the defense industry, who was largely out of Baibakov's control. Running this formidable power system was a collection of senior technocrats who had risen to power in the late 1930s. The chairman of the board, so to speak, was Leonid Brezhnev himself, who in the late 1950s had been the Central Committee secretary in charge of heavy industry, the military, and the space program.

In a system so tightly organized along vertical lines, the chronic weakness was horizontal coordination. Each ministry and each region strove to meet its targets by being as autonomous as possible. The chief function of the Party apparatus was to try to offset this bias by forcing the ministries to work together, but its efforts were never more than partly successful. The various bureaucracies had their own power structures, goals, and cultures. In response to this chronic problem, the system responded by putting senior managers in charge—"tsars," in the American parlance—who were given total command of a sector and total responsibility. Aleksey Kortunov, the father of the early Soviet gas industry, was one of those.

Yet Kortunov and gas sector paid a price for their partnership with the oil industry. Until the mid-1950s the discovery of gas reserves took place only as a by-product of exploration for oil, and consequently identified gas reserves grew only slowly. This turned out to be important in West Siberia, where the oil-rich region is located in the middle of the Ob River basin, whereas the gas-rich part of the province is located well to the north, where the climatic and logistical difficulties are of a whole different order. Bokserman, in his 1958 book on the gas industry, did not even include northern West Siberia in a list of the ten regions with the greatest potential for gas development.[56]

It is perhaps for this reason that when Soviet geologists finally did begin to uncover vast gas reserves in northern West Siberia, Baibakov grew cautious. In the course of several visits to the region, Baibakov took stock of the tremendous challenges involved, and he hesitated. This in turn caused a growing coolness between him and his protégé, Kortunov, and their gentleman's agreement came under strain as Kortunov campaigned loudly for early development of West Siberian gas. But Kortunov, to his frustration, was unable to persuade Baibakov to take the turn to West Siberian gas.

Baibakov's caution—matched by that of much of the Soviet establishment—caused the development of West Siberian gas to be delayed by a decade. This had two far-reaching consequences. One was the overuse of Ukrainian gas. The second was a long hesitation in organizing for exports, even though the gas sector badly needed imported pipe and technology. In Chapter 2, we explore how these factors influenced the first Soviet gas deals with the West.

CHAPTER 2

The Beginnings of the Gas Bridge

About thirty kilometers southwest of Vienna, nestled in the rolling hills that are the first sign of the Alps, stands a small castle built in the style the Austrians call English Gothic. Once a hunting lodge of the Habsburgs, Schloss Hernstein Bergdorf is a romantic spot: on a quiet lake and surrounded by woods, it is popular today for weddings and business conferences. It was here, in the late summer of 1967, that the Austrians hosted a delegation of eleven Soviet officials for the first concrete discussions about a gas contract. Over two weeks of talks and walks, the two sides hammered out the concepts that were to lead, two years later, to the first Soviet gas exports to the West. From that point on, the gas contracts multiplied and volumes grew rapidly. In retrospect, the meeting at Schloss Hernstein marks the historic beginning of the gas bridge between Russia and Europe.

The meeting had been arranged hurriedly at the request of the Soviets—who, after a summer's hiatus (partly connected to the diplomatic waves caused that spring by the Six-Day War in the Middle East), unexpectedly signaled that they wanted to talk. At that point, Soviet relations with Germany were as chilly as they had ever been, the French seemed more interested in doing business with the Algerians and the Dutch than the Russians, and early exploratory talks with the Italians had gone nowhere. The Austrians scented an opportunity, and they had a plan.

The group that gathered at the Schloss included many of the key players and personalities that would figure prominently in Soviet-European gas talks over the next decade. Representatives of OMV and VÖEST, the two state-owned Austrian gas and steel companies, and of Mannesmann and Thyssen, the two largest German steel companies, were there. The Soviet side was led by two men who were rapidly becoming familiar personalities in the European gas world and who would lead the Soviet side in many negotiations over the next twenty years: Nikolay Osipov, deputy minister of foreign trade, and Anatoliy Sorokin, deputy gas minister. Sorokin, who had just been elected president of the International Gas Union that spring, was known to be a confidant of the gas minister, Aleksey Kortunov.

Every proper castle needs a ghost; and over Schloss Hernstein there hovered, if not quite the ghost, at least the spirit of Rudolf Lukesch, the business director of VÖEST who had first masterminded the combination of gas, steel, and finance that could make a deal across the Iron Curtain possible and had persuaded the Austrian and German players to come together behind his plan. Four months before, Lukesch had died in an automobile accident. Yet it was his vision that had brought the players together and his plan that drove the meeting.

But here we are getting ahead of our story. Five years before, in 1962, gas exports from the Soviet Union to Western Europe were still hardly more than a vague idea. Since then the road that led to Schloss Hernstein had meandered through Rome and Bratislava and Munich, as well as Moscow and West Siberia, before coming back to Vienna. How the idea traveled and grew into a viable business plan with a coalition of governments and companies behind it—overcoming many obstacles along the way—is the subject of this chapter. In the first half we look at the origins of Soviet gas exports to the West; in the second half we turn to the Soviets' exports to their own bloc, showing the similarities and the differences.

The foundations of the gas bridge between Russia and Europe were laid over less than one decade, from the mid-1960s to the early 1970s. These were years of transformation on both sides of the Iron Curtain. Natural gas came of age in both East and West, as one immense discovery after another—especially in West Siberia—drove a hundredfold increase in gas reserves over the period, while advances in a dozen fields of technology, from steel to compressors to computers, enabled the construction of large-diameter pipelines operating at high pressures, allowing gas to be economically transported over thousands of kilometers to industries and households and displacing oil and especially coal in Europe's cities. The economic and environmental benefits were immediate—and highly popular. From Leyden to Leningrad, the air began to clear, and London's trademark pea-soup fogs became a thing of the past.[1]

In the first half of the 1960s, as reserves mounted in the Soviet Union and demand for natural gas surged in Europe, it was not long before gas people on both sides grasped the potential fit. Western Europe needed more gas, and from more diverse sources—while the still-nascent Soviet gas industry needed pipe, compressors, and finance. But how to turn potential into reality was far from obvious. The extent of the Soviet reserves were still highly uncertain, even to the Soviets, while the Europeans were uncertain how much gas they actually needed. But the fundamental obstacle was the Iron Curtain. Selling gas is not like selling oil; it is traditionally a business of relationships, and in the early 1960s those relationships were few and fraught. Communication was poor, and trust was minimal.

Yet the Iron Curtain contained a small loophole. By the late 1950s Austria—recognized as a neutral country only a few years before—had begun importing natural gas on a small scale from Czechoslovakia, largely in exchange for steel pipe. Through this modest exchange the Austrians and the Czechoslovaks had come to know and trust one another, and both sides had begun to think in larger terms, which a few years later led directly to the first export contract for Soviet gas into Western Europe. Thus, it was that the first span of the gas bridge was laid between two of the smallest countries on either side. The initial volumes were insignificant. But what was achieved at the Austrian-Czechoslovak border demonstrated that trading gas across the Iron Curtain was feasible. This was crucial for what followed, as both sides began to know one another and jointly learned the gas trade.

But the Soviet gas men were after bigger game. By this time, gas had been discovered in West Siberia, and they were anxious to develop it. For that they needed pipe, compressors—and money. Yet the Soviet leadership was far from persuaded. We turn first to the story of West Siberian gas, before returning to Schloss Hernstein.

The Battle for West Siberia

"Once they've sent you to Siberia," Siberians like to say, "they can't send you any farther." But actually they could—to Tyumen Province.[2] In those days it was the ultimate backwater, even within Siberia: other Siberian regions and resources, such as the coal basin of Kuznetsk, the powerful hydropower dams on the East Siberian rivers, and the naval shipyards of the Russian Far East, had greater priority, and Tyumen lagged far behind. Electricity was so scarce that the citizens of the provincial capital were forbidden to use electric space heaters, and most of the city's streets had no lighting.[3]

The defining event of the 1960s, on the Soviet side, was the discovery of the immense oil and gas reserves of West Siberia. Most of them happened to be located in Tyumen, in the vast basin of the Ob River, which runs through the heart of the province on its way to the Arctic Ocean. Soviet geologists, venturing farther and farther north into the frozen wasteland, found one supergiant field after another, with the main gas reserves found in the northern part of the province in the area drained by the smaller Taz and Nadym Rivers, in thick layers of prolific sandstone (see Map 2.1). They lay close to the surface in immense structures, largely undisturbed for millions of years. Yet because of their remoteness, the discoveries touched off intense political battles in Moscow over what to do with them. How much priority should be given to West Siberia, and particularly to gas? Of these conflicts the European negotiators were at first largely unaware, and their Soviet counterparts took care not to inform them.

The importance of the discoveries was dismissed at first. Planners and decision makers in the oil and gas business everywhere have learned to be wary of the optimism of geologists—an old joke in the industry is that geologists have successfully discovered ten of the last two supergiants—and Soviet geologists were no exception. In the years after World War II it took a lively imagination to believe that under the permanently frozen ground of West Siberia, an expanse of arctic wasteland the size of France,

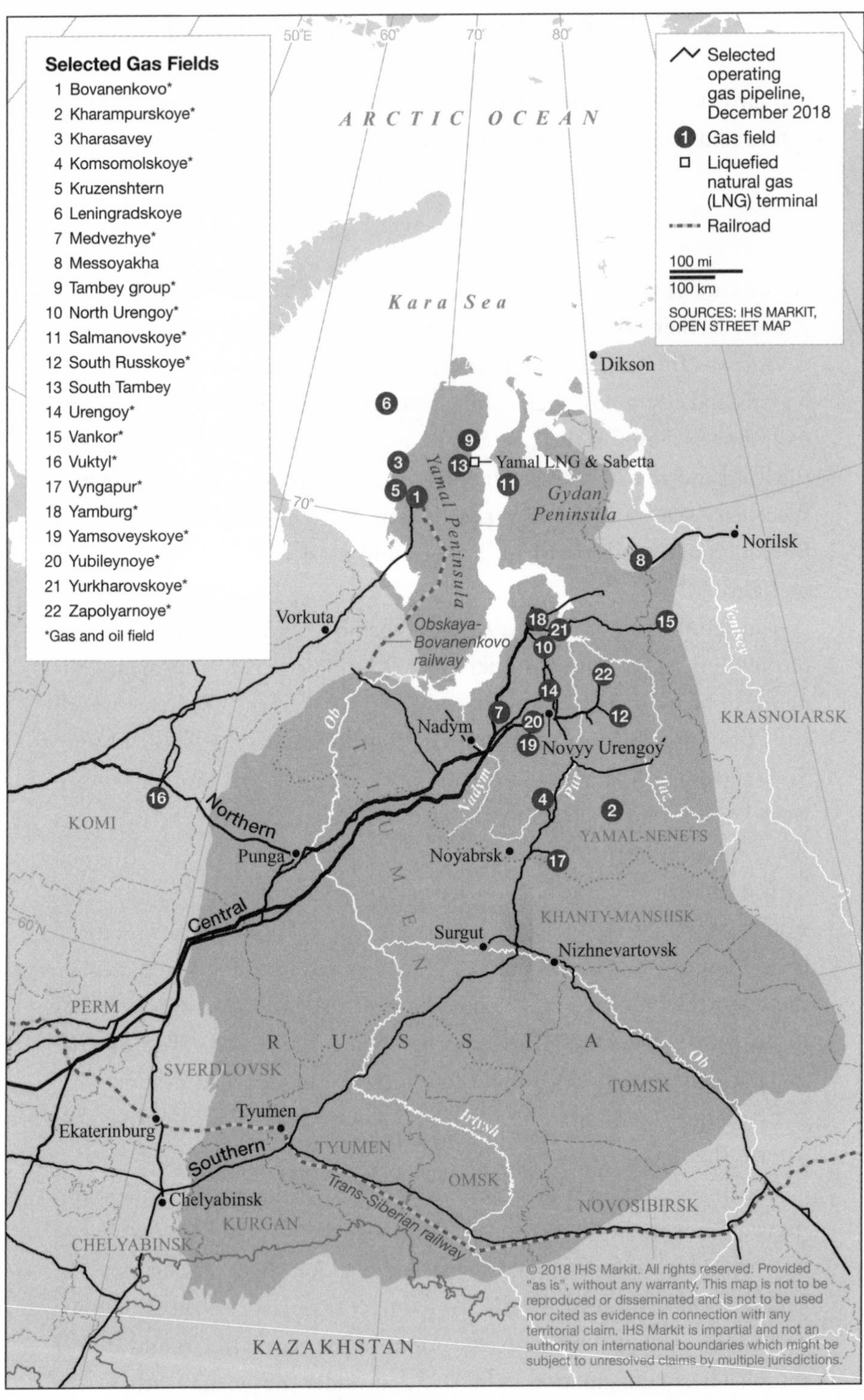

Map 2.1 Primary West Siberian gas fields and pipelines.

there could lie a treasure trove of hydrocarbons equal to that in the Middle East. Even the geological establishment in Moscow was skeptical: in the 1950s, local geologists who began exploring the Ob River basin in West Siberia were accused by Moscow of "squandering the people's money."[4]

West Siberian *gazoviki* ("gas men") like to point out that the very first major flow of hydrocarbons in West Siberia was not oil, but gas—in 1953, at Berezovo. It wasn't until 1960 that the first significant discovery of West Siberian oil began. But since there were as yet no gas pipelines, the oil was shipped out first, by tanker and barge, down the Ob River to the Arctic Ocean. For the time being, there was no similar outlet for the gas. A first small pipeline began operation in 1966, bringing gas over the Ural Mountains to industrial users in the Urals region. But the first real breakthrough did not come until six years later, in 1972, when an 800-kilometer pipeline from the Medvezh'e field delivered gas to the Urals via Nadym and Punga through a pipeline that was the first in the world to use 56-inch (1,420-mm) pipe.

As the Soviet geologists continued north, downriver toward the Arctic Ocean and the vast expanse the Russians call the Yamal, the great discoveries continued—but in the subsurface oil increasingly gave way to gas. This was a mixed blessing, to say the least. The challenges of West Siberian gas were of a whole different order of magnitude than those of oil. In northern Tyumen, where the giant gas fields lay, everything had to be learned the hard way. Winter was bad enough, with temperatures plunging to −40°F and lower, but the toughest test was the summer, when the deep snow and ice turned to impassable swamp, and clouds of mosquitoes and midges hovered over the construction crews. The midges (called *gnusy* by the locals) were by far the worst. In the words of one Western traveler in the early 1960s, "The Siberian midges rising like smoke out of the taiga, stinging, biting . . . were terrible things. And Moscow, true to its form, had forgotten to supply the workers with nets! Work came to a standstill."[5] But there were occasional compensations. One engineer from Ukraine, experiencing his first summer in Siberia, vividly recalls the day a senior official on an inspection trip from Moscow landed by helicopter next to their work site. The official jumped down onto the ground wearing nothing over his white shirt, mocking his fellow passengers for their timidity. Within minutes his white shirt had turned red with blood, and the loudmouth from Moscow slunk back into his helicopter without another word and took off. It was a byword among workers at the pipeline construction sites that no one could call himself a "Siberian" unless he had survived at least one winter—and one summer.[6]

Hydrophilia: How West Siberia's Oil and Gas Almost Ended Up under Water

Even before oil and gas were discovered, Tyumen appeared to have one great resource, the Ob River itself. To the Soviet planners of the 1950s it made eminent sense to harness the Ob for hydropower, to supply energy to the Tyumen region and the Urals industrial belt to the south and west. The politically powerful Soviet hydropower agency Gidroproekt, which in Beria's time had been an arm of the KGB's industrial empire and still retained informal ties to the security apparatus, conjured up a plan to build a string of ten hydropower plants along the entire course of the Ob. The largest one, the "Lower Ob Hydropower Station" (Nizhneobskaya Gidroelektricheskaia Stantsiia, or GES). would have put over 130,000 square kilometers, an area equal to one-third of the Baltic Sea, permanently underwater. In those days the environmental impact of such enormous projects hardly entered anyone's head, in Soviet Russia or elsewhere. By 1961, planning for the Nizhneobskaya GES was so far advanced that the Party leadership in Moscow appointed as *obkom* first secretary (i.e., first secretary of the province committee of the Communist Party) of Tyumen—Moscow's viceroy in the region—a man named Boris Shcherbina, who had already acquired a reputation as an effective Party manager for having successfully overseen the first massive hydropower projects in East Siberia.

But local geologists had come to believe that the Ob River basin might hold a great deal of oil and gas. The first major discovery in the lower Ob led them to petition the Party authorities to postpone the dam. Initially they got nowhere against the entrenched hydropower lobby. But by the early 1960s news of the first massive oil discoveries had started to come in, and the geologists pushed harder. A key turning point in the debate came when the new *obkom* first secretary, Shcherbina, became a convert to their cause, and from that point led the fight for West Siberian oil and gas.[7] Yet Gidroproekt was not to be denied. Clinging to its project, it argued that the vast flooding caused by the dam would actually make extracting the oil and gas more economical, since it would be easier to access the oil and gas fields by boat than to fight through swamps.

For Nikolai Baibakov, then in charge of planning for the oil and gas industry, this was too much. As a native of Baku, he had overseen the first offshore oil fields in the Caspian, and he knew how difficult and costly they had been (even though they were hardly more than trestles leading into shallow water). Thus, he too turned against the dam. Step by step, Gidroproekt was forced onto the defensive. Yet the battle was not over until the end of 1963, when a government decree ruled in favor of oil and gas over hydropower as the main priority in West Siberia. Even then, Gidroproekt soldiered on behind the scenes, using its historically close ties with the State Construction Committee (Gosstroi) to impede funding and supplies for oil and gas in Tyumen.

This was typical of battles over natural resources in the supposedly totalitarian and centrally planned Soviet system.[8] The ten-year conflict over the lower Ob had lasting consequences for our story in two respects. First, it puts in context the long hesitation of central policymakers in Moscow over the prospects and priority for West Siberian gas, which ultimately caused major delays in the development of the gas bridge to Europe. Second, it accounts for the special entrepreneurial energy—indeed, passion—of the West Siberian leadership in promoting the cause of oil and gas in Tyumen Province. Those who had launched their careers in the early 1960s in the successful battle against hydropower went on to become leaders of the Soviet oil and gas establishment for the next thirty years. Scherbina ultimately became deputy prime minister. His successor as *obkom* first secretary in Tyumen, the geologist Gennadii Bogomiakov, became the strongman of the West Siberian oil and gas industry for the next quarter-century, until he was overthrown in Mikhail Gorbachev's reforms of the late 1980s.[9]

It was people like Shcherbina and Bogomiakov—migrants from other regions who had staked their early careers on a total wilderness—who became the core of the emerging West Siberian power structure and the indispensable allies of Kortunov in the early days of the West Siberian gas industry. They looked back on their early battles with Moscow with pride. Bogomiakov recalled that the key moment of his career came when, as an unknown geologist, he traveled to Moscow to lobby against the lower Ob project, arguing that it would cause billions of rubles in damage: "You won't find those numbers in any institute. We came up with them one evening at the 'Moscow' Hotel. And the methodology remains on our conscience. But we knew that our cause was just."[10] In reality the battle was just beginning. Moscow had to be persuaded that the only way to develop West Siberian gas was to export it.

Kortunov's Strategy for West Siberian Exports: Reconnaissance, Stealth, and Assault

The many challenges of the Siberian north ultimately came down to one thing: pipelines. It was Kortunov and his small team at Glavgaz[11] who first grasped the essential connections between the growing gas potential of West Siberia, the advances in gas technology and engineering in the United States, and the growing popularity of gas in Western Europe. The challenge came down to a simple logical triangle: without pipelines, there could be no gas; without money, no pipelines; and without gas, no money. Each depended on the others. The conclusion was straightforward: the key to the development of West Siberian gas lay in Western Europe. Otherwise, it would not happen. This was the strategic vision that Kortunov sold to the Soviet leadership. Even today, it is the foundation of Gazprom.

To an unusual degree for a high Soviet official of the time, Kortunov was well acquainted with developments in the Western gas industry. He had made a long tour of Canada in 1958 to inspect the TransCanada pipeline, then under construction, and an equally extensive visit to the US gas industry in 1962 (a grainy black-and-white photograph shows a smiling Kortunov posing on Park Avenue, a high Party official alongside). As an experienced engineer, and especially as a manager, he could appreciate the achievements of the North American pipeline industry, as well as their implications for the Soviets. He published a book on each of those two trips, containing his detailed observations on the techniques he had witnessed—both the management and the engineering. These books were then diffused widely among Soviet decision makers as part of Kortunov's wide-ranging propaganda campaign for the emerging gas industry and his fledgling agency. He sponsored a similar book on France in the late 1960s, following his visit there—during which he explored with French officials and companies the purchase of a French license for liquefaction technology to support liquefied natural gas (LNG) export complexes on the Baltic and the Black Sea,[12] an idea to which Gazprom did not return until more than three decades later.

Kortunov was also an experienced military commander. In 1944, he was awarded the medal of "Hero of the Soviet Union" for having successfully commanded the troops that crossed the Vistula River. (The award of the medal in the field called for a ritual in which the medal was placed at the bottom of a large glass full of vodka. The recipient was required to drain the glass, recover the medal, and pronounce, as clearly as possible, "I serve the Soviet Union!") After the war, Kortunov was assigned to Weimar to head the military section of the Soviet occupation authority. His job was to dismantle German industry (it was called "German war potential") and send it back to Russia to rebuild Soviet industry destroyed during the war. By a neat symmetry, a quarter-century later, his mission had become the opposite: to assist in building a modern economy in Germany (both east and west) by supplying it and the rest of Europe with natural gas.

Kortunov's vision of developing West Siberia had taken shape in his mind almost as soon as news of the first major Siberian discoveries reached Moscow—especially that of the supergiant Urengoy in 1966.[13] For him it was obvious that West Siberia was the future of the gas industry, and almost immediately he began organizing for it, moving the center of gravity of the ministry to Tyumen Province.[14] But gas pipelines are much more demanding than oil lines, and this was especially true of West Siberia. Shipping West Siberian gas over thousands of kilometers and doing so economically required larger-diameter pipe and much higher pressures—and therefore thicker walls—than the Soviet gas industry had ever used before. Higher pressures meant stronger steels and larger compressors. These larger pipelines weighed more and were therefore more difficult to deliver to the site. But the challenges did not end there. Much of the ground consisted of frozen sand (called permafrost in English, but in

Russian, more lyrically, "eternal frost"), but the gas inside the pipelines was hot. Therefore, the pipe had to be cooled and raised on struts—hence more weight and more logistical headaches.

Yet here the Soviets were deficient—at least in the civilian sector. The military-industrial complex had a lock on the best domestically made alloys, and the lack of specialty steels for pipe had held back the gas industry from the beginning.[15] Turbines were primarily intended for jet bombers and naval vessels, and the military had first call. When the oil industry had faced similar problems a decade before, the Soviet leaders had turned to the West, trading oil for pipe and drilling technology. To extend this precedent to West Siberian gas was just a step, but it was one that the Soviet leaders and planners, suspicious of Kortunov's enthusiasm and that of the local promoters, were initially unwilling to take. Even Gosplan's chairman, Baibakov—who had been Kortunov's patron for the previous two decades and was otherwise favorably disposed to natural gas—was opposed to giving priority to West Siberian gas, as we saw in the last chapter.

There was only one answer: stealth. In early 1966, acting on his own initiative, Kortunov sent a team to Italy to explore with ENI a path-breaking concept—an export pipeline from West Siberia through Hungary and Yugoslavia to Italy. To head the team, Kortunov picked a thirty-four-year-old pipeline specialist named Stepan Derezhov. Derezhov had no prior background in international trade or external relations, and Kortunov had promoted him only a few months before to head a new division devoted to technology for gas production, storage, and transportation.[16] The Italians were initially as skeptical as the Soviet bureaucrats. As Derezhov recalled thirty years later: "The Italians reacted at first with great skepticism. Just think—they said—Where's Tyumen, and where's Rome? That's 4,500–5,000 kilometers, with no pipelines!"[17] Derezhov went on:

> When the delegation returned to Russia, Kortunov unexpectedly summoned me to his office for a private discussion. Once he had made sure I was on board with his idea, he proposed that I take charge of building the case and doing the numbers—on a confidential basis—for a memorandum to the government, arguing the absolute

> necessity of long-distance gas pipelines from Tyumen Province to Western Europe.[18]

Derezhov was a one-man team on what amounted to an undercover assignment. Years later he referred to it as a "conspiracy."[19] "It was an enormous job," said Derezhov. "I did it partly at work, partly at home on my own time." He then took a draft of the memorandum to Kortunov, who forwarded it to Prime Minister Aleksey Kosygin and other members of the government.

Kortunov's memorandum was roundly opposed, especially by Gosplan and Baibakov. As Kortunov's biographer writes drily, "Kortunov's arguments penetrated the minds of the top political government officials only with the greatest difficulty."[20] The memorandum and calculations from the gas ministry came back again and again with demands for revisions and further peer review. It took Kortunov months of determined lobbying, but step by step he made headway. The breakthrough came when Kosygin, the de facto second in command and the official responsible for the economy, was finally persuaded. In June 1966 he issued a decree endorsing the concept of an export pipeline based on West Siberian gas. It was a decisive conversion, the key to everything that followed. For the next decade and a half, Kosygin was Kortunov's most powerful and reliable backer.

Considered from the broad standpoint of Soviet interests, Kortunov's plan was the perfect win-win situation: it served the development of the West Siberian frontier; brought in Western finance, equipment, and technology; delivered a convenient and clean fuel to domestic industry and consumers, especially in the energy-poor western regions of the Soviet Union; consolidated the Soviet empire in Eastern Europe; and advanced the interests of Soviet diplomacy in Central and Southern Europe. The one leader who was the first to be persuaded and whose support proved crucial—Premier Kosygin—was the most reform-minded and the least ideological, as well as the most practical economic manager of the entire ruling coalition that succeeded Nikita Khrushchev. But the contrast with Khrushchev was striking: whereas Khrushchev was immediately charmed by the propaganda aspect of natural gas—trumpeting it as evidence of Soviet superiority over the United States before the Soviet gas industry even existed—Kosygin was resolutely down-to-earth and cautious.

Kosygin was already familiar with the fast-growing gas industry in Europe. In June 1962, as deputy prime minister, he had made a semi-official visit to Italy and been taken on a grand tour of northern Italian industry by a group of CEOs including Enrico Mattei, the highly entrepreneurial head of ENI. By that time Mattei and ENI were already importing over two million tons of Soviet oil a year, much of it in exchange for Italian oil pipe. But the balance of trade was strongly negative in the Soviets' favor. As Kosygin observed to his hosts, the challenge for the Italians was to find things that the Soviets wanted to buy and credits to pay for them.[21] It would be surprising indeed if the subject of trading pipe for gas did not come up during that visit, although the Soviets' official stance—reflected in a statement from Kortunov in the same month—was that the Soviet Union did not yet have any plans to sell gas to capitalist countries.[22] But the idea had been born, and Mattei's untimely death in a plane crash over Milan only four months later did not hold back ENI's growing interest.

The *Gazoviki* and the Outside World

How much did the first Russian *gazoviki* know about the outside world in the early days of the Soviet gas industry? We have already commented on Kortunov and Kosygin, but they were relative exceptions. The Soviet elite whose members advanced to positions of power in the late 1930s and led the Soviet Union through World War II and the postwar reconstruction was above all made up of technocrats—engineers by training and specialists in narrow technical branches,[23] who then rose by dint of administrative skill and key support from powerful patrons. They had no background in economics, finance, or law, and they needed none. Moreover, until the late 1950s, when the Soviet Union began to open up very slightly to the outside world, they had had no exposure to it, reflecting the extreme isolation of the Soviet system in the Stalin years. Some might remember bits of schoolboy German; virtually none would have studied English.

All contacts with the outside world were closely monitored, and foreign trade was tightly controlled. By the 1970s, as money from oil and gas exports began to flow in, the system loosened a bit. By 1971, an up-and-coming regional official, Mikhail Gorbachev (then Party boss of Stavropol' Province, in the south of Russia) was able to take his wife, Raisa, on a personal vacation to Italy, during which they rented a car and toured the sites like private people—although all at Party expense. But that was highly exceptional: Gorbachev had already been earmarked by the senior leadership and was on a fast track to the top.

For ordinary mortals, even fairly high-ranking ones, access to foreign travel and foreign currency was so tightly controlled that only intervention from the very top could loosen the system even slightly. One episode speaks volumes. Just a year or two after Gorbachev's Italian vacation, a fire broke out in one of the Rolls-Royce compressors powering the export gas pipeline. The gas minister, by then Baibakov's friend Sabit Orudzhev, decided to send his top foreign-relations expert (Derezhov, as it happened) to England to buy a new compressor on an emergency basis. But to authorize the trip on short notice—even to obtain a passport and the authority to negotiate a foreign purchase—Orudzhev had to call Prime Minister Kosygin personally on the closed Kremlin phone line to ask for approval. Kosygin, in turn, made a personal call to the foreign minister, Andrey Gromyko, asking him to issue a passport and the necessary travel orders. The burned-out compressor was quickly replaced.[24]

The only exceptions to this general pattern were the specialists—people whose jobs consisted of working with foreigners in foreign postings. These people were an entirely different sort: spies and counterspies, diplomats, international bankers, foreign trade specialists, plus the odd journalist. They were educated in much the same special schools together, and their specialties frequently overlapped. A KGB operative, for example, might have a cover as a banker or a foreign trade representative, and his day job might be as real as his night job. These were frequently sophisticated people with extensive foreign experience and deep knowledge of their official fields, and they often spoke perfect English. By the Gorbachev years, as the Soviet system loosened further, they were well positioned to parlay their skills into private wealth.

But in the Soviet system these people did not make the decisions: they were on tap but not on top. As Western companies first began discussing energy trade deals with the Russians in the late 1950s, they found that, as a general rule, the more senior the Soviet figure on the other side of the table, the less he tended to know about the outside world, foreign trade, or economics—although his command of science and the engineering details might surpass that of his Western counterpart. As a result, the Westerners sometimes underestimated the people they were dealing with. As a company history of Gasunie describes the first contacts between Shell's gas people and the Russian *gazoviki:*

> When the Russians got into the natural gas game and needed rights for transmission through their satellite countries, they approached [Shell's International Gas Transport Company in The Hague] for advice on setting up a rate structure. . . . The Russians were completely ignorant of business economics and financial figuring Western-style and sat dumbfounded while an American economist gave a course in rates and tariffs.[25]

But the Russians learned fast. The oil trade was an early source of expertise and experience, which then carried over directly into the gas trade. As early as the 1950s the Ministry of Foreign Trade had created a specialized agency for oil trade, Soiuznefteksport. When gas discussions first began between the Soviet Union and Austria, the way was smoothed by the fact that the Austrians had already been importing oil from the Soviet Union and were familiar with Soiuznefteksport and its chairman, Iurii Baranovskii (about whom more below). When the ministry in 1973 created a separate export agency for natural gas, Baranovskii moved over as chairman, where he remained for the next three decades.[26] Over the years, this cast of characters became very familiar to a whole generation of Western businessmen: when the time came to close a deal, the gas minister—first Kortunov and then Orudzhev—sat in the chair and led the final talks, but with Baranovskii and his foreign trade colleagues sitting at their side, having done the homework.[27]

This was the background, then, that the Soviet officials brought with them to Schloss Hernstein. The story of West Siberian gas and the politics behind it were unknown to the Western Europeans. Indeed, it is not certain how many of the Soviet delegates knew much of it. But what was striking to the Europeans was the conviction of the Soviet gas men that they were onto something important.

The Significance of Austria: The Logic of Small Beginnings

The volumes were small, but their significance was very large. The role of Austria as the pathbreaker was crucial, because it showed for the first time—as a custom-built prototype, as it were—that gas exports across the Iron Curtain, despite ideological differences and recurring diplomatic tensions, could be conducted reliably on straightforward commercial principles. Not the least important result was the beginning of an informal community of people (gas producers and shippers, foreign trade officials, steelmakers, and bankers) who came to know and trust one another as the gas bridge expanded over the following decades.

But why Austria? Austria brought together a unique combination of favorable features.[28] To begin with, Austria was neutral, and as such it had already been a bridge between East and West for nearly two decades. More than any other European country, Austria had extensive prior experience in dealing with the other side of the Iron Curtain. OMV, the Austrian state-owned monopoly oil and gas company that had been created by the Nazis after the *Anschluss* (the Nazi annexation of Austria in 1938) had been taken over by the Soviets in 1945 and managed by Russian personnel until 1955, the year of the conclusion of the Austrian State Treaty—at which point the Russians went home, and the company was nationalized by the Austrian government. Yet the legacy of this period was that the Austrians, more than any other Westerners at the time, were familiar with the Eastern bloc and comfortable in dealing with it. OMV had particularly close ties with its counterpart in neighboring Czechoslovakia. A sizable gas field called Zwerndorf happened to lie across the common border. The two companies formed a joint venture to operate it together, and through an ingenious swap arrangement, part of the Czechoslovak production was credited as an export to Austria.

Thus, OMV knew the Soviet bloc well and kept close tabs on developments there. As Austria's own modest gas reserves declined and it became clear that domestic production could not meet the country's growing demand for gas, the Austrians followed with interest Soviet plans to build a pipeline to Eastern Europe, the so-called Bratstvo (Brotherhood) line. The Czechoslovak terminus of the line in Bratislava lay only fifty-five kilometers from Vienna, and the two countries were soon holding talks about building a connector across the border. In addition, OMV had one more powerful motive to look east: gas distributors inside Austria, which had hitherto been OMV's customers but had long chafed under its monopoly, were starting to look at alternative sources—Dutch gas, in particular, which by this time had reached northern Germany. OMV feared it would lose control if it did not act quickly.

Business was business, but Cold War politics were never far away. Commercial motives interacted with diplomatic considerations to put Austria at center stage. The Austrian State Treaty required that Austria remain neutral, and the Soviets wanted to keep it that way. In the mid-1960s they watched uneasily as Austria increasingly turned toward the emerging European Economic Community (EEC). The Kremlin saw gas exports to Austria as one way of countering the growing attraction of the EEC; and as plans for exports to Italy and France developed, Austria's very neutrality made it potentially attractive as a hub. The fact that such geostrategic issues were on the minds of the top Soviet leaders could be seen from the fact that Nikolai Podgorny, then the head of the Soviet state, led the Soviet delegation that traveled to Austria in November 1966 and announced the Soviet plan for an export deal to Italy with Austria as the transit country.[29]

But while a senior figure like Podgorny or the Soviet foreign trade chief Nikolay Patolichev[30] might represent be the official face of Soviet policy, negotiating and closing actual deals required teams of specialized experts. In the mid-1960s the gas industry was still new to all the players, but especially to the Soviets. Yet by the time the Soviet and Austrian delegations came together at Schloss Hernstein, a core of Soviet expertise (the Soviets would have said a "cadre") had already begun to emerge. Who were the Soviets experts who came to Schloss Hernstein, and what did they bring to the table?

Organizing for Gas Exports: Building the Team, Learning the Language of Gas

A stock theme in the history of Soviet foreign economic policy was the fraught relationship between two very different groups of people, the foreign trade experts in the Ministry of Foreign Trade and the technical specialists in the industrial ministries. The Ministry of Foreign Trade had sole authority to negotiate foreign trade contracts, but the orders for foreign goods ultimately came from industrial ministries and enterprises. Thus, the two groups were bound together like Siamese twins. Yet they represented two very different cultures—the one commercial and international, the other focused on technical issues and internal production plans—and they were frequently at odds.

In the case of gas, there were additional complications. The gas ministry was interested not solely in exporting gas but also in importing a wide range of machinery and technology, but these were the responsibility of several different specialized industrial ministries and foreign trade organizations, each with its own personnel and subculture. For example, there was a foreign trade organization for machinery, called Machinoimport. If credits were involved, yet another bureaucracy came into play—the Foreign Economic Bank, a subsidiary of the USSR State Bank. Finally, inside the Soviet bloc different rules and players prevailed. East Germany, for example, was treated as an extension of the Soviet Union, in the sense that trade relations were conducted by the Soviet planning authorities, not the Ministry of Foreign Trade. As a result, when it came to foreign trade, the usual problems of coordination on the Soviet side were compounded, and the result at times was cacophony, as the Europeans soon became aware.

The Soviet gas industry had only recently grown out of the oil industry, and the only foreign trade group with even remotely relevant expertise was Soiuznefteksport, which handled oil exports. As it became clear that the initial talks with the Europeans would soon lead to actual contracts, more specialized gas expertise was needed. A "gas office" *(gazovaia kontora)* was created within the oil export agency Soiuznefteeksport, and

Baranovskii, an experienced oil trader, was named to head it. Like most of the early Soviet negotiators, Baranovskii was originally a petroleum engineer. He was a graduate of the prestigious Gubkin Institute of Oil and Gas, the leading technical university in the field, with a specialty in oil refining. When the Soviet Union resumed oil exports in 1958, Baranovskii was recruited into oil trading, and for the next several years he negotiated oil export contracts—notably with England and Japan, where he was posted. By the time of the first gas talks, Baranovskii was in his mid-thirties and already had nearly a decade of sophisticated experience in the West and foreign oil deals, but he had no background in gas.

Over the following decades there was a thinly veiled rivalry between the *gazoviki* (the gas professionals from the gas ministry) and the foreign trade specialists. Derezhov headed the former, Baranovskii the latter. The first group was made up mainly of engineers, for the most part specialists in pipeline construction; the second group was made up of traders, originating primarily from backgrounds in the oil trade. Although members of both groups were frequently alumni of the same institute—particularly the Gubkin Institute, which was known irreverently to generations of graduates as Kerosinka ("Kerosene U")—their subsequent careers diverged as they joined the two parallel bureaucracies.

Between the two sides there was a constant jockeying for power. Over the course of three decades the Ministry of Gas gradually gained the upper hand, but the struggle did not finally end until the late 1980s, when Gorbachev broke up the Ministry of Foreign Trade and abolished its monopoly—and the Ministry of Gas (by this time renamed Gazprom, as a "state concern") swooped in and swallowed its rival. Gazeksport, under the name of Gazexport, the descendant of Soiuzgazeksport, is now firmly part of Gazprom. Yet to this day there is a noticeable difference in culture and considerable internal rivalry.

But all that came later. The immediate challenge for the Soviets, once Kosygin's June 1966 decree endorsed the principle of gas exports based on West Siberian gas, was to find the right people to carry it out. Immediately following the June 1966 decree an interagency team (the Russians called it a *brigada*) was formed to make first contact with potential European partners, consisting of representatives from the

Ministry of Foreign Trade, Gosplan, the Foreign Trade Bank (Vneshekonombank), and the Ministry of Gas under the leadership of Osipov, deputy minister of foreign trade. The gas ministry's response—typical of Kortunov's drive to create a vertically integrated structure that could handle every aspect of the business—was to create its own international relations department, which paralleled the oil-trading arm of the Ministry of Foreign Trade. Kortunov put Derezhov in charge, and Derezhov recruited a team of young engineers who became the core of the future Gazprom's expertise in foreign trade and relations.

Interestingly, although Kortunov's first overtures had been to the Italians, the Soviet delegation began its first European road trip by approaching the German gas company Ruhrgas. Soviet-German relations were still frosty, and Ruhrgas's reaction, although one of curiosity, was understandably cautious. As Derezhov relates:

> At first their representatives viewed us with mistrust. They were not even willing to meet us in Germany. So we suggested they meet us in the Austrian capital, Vienna, in the office of the Soviet trade representative. The chief executive of Ruhrgas, Mr. Schelberger, came, together with some of his aides. The Soviet side offered to sell 5 Bcm [billion cubic meters] a year, but the cautious Germans in the end agreed only to 3 Bcm.[31]

At Schloss Hernstein, two people were key: Nikolai Osipov, the deputy minister of foreign trade, and Aleksey Sorokin, the deputy minister of gas and a close associate of the gas minister, Kortunov. Osipov was the foreign trade man; Sorokin represented Kortunov. Osipov was already an old hand in the Soviet foreign trade system, with experience in negotiating oil exports. Newly promoted to deputy minister the year before, Osipov continued to negotiate oil-export deals and led every Soviet gas negotiating team over the following twenty years, until the final breakup of the Ministry of Foreign Trade in 1987.[32]

Sorokin's role was somewhat different. A rotund, cheerful man, Sorokin had been an unofficial goodwill ambassador for Soviet gas ever since the late 1950s. He first appeared on the international scene shortly after

Glavgaz joined the International Gas Union, sometime after 1957. He typically led Glavgaz's international delegations in those years, as the head of Glavgaz's international relations department. In 1961 Sorokin was elected chairman of the union.

By all accounts, whatever tensions there were between their two houses, Osipov and Sorokin got on well and formed a smooth team. But as negotiations grew in scale and complexity in the early 1970s, Sorokin faded from view, and a new generation of *gazoviki,* led by Derezhov, came to the fore.[33] At the same time, there was a greater effort on the Soviet side to coordinate the different moving parts—the gas, pipe, compressors, and finance. The coordination came from the top, and particularly from the quartet of Osipov, Baranovskii, and the representatives of Machinoimport and Vneshekonombank, Stanislav Volchkov and a foreign trade banker named Ivanov,[34] under the overall leadership of Osipov—who reported directly to Patolichev, the minister of foreign trade.[35] This team managed the whole process for over a quarter-century.

The meeting at Schloss Hernstein did not yet close the deal. The Soviets and the Austrians were far apart on price, and talks went on through the autumn and winter of 1967. One difficulty was that they did not yet have a common language for key commercial concepts. The notion of the "netback price," based on the cost of competing fuels in Austria, had not yet been perfected. Instead, they traded reference prices based on what was known about other deals at the time, such as "Dutch prices plus transportation costs to Austria." The principle of *Anlegbarkeit* (a price based on alternative fuels at the destination) had not yet been accepted by the two sides as a workable convention. In the end, the Austrians, under pressure to close the deal, finally accepted a price that was far higher than they had wanted and probably considerably higher than the oil-referenced netback price they might have gotten a few years later.[36] But once agreement was reached, the deal was duly signed in June 1968, and deliveries began on schedule on September 1, 1968—with Kortunov and Ludwig Weiss, the Austrian minister of transportation, cordially presiding. It is striking that this occurred despite the invasion of Czechoslovakia by the Warsaw Pact forces ten days earlier. Thus was the East-West bridge born, Cold War or no.[37]

The Emerging Roles of the Two Germanies

Meanwhile, what of Germany? West German steel companies were present as interested observers at Schloss Hernstein, and the steel for the pipe in the deal came largely from them. Thus, one could say that the West Germans were involved from the beginning. But the first Germans to sign an actual gas contract with the Soviet Union were not West, but East, Germans.

Controlling one's puppets is always a challenge. Relations between Moscow and the German Democratic Republic (GDR) were notoriously difficult and were not helped by what was perceived by the Russians as East German arrogance and unreasonable demands for economic assistance. Relations between Khrushchev and Walter Ulbricht, the Party boss in East Germany, were especially bad.

Ulbricht constantly demanded large-scale help from the USSR. The gap between the Federal Republic of Germany (FRG) and the GDR was widening, he said. GDP growth in 1960 was 12–13 percent in the FRG and only 8 percent in the GDR. The GDR needed raw materials, semifinished goods, and basic foodstuffs, and was beginning to import them from the FRG. Khrushchev seems not to have realized how dependent the GDR had become on West Germany. "We didn't know that the GDR was so vulnerable to West Germany," he told Ulbricht.[38] Already the Russians' exasperation with the East Germans is evident. As Khrushchev said to Ulbricht in the same conversation, "You should have learned how to walk on your own two feet, instead of leaning on us all the time."[39]

Ulbricht constantly reminded the Soviets that one important reason why the GDR was lagging behind economically was the Soviets' punitive policy of reparations between 1945 and 1954. In a letter to Khrushchev in January 1961, Ulbricht wrote: "While in the first ten postwar years we paid reparations by the withdrawal of existing plants and from current production, West Germany . . . received large credits from the USA to save the monopoly capitalist system and German militarism. . . . This is the main reason that we have remained so far behind West Germany in labor productivity and standard of living."[40]

The GDR kept raising its demands for Soviet aid, particularly raw materials and oil, while the Soviets resisted. Ulbricht bombarded

Khrushchev with letters and memoranda, which grew increasingly bitter and sarcastic as time went on. Natural gas featured prominently from an early date: in 1964 Ulbricht demanded that Khrushchev commit to building a pipeline to the GDR and to begin gas deliveries by 1969. In a tense meeting with Soviet economic officials, Ulbricht charged that the Soviets were already building a gas pipeline to Poland and Czechoslovakia—why not build one to the GDR? The Soviets replied that this was only an expansion of the line that had been built before the war by a German company. To which Ulbricht retorted, only half in jest: "What you're talking about is the old pipeline that the Nazis built. The subject here is [your new] pipeline to Silesia. We'll sniff the gas at the border; we'll learn about it from the smell. The only thing that interests us is the answer to the question, 'Is the Soviet Union supplying gas?'"[41]

Yet as acerbic as their relationship was, Khrushchev in the end was prepared to go along with Ulbricht's demands, because he believed that the prestige of the socialist cause depended on helping the GDR thrive. As Khrushchev wrote in a memo to the Politburo:

> The GDR's needs are also our needs. We can't permit it that they come to us in such a state that either they sink or we throw them a rope. Let's stop playing games about this question. . . . We cannot be blind money-counters and every time construct our trade around whether to give or not to give 1,000 tons more. Malenkov and Beria wanted to liquidate the GDR, but we fired one and shot the other and said that we supported a socialist Germany.[42]

But there were limits to Khrushchev's generosity. In one conversation, when Ulbricht asked for gold, Khrushchev exploded: "You ask for 68 tons of gold. This is inconceivable. We can't have a situation where you buy goods and we must pay for them. We don't have much gold, and we must keep it for an emergency."[43]

The essence of the Soviet view of the importance of the GDR was expressed by Anastas Mikoyan, then deputy premier, in a closed conversation with Bruno Leuschner, the GDR planning chief, during his visit to Berlin in June 1961:

> The GDR, Germany, is the country in which it must be decided that Marxism-Leninism is correct, that communism is also the higher, better, social order for industrial states. Marxism was born in Germany and it must prove its correctness and value here in a highly developed industrial state. We must do everything so that your development constantly and steadily goes forward. You cannot do this alone. The Soviet Union must and will help with this. . . . If socialism does not win in the GDR, if communism does not prove itself as superior and vital here, then we have not won.[44]

As Khrushchev wrote in his memoir, *Khrushchev Remembers: The Glasnost Tapes,* "It was my dream to create such conditions in Germany that the GDR would become a showcase of moral, political, and material achievement—all attractively displayed for the Western world to see and admire."[45]

But Khrushchev's apparent favoritism toward the GDR, not surprisingly, exasperated the other satellite countries. To Khrushchev's evident frustration, the East European socialist leaders delivered a battery of polite yet firm and detailed protests, citing their own economic problems.

The fall of Khrushchev marked a turning point. By the mid-1960s, his successors had become impatient not just with Ulbricht, but with the East Germans in general. As Leonid Brezhnev put it to Erich Honecker, shortly before the two of them engineered the ouster of Ulbricht in 1971: "There is a certain superiority of [the East Germans] with regard to the other socialist countries, your experiences, methods of leadership, etc. This is also directed toward us. This also upsets us. This must be changed. The SED [Sozialistische Einheitspartei Deutschlands] Politburo, you, must change this."[46]

But it did not change, even after Honecker replaced Ulbricht. And as the 1960s went on, the increasingly exasperated Soviet leaders grew tougher on the subject of economic aid to the GDR. From the moment they took over in 1964, Brezhnev and his team were less tolerant, but it was especially Kosygin and Baibakov who were inclined toward a hard line. Whereas Khrushchev still hoped that East Germany could be a showcase, his successors became increasingly resigned to the fact that East

Germany was underperforming relative to the West and probably always would.

By the late 1960s, for the Soviets West Germany increasingly became the key and the prize—for pipe, compressors, finance, and trade generally. The socialist countries were welcome to gas at subsidized prices, but only if they pitched in. Getting the East Germans to cooperate, however, turned out to be difficult. They did not like taking gas for oil, they did not like building pipelines for it, and above all they did not like the West Germans receiving Soviet gas.

East German Prelude

Beginning in 1968, the Soviets began proposing gas exports to Eastern Europe as a partial substitute for oil. The East Europeans did not take this well: what they wanted was oil, not gas. But the Soviets were increasingly insistent. At the beginning of 1968—a year before the first serious gas discussions between the Soviets and the West Germans but in the middle of the talks with the Austrians—the Soviets began talks with the East Germans and the Poles about running a gas pipeline along the same route as the existing "Friendship" oil pipeline that ran through Poland to East Germany. The Soviets initially proposed a small line of only 600 millimeters in diameter, which would have been adequate for only 3 billion cubic meters of gas per year. But as the talks proceeded, the conversation shifted to a line 900 millimeters in diameter, large enough to carry up to 7 billion cubic meters per year. The Soviets quickly endorsed the idea of the larger diameter, suggesting that what they had in mind could also be used for exports of Soviet gas to West Germany, including West Berlin.[47]

The idea of supplying gas to West Germany—let alone West Berlin—brought fierce opposition from the East Germans, and especially from Ulbricht. At this time relations between the two Germanies were still in a deep freeze. In the end, the Soviet and East German negotiators could agree only to postpone the issue. Thus, the first gas contract between the Soviet Union and the GDR, signed in May 1968, provided for only 3 billion cubic meters per year (although the Soviets insisted on the larger diameter pipe, suggesting that they still had the West in mind as the ultimate market). Since the agreement was a "certain compensation" for the fact that the Soviets were not shipping as much oil as the GDR would have liked, the Soviets dropped their initial demand that the East Germans help finance the pipeline and agreed to take the equivalent in industrial and consumer goods instead.[48]

This early episode shows that the Soviets already had a larger goal in mind—to transit gas through the satellite countries to Western Europe. This became explicit later in 1968 when Baibakov, the head of Gosplan, met with his East German counterpart, Gerhard Schürer, and insisted to him in no uncertain terms that in any future gas contracts the GDR would have to contribute to the costs of gas development. But, Baibakov went on, West German pipe suppliers would play an important part. Future gas exports to West Germany and France, he assured Schürer, would include branching off some of the gas to the GDR. The East Germans remained unenthusiastic, to say the least. Schürer recorded that he didn't go further with this subject with Baibakov, except to reassert the East German position that they could not replace oil with gas, or at least only in limited quantity.[49]

Almost immediately problems cropped up between the East Germans and the Poles. It had originally been agreed that the GDR would grant credits to the Poles for the construction of the Polish segment of the pipeline. The East Germans promised to deliver 725 kilometers of pipe, compressors, and all ancillary equipment up to a value of one billion *valutamarken* (hard-currency marks). The actual construction work would be done by Polish enterprises.[50] The heads of the Joint Polish-German Economic Commission signed a *Generalschema* in May 1969, on the strength of which the East Germans made initial deliveries of ten million transferable rubles' worth of pipe and machinery, even though no actual contract had yet been signed.

But from the first there were disagreements over the size and terms of the credits, and above all over transit tariffs. The two East European leaders, Ulbricht and Wladyslaw Gomulka, had agreed verbally that the transit tariff would be set according to "global market criteria," but the two sides could not agree on what those were. The Poles wanted 3.65 rubles per thousand cubic meters, while the Germans refused to go above 2.02.[51] By the end of 1969 this had led to a breakdown of the project.

In July 1970 the East Germans and the Poles canceled their plan. The East Germans paid a one-time settlement to the Poles for the work already done by the latter and took back the pipe and equipment that had already been delivered to Poland. It was not until the 1990s, a generation later, that a gas pipeline through Poland finally reached Germany, and when it did, it was the work of a joint venture that would have seemed inconceivable in the late 1960s—between a Russian company (Gazprom) and a leading private-sector player in a now-reunited Germany (Wintershall).

But back to 1970: as the East German–Polish project ground to a halt, the Soviets made it clear that they would not allow the setback to stand in the way of their immediate goal, to sell gas to the West, and now they had an alternative route in prospect, through Czechoslovakia. In February 1970 the Soviets informed the East Germans that they were welcome to join in the construction of the pipeline through Czechoslovakia and advised them not to agree to the Poles' demands. The East Germans hurriedly began talks with the Czechoslovaks. They met with a friendly reception, and the transit tariffs through Czechoslovakia were only half of those the Poles were demanding. In July 1971 the GDR and Czechoslovak governments signed an intergovernmental agreement to cooperate in building a 1,000-kilometer line from the Soviet-Czechoslovak border to the GDR border. The Germans agreed to supply pipe and equipment, as well as a trade credit of 115 million transferable rubles. Construction began in September 1972, and the first Soviet gas flowed in May 1973. Thus, Czechoslovakia became the main route for the export of Soviet gas to both halves of Germany.[52] But by this time the West German market was clearly the one the Soviets were after, and the East Germans had to resign themselves to playing second fiddle.

Willy Brandt and Soviet Gas to West Germany

Meanwhile, major changes were under way in West Germany. In 1966 a charismatic politician, Willy Brandt, became foreign minister of the FRG. He launched a radical new policy toward the Soviet Union, which he dubbed *Ostpolitik*, the "Eastern policy." At its center was trade in gas.

Willy Brandt was mayor of West Berlin for nearly a decade, from 1957 to 1966, twice as long as his tenure as chancellor.[53] He has gone down in history as the architect of *Ostpolitik.* But long before that, as mayor of West Berlin, he was the symbol of the successful defense of the city as the outpost of the West behind the Iron Curtain. In November 1958, one month after Brandt was named mayor, Khrushchev delivered an ultimatum: he gave the Allies six months to terminate the occupation and leave the city. Over the following four years, to the building of the Berlin Wall in 1961 and the tense confrontation that followed, Brandt was the visible face of the West's determination to resist Khrushchev's demands.

Brandt quickly became an international figure. He was perfectly cast for the part: his tall and athletic physique, strong jaw, and broad face made him the very portrait of tough determination. In those years, he was known as a hard-liner, as far to the right as you could get and still be a member of the Social Democratic Party (SPD). He was a favorite of Axel Springer, the powerful conservative publisher of the daily *Die Welt* and the popular tabloid *Bild Zeitung.* In May 1959 Brandt posed alongside Springer at the ground breaking for the new Berlin headquarters of the Springer Group on the Kochstrasse, located in a deliberate provocation at the edge of the Soviet zone. Springer controlled 80 percent of the press in West Berlin and did much to promote Brandt's image as a future chancellor.[54] It came as something of a surprise, therefore—and not least to Springer, who subsequently turned bitterly against Brandt—that as early as 1963 the mayor of West Berlin began to speak in private of détente.

West Berlin prospered under Mayor Brandt. After 1958 the economy of West Berlin grew faster than that of the rest of West Germany (although mainly, to be sure, because it was starting from a lower point). It became a visible symbol of West Germany's strength and East Germany's weakness. In 1964, toward the end of Brandt's tenure as mayor, 89 percent of West Berliners polled pronounced themselves satisfied with both the man and his policies. In his memoirs, Brandt commented that West Berlin was the place where he had "learned his trade" as a politician.[55] Yet much of the city's economic success had little to do with Brandt's management. Revealingly, in his memoirs Brandt devotes almost no space to his day-to-day running of the city. Brandt was a rising star who spent more and more time on the national and international stages and less and less in Berlin. As early as May 1959, Brandt had made the cover of *Time* magazine, the first of several covers to come.

He had little control over the city's revenues, much of which came from subsidies from Bonn[56] and went to support the construction of desperately needed housing and essential services such as water and sewage. Twenty thousand new dwellings opened each year during Brandt's tenure as mayor; new hotels and office blocks appeared; and portions of a new inner-ring highway, the *Bundesautobahn 100,* known to two generations of Berliners as the *Stadtring* or more often as the infamous "BAB 100" because of its traffic jams, were inaugurated under Mayor Brandt, paralleling the tracks of the existing S-Bahn. Yet West Berlin did not receive natural gas until many years after Brandt had left. The city remained dependent on prewar coal gas and increasingly on imported oil, supplied through pipelines built by the Allied powers. The city was chronically short of energy. One resident recalls that in her childhood, "during the war and after, the trees in the Tiergarten and elsewhere had been cut down for firewood; fuel for heat was short, even into the 1960s. One of the biggest items in the cargo holds of the Allied planes during the Berlin airlift was coal."[57]

But from the moment Brandt reached the national stage as foreign minister in 1966, he took an active interest in the possibilities of rapprochement, and natural gas soon played a central role. Ironically, a key turning point in West German policy was the Soviet intervention in Czechoslovakia in August 1968, which stimulated governments on both sides to seek new avenues for détente. As Angela Stent writes in her history of Soviet-German relations, "The invasion emphasized for Brezhnev the need to secure an agreement with the West that would recognize the legitimacy of Soviet influence in Eastern Europe and thereby lessen the prospect of another Czechoslovakia." Interest on the West German side was equally strong. As Stent comments,

> Brandt's Ostpolitik was primarily *defensive,* "to maintain the substance of the nation," that is, the common ties that existed between the two Germanies. The key determinant of this Ostpolitik was the desire to improve intra-German relations. . . . Bonn had learned the main lesson of the Czechoslovak invasion, namely, that the key to any future settlement of the German issue lay with Moscow and with no one else.[58]

Natural gas was a perfect candidate for both sides, as an economic commodity that was clearly beneficial to both and served as a symbol of détente. The first expressions of interest came from the Soviets, as Kortunov's top deputy, Sorokin, approached a German delegation at the International Gas Union's annual meeting shortly after the Czechoslovak intervention. The Germans reciprocated, initially cautiously but soon with increasing enthusiasm. Progress from that point was remarkably rapid. The key breakthrough in the talks came in November 1969, only just a year and a half after the first informal conversations.

In retrospect, the negotiations were noteworthy for two things. The first was the increasingly strong role played by the West German government as the talks went on. The turning point was the parliamentary election of September 1969, which led, for the first time, to a coalition headed by the SPD without the participation of the Christian Democratic Union, which had been the senior member of every coalition in the past. As soon as Brandt became chancellor, he began exploring the prospects for an intergovernmental treaty with the Soviet Union, in which natural gas would play a central role. As talks between the governments progressed, the West German ministries of economic and foreign affairs applied growing pressure on the gas negotiators to conclude an agreement.

Second, most of the disagreements that repeatedly threatened to derail the talks had to do with differences over price, not over politics. The Soviets and the West Germans put forward several different concepts along the way, and the result was a joint process of exploration and learning in a business that was still relatively new to both. There were at least five concepts in play: (1) the prices of other contracts and other ongoing negotiations elsewhere and broad price trends in the gas industry, as filtered through changes in exchange rates; (2) the concept of the marginal cost of competing fuels; (3) the costs of transit and pipeline construction; (4) anticipated future levels of gas demand; and (5) more subtle factors affecting cost, such as chemical composition, reliability, and seasonal variation. The latter, in particular, led to discussions of various levels of flexibility around the buyers' take or pay commitment.[59] Price, in other words, was the place where all the swirling elements came together.

During this period the price of Dutch gas was falling as new reserves were being discovered, and the West German negotiators hoped to take advantage of this point. Two men—Herbert Schelberger, the chairman of Ruhrgas, and Osipov, the Soviet deputy minister of foreign trade—played the central roles as the two sides maneuvered to reach a basis for the gas price. Along the way, for the first time the principle of *Anlegbarkeit* made its appearance, as Schelberger successfully argued that the prices of competing fuels in the north German market would determine the price at which Ruhrgas would be able to sell the Soviet gas. In the end, the breakthrough—after a standoff that lasted nearly a year—came when Osipov finally accepted the Dutch border price as the price standard. By this time, the political authorities of both countries were practically shouting for the gas talks to be concluded so that the final arrangements for the larger comprehensive treaty could proceed.[60]

The first Soviet–West German gas deal, like the previous Soviet-Austrian deal, was noteworthy for the role played by key individuals. The gas deal itself would not have happened without the expert knowledge and skill of the Soviet and West German negotiators who put it together. Schelberger, in particular, was just the first of a long line of chairmen of Ruhrgas who acted as overseers for the Russian-German gas relationship down through the years.[61] Government officials played key roles as well. Norbert Plesser, a gas expert from the Federal Economics Ministry, was officially only an observer, but he spoke up at a key moment when the two sides were deadlocked over price. He urged the Soviets to accept the fact that Dutch gas prices were heading downward, and that if the Soviets wanted their gas to be competitive in Germany, they needed to price their own gas accordingly. The argument proved decisive with Osipov, and it led directly to the final breakthrough in the talks.[62] Other leading officials of the German economic and foreign ministries played significant roles as well. As the talks proceeded in the summer and autumn of 1969, there were side channels and back channels as various officials met to emphasize the keen interest of both sides in reaching an agreement. There was a growing buzz of activity and messages, official and unofficial. The commercial and diplomatic efforts advanced in lockstep.

But without Brandt and his key foreign-policy advisor, Egon Bahr, the gas deal would not have happened. *Ostpolitik* remained the foundation of West German policy over the following decades (an updated variation on this has been a guiding policy of a reunited Germany ever since), and it provided the all-important political framework for the gas bridge. It was no accident that signing the gas contract was the first major event in Brandt's *Ostpolitik,* a key starting point in the negotiations between Bahr and his Soviet counterpart, Andrei Gromyko, over the renunciation of force. As Stent observes in her landmark study, "The conclusion of the deal under favorable economic conditions for the Soviets . . . was a factor that contributed to the improvement of German-Soviet political relations."[63]

The Role of *Ostpolitik* as Catalyst

It was a moment of high drama and ceremony. On Sunday, February 1, 1970, the first gas export contract between the Soviet Union and West Germany was signed at the historic Hotel Kaiserhof, a former Capucin convent in Essen, the home city of Ruhrgas.[64] The mood was festive as a large group of Russian and German government and company officials celebrated the outcome of nine months of intense negotiations. Presiding over the ceremony were Karl Schiller, the West German minister of the economy and a close ally of Brandt, the newly named chancellor; and Patolichev, the veteran Soviet minister of foreign trade. But the stars of the day were the three men who had conducted most of the negotiations: the duo of Osipov and Sorokin, who had led the Soviet campaign for gas since its beginning in the mid-1960s and who had become such familiar figures on the European gas scene that they were known as Castor and Pollux,[65] and Schelberger, the president of Ruhrgas and leader of the German side. Schelberger, savoring the moment on his own home ground, gave the welcoming address. But it was Osipov who captured the symbolic importance of the moment, when he stressed the wider economic implications and, above all, the political significance of the contract:

> The concluded agreement . . . is a good example of economic cooperation between Europe's countries. This example also indicates that other economic problems in Europe can be successfully solved, under the condition that all European countries manifest their good will. . . . I can add, gentlemen, that what I have said also fully concerns the

> solution of all political issues in Europe. It is precisely from this perspective that we view the signed contract.[66]

The Soviet–West German gas agreement of February 1970 marked the cornerstone, and the single most important moment in the creation, of the gas bridge. The Soviet-Austrian contract of 1968 had shown that an East-West gas deal was feasible, but what had made it possible was the exceptional circumstances of the Austro-Czechoslovak relationship—in effect, a loophole. But if the Soviet-Austrian deal was a loophole, the Soviet–West German contract was a breakthrough, the beginning of a breach in the Iron Curtain that would only grow larger over the following decades. To this day, Germany remains the largest buyer of Russian gas in Europe (see Table 2.1) and one of Russia's most important trade partners, in a many-sided commercial and political relationship in which gas remains the bedrock.

A sign of the high political importance attached to the deal by the German side—and of the key importance of the steel component of the deal—was the generous credit terms extended by the government: 85 percent of the value of the German pipe and equipment exports, with the federal government guaranteeing half of the total. This was much criticized by other EEC member governments.[67] The German opposition also perceived the deal as too generous to the Soviets and "demanded to know what political price the Soviets would pay for such a generous economic deal."[68]

At its outset in the mid-1960s, the commercial case for importing Soviet gas to Germany (at least on the German side) had seemed less than compelling. The role of natural gas in the West German economy was still modest, and its future needs appeared amply covered by Dutch gas. Only in south Germany, particularly Bavaria, was there a perceived urgent need for additional gas supplies, but alternative sources, notably Algeria, seemed available. Moreover, gas imports from the Soviet Union were strongly opposed by powerful business interests—in particular, the north German coal industry and the major oil companies, Esso and Shell, which were the chief producers of Dutch gas. Finally, Ruhrgas itself was initially reluctant. As Per Högselius recounts in detail, only the threat of a separate deal by Bavaria and the strong urging of the government caused Ruhrgas to move.[69]

Table 2.1 Soviet natural gas exports by country, 1970–1990 (billions of cubic meters)

	1970	1975	1980	1985	1990
USSR total	3.3	19.4	55.0	69.8	109.0
CMEA or former CMEA	2.4	11.4	31.4	38.3	46.1
Czechoslovakia	1.4	3.8	8.3	10.8	12.6
Poland	1.0	2.5	5.3	5.9	8.4
Romania	—[a]	—[a]	1.6	1.8	7.3
Bulgaria	—[a]	1.2	3.8	5.5	6.8
Hungary	—[a]	0.6	3.8	4.0	6.5
Yugoslavia	—[a]	—[a]	2.0	4.1	4.5
East Germany	—[a]	3.3	6.4	6.2	—[a]
Other	1.0	8.0	23.6	31.6	63.0
West Germany[b] (and reunified Germany in 1990)	—[a]	3.1	9.3	13.4	26.6
Italy	—[a]	2.3	6.5	6.0	14.3
France	—[a]	—[a]	3.8	6.8	10.6
Austria	1.0	1.9	2.7	4.5	5.1
Turkey	—[a]	—[a]	—[a]	—[a]	3.3
Finland	—[a]	0.7	1.2	0.8	2.7
Switzerland	—[a]	—[a]	—[a]	—[a]	0.3

Data sources: Matthew J. Sagers et al. 1988. "Prospects for Soviet Gas Exports: Opportunities and Constraints." *Soviet Geography*, Vol. XXIX, No. 10, p. 885 (for 1970–1985); "News Notes," *Soviet Geography*, November 1987, p. 697; *Vneshnyaya torgovlya SSSR 1987* and other selected years; *Soviet and East European Energy Databank* (PlanEcon, Inc., 1987); *Vneshniye ekonomicheskiye svyazi SSSR v 1990 g.* (for 1990).

Notes: Numbers may not sum to totals due to rounding. The table uses the Russian measure of 8,850 kilocalories per cubic meter (in gross calorific value). CMEA is the Council for Mutual Economic Assistance (Comecon).

[a]Neglible or none.

[b]Includes West Berlin.

Yet even as the talks proceeded, the energy balance was shifting rapidly, as gas demand boomed in Germany in the late 1960s. Having contracted to buy Dutch gas and laid a pipeline from the Netherlands to Germany—the so-called Brummer line—Thyssengas and Ruhrgas now had to sell it to German households. They pulled out all the stops in a massive publicity campaign. An advertising company was hired to create a cute mascot, "Trixie Sunshine," who became the symbol of the new fuel.[70] The campaign had its comical moments: one slogan ran, "Soon it will not be the gas man who is kissing your wife, but the natural-gas man."[71] But it worked, if not with German housewives, at least with local gas distributors. By the end of 1968 over half of the thirty-five local gas companies supplied by Thyssengas had been converted to natural gas. Even RWE, the dominant supplier of coking gas, found itself being rapidly squeezed out of its traditional market in the Ruhr.[72] At this point—on the eve of the first German-Soviet gas negotiations—demand for natural gas in Germany was growing at 30–60 percent a year.[73]

Once the 1970 contract was concluded, the direct involvement of the German government as a promoter and facilitator mostly ended. (In Chancellor Helmut Schmidt's memoirs, there is only one passing reference to German-Soviet energy trade and pipelines as a subject of conversation between him and Brezhnev.)[74] There were several reasons for this. First, German-Soviet relations entered a period of relative quiescence, as though both sides had achieved from *Ostpolitik* what they wanted. Second, the government team that had played such a key role dispersed, even before the end of the Brandt chancellorship in 1974. Third, the oil shock of 1973–1974 suddenly changed the whole context of energy policy, making it clear that, whatever the other consequences of the oil shock might be, Soviet gas now had an essential role to play in the European portfolio.[75]

As for West Berlin, it had to wait until the mid-1980s before it received natural gas—again by way of the gas bridge from the Soviet Union. Ruhrgas decided to deal directly with Soiuzgazeksport, signaling its interest in importing Soviet gas for West Berlin. According to Ruhrgas chairman Klaus Liesen, the East German authorities were reluctant, and the Soviets basically imposed a decision, while Ruhrgas acted as a mediator. The deal ended up being negotiated on the East German side by the Commercial Coordination Department (known as Ko-Ko) of the GDR Foreign Trade Ministry, led by the notorious Alexander Schalck-Golodkowski, who was in charge of the many shadowy arrangements under which the West Germans financed East German exports (including traded spies and refugees) for hard currency and traded in embargoed goods.

VEB Gaskombinat, the East German gas company, built a new pipeline from the Czechoslovak border to the city limits of West Berlin, but at this point came the hardest part of the project—the actual crossing into the West, which required taking down a twenty-meter segment of the Berlin Wall for several days and entrusting the job to a specially cleared team of workers—the so-called Travel Cadres Category I, the highest classification of reliable workers authorized for foreign travel. The pipeline was finally inaugurated in October 1985. The Soviet and East German sides were discreetly absent from the ceremonies, since it was a nonstate deal between two companies. This was a German-German deal at a time when the three occupying Western powers in principle still had to approve transit across the Soviet zone (GDR) to West Berlin. Gas traveled where planes and trains could not. For the postwar generation of Germans, this was a big deal. From that moment on, Soviet gas supplied up to 90 percent of the gas used in West Berlin.[76]

When German reunification took place in 1990 and Berlin became one city again, it found itself in the absurd situation of having two completely separate gas networks—both of them supplied with Soviet gas. As far as the West Berlin gas company, GasAG, was concerned, integrating the two systems was a straightforward engineering problem, and it immediately set about drawing up technical plans. But for the gas people in East Berlin matters were more complicated. At stake were jobs, benefits, and above all independence.[77] The East Berliners fought hard to remain independent by building what amounted to a new Berlin Wall, this one made of legal paper and corporate defenses. For a brief time, gas in East Berlin was supplied by something known as Berliner Erdgas AG, which featured a brand-new logo but had the same old management. It took three more years and the vigorous intervention of the Berlin government before the resistance of the East Berliners was overcome and Berlin's gas supply was unified under a single company.[78]

The first span of the Russian-European gas bridge was built under conditions very different from those of today. It was a countertrade deal, in which the primary objective on both sides was to trade goods and not money. The Soviets needed steel, technology, and finance; the East Europeans needed energy; and the West Europeans needed export markets. For all sides, gas was a currency, a substitute for money. This had one important side effect: since the pricing of the goods exchanged tended to be arbitrary, none of the parties really knew what the actual price of the gas was. To be sure, the contracts contained pricing formulas, but these had little meaning if the gas bill was then settled in arbitrarily priced goods and politically priced credit. This remained a central feature of the Soviet-German gas trade through the 1980s.

Russian gas posed a whole new level of challenge in defining value and risk, for multiple reasons. First, there were the long distances and heavy capital costs (and hence long payout periods). Second, they were essentially barter deals—steel and equipment for gas. And third, there was a lack of information about risks, especially from the Russian side. For example, the Germans were unaware of problems the Russians were having in sticking to the agreed export schedules with the Austrians; they learned about them from OMV only after the first Soviet talks were well under way. The Germans were also largely unaware of the complex politics surrounding West Siberian gas. If they had known more, they might have bargained harder.

In the background of the negotiations was the fact that in the 1960s economics was just being revived as a legitimate science in the post-Stalin USSR, having been driven essentially underground in the Stalin years.[79] Soviet foreign trade officials, because of their exposure to the world economy, should have been better versed than their counterparts in the gas ministry. But as we have seen, most of them were recent transfers from the oil-export business and were mostly engineers by education.

There was much anxiety in Washington and the European capitals about whether the Soviets would use gas as a weapon in the Cold War. But the key point of the early history is that for the Soviets, countertrade was the key to the entire development of West Siberia's gas reserves: no Western pipe and equipment and finance, no West Siberian gas. It was that simple. And as long as countertrade remained at the center of the Soviet motivation, it was nearly guaranteed that the Soviets would stick to business. This remained true for the next thirty years, up to the end of the Soviet period—at which point the entire context changed, as we shall see.

CHAPTER 3

From Optimism to Anxiety

The period from 1970 to the late 1980s marked the end of the age of optimism that had prevailed since the end of World War II. But the reasons were different on the two sides of the Iron Curtain. In Western Europe, the first oil shock of 1973–1974, followed by the second shock of 1979, brought an abrupt end to the sunny optimism of the thirty most prosperous years in West European history. Suddenly, growth gave way to stagnation, rising unemployment, and inflation—a combination that economic theory had previously thought impossible—which was quickly dubbed stagflation. On the whole, it is not most historians' favorite period. In his history of postwar Europe, Tony Judt titles his chapter on the 1970s and early 1980s "Diminished Expectations."[1] Mark Mazower, whose book is revealingly enough called *Dark Continent,* names this period "The Social Contract in Crisis,"[2] while Daniel Yergin's history of the oil industry captures the mood of the later 1970s and early 1980s as "The Great Panic."[3] As Yergin concludes, "The decade of the 1970s was, for Hydrocarbon Man and for the industrial world as a whole, a time of rancor, tension, unease, and gritty pessimism."[4] The first half of the 1980s offered more of the same. The key word of the period was "anxiety."

Behind the Iron Curtain, the Soviet Union in the late 1960s and early 1970s was at the height of its power and self-confidence, as yet unshaken.[5] We now know, of course, with history's 20 / 20 hindsight, that the Soviet

Union was ultimately headed for failure, but that was far from obvious in the 1960s, least of all to the Soviets themselves. When Nikita Khrushchev first began tinkering with the structure of the economy in the late 1950s, his aim was not to make it grow faster (the Soviet Union at that time was the world's fastest-growing economy) but rather to improve the quality of growth, particularly by dealing with the problems of agriculture and consumer-goods production. The massive failure of his experiment with radical decentralization (the so-called *sovnarkhozy*) could be dismissed by his successors as an aberration, one of Khrushchev's notorious "harebrained schemes." They too sought to address the problems of agriculture and light industry, this time through the more carefully designed reforms of Premier Aleksey Kosygin, which aimed at making enterprise directors more efficient by giving them more of a profit motive.

It was only in the second half of the 1960s and the early 1970s, when the Kosygin reforms failed as well and GDP growth began to slow, that the more perceptive Soviet economists and planners began seriously to worry about the long-term competitiveness and even the viability of the Soviet model. But on the eve of the first oil shock, most Soviet leaders, typified by Kosygin, still believed that the command economy, whatever its defects, was still the surest—indeed, the inevitable—way to the future. The Soviet Union at the beginning of the 1970s was still largely a closed system, isolated from the outside world and not essentially different from the autarkic system that Stalin had left to his successors at his death in 1953. People like Nikolai Baibakov, chairman of Gosplan, who had helped build the system before the war and rebuilt it afterward, liked it that way and saw no reason to change. The Brezhnev leadership disagreed, and reacted to the oil shocks by increasing oil exports. The oil shocks brought Moscow a welcome windfall: augmented by higher prices, Soviet revenues from oil exports mushroomed from a few hundred million dollars a year in the 1960s to nearly $10 billion in 1975 and over $25 billion a year in the first half of the 1980s.[6]

Thus the oil shocks of the 1970s, by bringing a massive infusion of oil revenues, began a fateful dependence that only increased over the following decades. Yet paradoxically, these did not save the system, but only masked its decay. There were repeated crop failures, and in bad years food grew scarce. Despite strenuous efforts early in the period to reform the command economy,[7] especially in agriculture and light industry, the loss of momentum in GDP growth could not be halted, and the Soviet Union began to import massive quantities of grain and consumer goods. Through 1974 these were paid for chiefly in gold (here the Soviet Union enjoyed another windfall, thanks to the decision of the administration of President Richard Nixon to leave the gold standard and devalue the dollar, with world gold prices tripling in dollar terms over the next two years).[8] But beginning in the mid-1970s the Soviet leadership was paying for imported food and consumer goods by exporting steadily growing amounts of oil to the capitalist West. By the 1980s, the Kremlin was hooked on oil exports.[9] At the same time, the costs of all domestic inputs, especially labor, began to climb. From 1970 on, the Soviet Union was experiencing its own version of stagflation,[10] and the populations of the Soviet bloc became increasingly demoralized.

Thus the period from 1970 to the mid-1980s is a key turning point in our story. By the second half of the 1980s, Europe and Russia were on the verge of revolutionary changes in economic and political structures and modes of thought, for which the way had been largely prepared during the preceding decade and a half. In this chapter we focus on Western Europe before returning to the Soviet Union in Chapter 5.

Three changes are especially significant: the turn to nuclear power, especially in France and Germany; the rise of the antinuclear protest movement in both countries; and its connection to growing environmental awareness and anxiety. We then home in on Germany, where the political responses to the anxiety. of the age touched off far-reaching changes that set the stage for today's *Energiewende* (energy transition) program, with radical consequences for the future of gas.

Meanwhile, in the background of these events, natural gas (based largely on imports) was gradually penetrating the energy fabric of Europe. "Despite a lack of public attention, often verging on indifference," wrote the gas expert Jonathan Stern at the end of the 1980s, "the natural gas industries in the major European countries developed an extremely powerful position . . . without the publicity accorded to most of the other fuel industries. . . . Gas has been seen as useful, even necessary, but boring."[11]

Yet the story is hardly boring. In the North Sea, production began as the first big fields were discovered and undersea pipelines built (see Chapter 4). In the Soviet Union, faced with troubles in the oil fields, the leadership launched a top-priority gas program (see Chapter 5). In Algeria, after the war of independence the gas sector grew strongly, chiefly thanks to the early adoption of liquefied natural gas (LNG) technology for exports. And for the men who built those industries, especially those involved in the Soviet-European gas trade, there was not only the satisfaction of developing and delivering their major engineering projects, but also the fascination of being personally involved in a political adventure across a frontier (the Iron Curtain) that many thought was unbridgeable. It was a slow and partial penetration, held back by both physical and regulatory constraints. Still, throughout that period, the public image of gas remained highly positive. Gas was the virtuous fuel in the popular mind, bringing the benefits of cleanliness, comfort, and convenience—not to mention health.

Yet natural gas had a formidable competitor: nuclear power. Nuclear power at the time was expected to make electricity so widely and inexpensively available that it would be "too cheap to meter."[12] And in such a world, gas would have little or no value. In France nuclear power in the 1970s and 1980s came to occupy the niche that might have been occupied by natural gas, once an efficient gas-burning technology (combined-cycle gas turbines) became widely available at the end of the 1980s. It is only recently that the French national consensus in favor of nuclear power has been questioned. In Germany, however, opposition to nuclear power catalyzed a broad environmental movement that culminated in the rise of the Greens in Germany, which in turn led eventually to the *Energiewende*.[13] The social movement that originated in the anti-nuclear protests of the 1970s and 1980s has matured and gained power, to the point where the future of nuclear power is sealed in Germany. In the process, it has now reopened the door to the gas question in Europe, in the form of the following question: Once nuclear power has been shut down (as in Germany) or increasingly challenged (as in France), what will replace it, if not natural gas? Or will it be renewable sources, which have emerged more recently as powerful competitors, notably solar and wind?

The Impact of the Oil Shocks

The oil shocks of the 1970s were a classic case of the price cycle that rules all commodities. Booms and crashes in commodity prices follow one another like the tides of the sea. Most commodities rise and fall together, the result of the tendency of investors to overinvest when prices are high, creating excess capacity which then causes prices to crash, followed by a drying up of investment and a price recovery, starting the cycle all over again.[14] But each price cycle does more than affect prices. It brings sweeping changes in entire economies and societies. Long-established firms go bankrupt, along with innumerable start-ups. Price cycles are destructive, but the destruction they bring is also creative. New technologies are created and diffused into new applications. New companies spring up, with new ideas and business models. The efficiency of resource use is increased.[15]

Energy is a classic commodity, but with a difference. Energy presents decision makers with three basic challenges: high capital intensity, long lead times and lifetimes, and wide economic and political impact. Energy projects can take as long as twenty years from conception to production. This aggravates the swings in commodity cycles, and it causes a lag between policymakers' decisions and the evolving market. Decisions made when prices are high and supplies are insecure do not take effect until prices have dropped and supplies are abundant. As a result, policymakers are always addressing the last energy crisis—always half a turn of the wheel behind—and having to live with the results of their predecessors' decisions.

So it was with the two great oil shocks of 1973–1974 and 1979 and their aftermath. In the 1970s, following the first shock, energy policy was dominated by fears of scarcity, as energy prices spiraled upward. Security was decision makers' main concern. This led to a powerful wave of energy investment and conservation. But in the 1980s, the setting flipped: energy scarcity was replaced by abundance, energy prices stagnated and then began to fall, and decision makers worried about cost and efficiency instead of security. Yet the policy decisions made during the decade of scarcity—and society's responses to them—lived on, shaping the energy balance and the political scene of Europe down to the present day.

On both sides of the Iron Curtain, the main initial response to the two oil shocks was a reinforcement of the prevailing statist models of decision making. In energy, the broad response was a combination of coal, conservation, and nuclear power—in Germany frequently referred to as the "CoCoNuke" policy. This took different forms in each country. Britain and Germany partly retreated to coal, while France went for nuclear power and stressed conservation. The two oil shocks of 1973 and 1979 caused a drive throughout Western Europe to displace oil. One of the chief consequences of the oil shocks was a permanent decline in the role of oil in Western Europe's energy mix. Over the next decade and a half, as the Organisation for Economic Co-operation and Development (OECD) countries sought to diversify their sources of energy, whole sectors were lost to oil—notably in power generation. In the Soviet Union, in contrast, the Brezhnev leadership redoubled its efforts to boost oil production (at considerable cost in lost long-term recovery) but also forced the pace of central investment in nuclear power, particularly in the western regions of the USSR and notably Ukraine. It was not until the 1980s that gas became a central part of the Soviet energy response, when Leonid Brezhnev launched the Soviet gas campaign, as we shall see in Chapter 5.

Yet statist solutions failed to restart growth. In both the West and the Soviet bloc, the inability of governments to find solutions to stagflation caused disillusionment with the traditional state-led approaches and stimulated a search for alternative models. Disappointment with the conventional narratives of state planning and growth also engendered skepticism about the side effects of economic growth and its impact on health and nature. One result—both in the West and in the East—was the rise of environmental protest. Thus the anxiety that characterized this period took two seemingly opposite forms: on the one hand, anxiety over the harm caused by slower growth; and on the other hand, increasingly, anxiety over the damage inflicted by growth itself.

The need to diversify energy sources had begun to dawn in some European minds well before the first oil shock of 1973. Nearly two decades before, the Suez Crisis of 1956 had been the first warning that dependence on Middle Eastern oil posed dangers. But public opinion did not view energy as an urgent problem, and the pace of change among politicians is best described as leisurely.[16] During the postwar decades, Western

Europe had shifted massively from coal to oil;[17] the same had happened in the Soviet Union[18] and, to a lesser degree, Eastern Europe. By the beginning of the 1970s, Europe had grown addicted to cheap imported oil. Western Europe depended on the Middle East, while Eastern Europe depended on Russia.

The decade and a half between 1970 and the mid-1980s can be divided into two acts: first, the shock of 1973–1974 and the initial responses to it; and then the second shock in 1979 and the overshoot that followed. The two phases were linked, as the responses of the 1970s both increased energy supply and curtailed demand, setting the stage for the final phase: a massive collapse in oil prices and commodity prices generally in the mid-1980s. As energy became increasingly abundant and prices dropped, there was a corresponding decline in anxiety over security. Instead the dominant concern of policymakers was efficiency (or, in one author's nice phrase, "sweating assets instead of creating them"[19]), while society increasingly worried about the environment—or, as it came to be called in the mid-1980s, sustainability.

Each national government, however, developed its own policy in near isolation from the others. The European Economic Community, exhausted by ferocious fights in the 1960s over membership and subsidies, was in one of its recurring periods of dormancy, from which it did not begin to emerge until the end of the 1970s.[20] (The only supranational body to play a significant role in European energy policy in the 1970s was the newly created International Energy Agency, which coordinated the West's responses to the oil shocks.) Each government's energy policy reflected its own unique circumstances and politics, and there was little communication among governments. At the beginning of the 1970s, each European state was effectively still an energy island unto itself.[21]

Natural Gas: From Evolution to Revolution

Meanwhile, what of natural gas? One might have supposed that the energy shocks would bring a turn to gas as one of the main policy responses in Western Europe, and they did, on the supply side at least. Growing discoveries offered the promise of expanding the gas bridge to Europe. The way had been opened in the 1960s by the first great finds in the

Netherlands, the North Sea, Algeria, and West Siberia and by technological advances in production and transportation—especially LNG and long-distance large-diameter pipelines. In the North Sea, Norway's Ekofisk field was discovered (1969), almost simultaneously with the first three great giants of the Soviet Yamal region.[22] In short, the 1960s and early 1970s had dramatically widened the circle of gas supply beyond the Continent to the periphery of Europe. As a result, Europe—both capitalist and socialist—became divided into two interdependent zones: a gas-rich periphery (Algeria, the Soviet Union, and the British and Norwegian sectors of the North Sea) and a core that, with the exception of Groningen in the Netherlands, was gas poor. That is essentially the pattern that remains to the present day, especially in view of the recent decline of Groningen.[23]

The security issue played out in two different ways. By the 1970s, to Europeans insecurity referred to the Middle East more than the Soviet bloc. The advent of East-West détente in Europe in the late 1960s and 1970s, against a backdrop of continued US-Soviet tensions, split the security issue in two: it accelerated the export of Soviet gas as Europeans sought alternatives to oil (despite Washington's opposition), while simultaneously stimulating the search for hydrocarbons from the North Sea (with Washington's support).[24] These twin concerns over security combined to increase the gas available from suppliers at the periphery of Europe.

Yet it took time before the increase in potential gas supply produced significant penetration of gas into the energy economy of Europe, and it was two decades before natural gas became a contender for primacy in European energy demand. The reasons are straightforward: On the supply side, for major field developments such as those in the Soviet Union or the Norwegian North Sea it could take more than a decade before licensing, exploration, commercial negotiation, and development yielded actual gas. In the so-called midstream, the penetration of gas into Europe was constrained by the lack of transportation and distribution capacity. In the early 1970s, the web of pipelines, big and small, that would eventually connect the new gas to its market was still largely confined to Dutch gas, which had been flowing just for a few years to Belgium and to the northern regions of France, Germany, and Italy.[25] The first Soviet gas had just begun to arrive in Europe, but for the moment it was flowing only to Austria and south Germany, as well as Eastern Europe. Throughout the 1970s and 1980s, the network of gas arteries and capillaries, like an embryonic circulatory system, grew steadily throughout the body of Europe, eventually reaching every major city and community.[26]

There were equally long lead times on the commercial side. The business framework for buying and selling gas across international borders had just been invented by the Dutch, led by Shell and Esso, to market their gas quickly to customers in the neighboring European countries, as we saw in Chapter 2. The principles developed for this trade were extended in the 1970s and 1980s to the much larger quantities of gas that the Soviet Union, Norway, and Algeria were able to offer. Central to the commercial arrangement was the deal by which a monopoly distributor in the purchasing country—state-owned entities like Gaz de France or Italy's SNAM, and private companies like Ruhrgas and Thyssengas in Germany (each with its regional monopoly protected under German law)—undertook to take or pay for the volumes they committed to buy. In exchange, the seller would commit to a price for the gas that would ensure its competitiveness with the energy products whose markets the distributor was intending to win.

This was the era when a generation of men emerged in the West European countries who specialized in the negotiation and renegotiation of these contracts on the other side of the table from the Soviet negotiators, whom we met in the last chapter. Burkhard Bergmann at Ruhrgas, Yves Cousin at Gaz de France, Domenico Dispenza at SNAM, James Alcock at British Gas, and their colleagues in smaller European companies such as OMV in Austria, Neste in Finland, and Distrigas in Belgium held a special place in their respective companies. Plenipotentiary in their roles as chief negotiators, they were instrumental in delivering commercial deals whose terms would commit their companies to spending billions of dollars—commitments that would last not just for years but for decades. Their relationships with their Soviet (and Norwegian and Algerian) counterparts took on a character that was molded by the trust and tension that were intrinsic in such enormous deals. And with that came the awareness that there was a vital political dimension to the deals they struck—above all to the importance of producing something that was commercially solid in a climate that was politically fragile.

These contracts took time to be negotiated. The principles developed for the Dutch exports proved to be remarkably robust and well suited to the need for commercial solidity in the context of political fragility. As time passed and gas flowed reliably in one direction and money flowed steadily in the other, the experience of success convinced both Soviet sellers and European buyers that their model of long-term take or pay contracts, with a negotiated base price indexed to monthly or quarterly changes in the price of oil products and subject to renegotiation every two or three years (with the right to exceptional recourse to renegotiation in the event of financial hardship for one of the parties) was perfectly adapted to the needs of the trade. It gave the buyers time to invest in extending the pipeline networks in their home countries and regions, knowing they would have the gas available to supply when consumers had signed up to use gas instead of coal or oil. It gave the sellers the ability and the financial resources to undertake the long-lead-time investments in developing fields and building pipelines to bring the gas to market. The contracts and the strong personal relationships that they engendered underwrote the confidence of banks that were invited to provide debt finance for the investments both upstream (in gas field development) and midstream (in long-distance transmission pipelines). By the late 1970s, major commercial banks like ABN Amro, Deutsche Bank, Barclays, and Société Générale were willing, even eager, to enter consortia that would offer debt finance of up to 85 percent of these multibillion-dollar investment programs. This testifies to the robustness of the commercial model that the Dutch had first developed.

So it was that the previously low-profile, "boring" state and private companies that had originally been small distributors of town gas grew into major players in international trade, rubbing shoulders with the international oil majors and the giants of the European coal industry. They campaigned to boost the sales of their product, with gas distributors in Great Britain, France, and Germany marshaling all the resources of modern advertising to promote it.

Technology was both a constraint and a stimulus. The presence of a mature nuclear defense industry in France, combined with an industrial policy oriented toward the kind of high-tech solutions that nuclear power symbolized, made nuclear power the option of choice in the French electricity sector. But in a broader sense the development of gas in Europe would have been impossible without a revolutionary process of innovation on a broad front. To cite just one example: the development of combined-cycle power plants, which bring together an advanced gas turbine with a steam-driven generator to use the waste heat from the turbine, was one of the breakthroughs of the 1980s and transformed the landscape in the 1990s. It could not have been realized without advances in the surface chemistry of turbine blades that enabled superhard surfaces with long service lives at high temperatures.

Gas "Too Valuable" for Power

Gas was slower to penetrate the power sector than others. There was a long-standing belief that gas was too scarce to waste in the power sector. Sales of gas for power generation actually began to decline after the first major oil price increase in 1973, "reinforced by a directive from the European Commission in 1975, which declared gas to be too valuable a fuel to be used under boilers to generate power."[27] This was very much the conventional wisdom at the time. The United States introduced a similar policy with the Fuel Use Act of 1977. In the European countries in the OECD as a whole, after an initial fast start, gas consumption in the power sector peaked at 32.5 billion cubic meters in 1975 and then declined for the next decade, bottoming out at 25 billion cubic meters in 1985.[28] Elsewhere, it stabilized at about 15 percent of total demand—in Italy and the Netherlands somewhat more, in Germany a little less. In France (because of the pro-nuclear policy), and in Great Britain (because of coal) the use of gas in power generation was practically zero. In contrast, industrial and residential demand for gas boomed throughout the European OECD countries, reaching about two-thirds of total energy demand in those sectors.[29]

The dramatic increase in efficiency brought about by combined-cycle gas turbine technology in the 1990s helped overcome the prejudice against using a high-value fuel such as gas for power generation and opened the way for the "dash to gas" that followed in the 1990s, especially in Great Britain. (There was a similar prejudice against using gas for power generation in the Soviet Union. Nevertheless, in the 1980s there took place a massive backing out of oil by gas in power plants because the displaced oil could then be exported and replaced with increasingly available gas.) Similarly, combined-cycle gas turbines did not arrive in Russia on any scale until after the collapse of the Soviet Union.)

In contrast, residential heating with gas grew rapidly in the 1970s and 1980s, starting from zero in 1960. For many people it was the first time that every room in their home could be heated through the winter. This transformed even the social life of many families, as evenings were no longer spent crowded into one heated room. For other, more prosperous families, who might already have had an oil-fired heating system for the whole house, conversion to gas-fired heating could liberate space in the basement (no longer needed for an oil tank or reservoir); reduce inconvenience (no need for a tanker delivery or two in winter); and eliminate the need to pay ahead for the winter's needs, as the gas was paid for after it was metered and used. These quality of life issues, as well as the commercial practice of keeping the price for gas just at or below what a homeowner would pay for oil, ensured the steady and sustained penetration of gas in the home heating market throughout Western Europe in these years—at least for those fortunate enough to live in the regions that had access to the pipelines that supplied it. Local political pressure for the extension of these regions soon built up, as word spread about the qualities of this new fuel.

But there were differences from country to country. There were state-driven obligations on monopoly suppliers in the Netherlands and Great Britain for them to make a gas connection, at their cost, to any householder who lived within twenty-two meters of a gas main. In urban areas, where there were local gas grids dating from Victorian times, supplying gas from the coke factories for street lighting, this meant just about every household in the street. But in Germany, where the industry was not nationalized and depended instead on cooperative ventures between independent municipalities and private companies, no such obligation existed. Accordingly, the pace of penetration of this market was slower in Germany than it was in other European countries.

The German Story: Nuclear Power and Coal

If anyone had asked in the 1950s and 1960s what source the world would rely on for power once the age of oil had passed (and it was increasingly perceived at the time that it would pass), the nearly universal answer would have been nuclear power. In the United States, by 1970 civilian nuclear power already accounted for a rapidly growing share of power generation. It enjoyed strong public support, and nothing, it seemed, would hold it back. In the Soviet Union as well, nuclear power appeared a successful alternative to oil and was firmly backed—especially by the scientific elite, whose members saw nuclear power and coal as the long-term energy future of the country.[30] Throughout Europe, nuclear power at first appeared to be the answer. In France, the commitment to nuclear power became total.

But in West Germany nuclear power was more complicated from the start. Prior to the oil shock there had been practically no attention given either to energy policy per se (only to individual fuels) or to the environment.[31] The government's main energy worry was the decline of the coal sector in power generation, which had become particularly serious from about 1957, together with a corresponding loss of jobs. The impending scarcity of oil was largely ignored by the German government, despite an early warning to the OECD from the US government in 1968. Very few were alert to the danger. One rare exception was State Secretary Ulf Lantzke, who subsequently became the first executive director of the International Energy Agency. He was first alerted by the Suez crisis in 1956. He later recalled:

> For me that was a triggering point. From that point onwards, I was trying to turn around energy policies in Germany. The issue was no longer how to resolve our coal problems, but how do we get security of supplies into our policies? It was very, very cumbersome. It took

> me five years to prepare the ground and convince people, so deep-seated was the political belief that energy supplies were no problem.[32]

But there were few like him. German energy policy through the 1960s consisted mainly of imposing high taxes on oil to protect the coal sector.[33] Yet the effort to defend coal failed, because imported oil was so cheap that it penetrated the West German economy in all sectors (in contrast to East Germany, which remained dependent on coal for both power and home heating). In 1970, the peak year, oil accounted for 53 percent of the primary energy used to meet demand in West Germany.[34]

The oil shocks brought sharp changes: by 1990, West German oil's share of primary consumption had been cut back to 40 percent and was essentially concentrated in the transportation sector and home heating.[35] Coal's share had collapsed under the previous onslaught of oil, going from nearly 75 percent in 1960 to 30 percent in 1980.[36]

As oil retreated, energy for residential use became a battlefield between gas and electricity, although home heating was still largely fueled by oil. Electricity and gas had been bitter competitors since the 1920s, but gas in those days meant coal gas, and electricity meant coal-fired power. Each had its own champion, RWE for coal-based electricity and Thyssengas for coal gas.[37] With the arrival of natural gas from the Netherlands in the late 1960s, the competition intensified. The electricity industry (still fueled mainly by coal at the time) mounted a determined defense against natural gas. One of the arguments used by RWE was that natural gas would run out after twenty to forty years, and therefore only a foolish householder would rely on it. Moreover, the heavy investment required for pipelines and other infrastructure meant that natural gas was bound to be expensive. Instead, in a vigorous ad campaign the power industry promoted the virtues of the all-electric household (using slogans such as "Sei auf Draht—mach's gleich elektrisch" [Get on the ball—go electric]), without which such newly introduced appliances as automatic washing machines and dishwashers would not be possible. In some of its market areas RWE pushed the campaign to the point of sending electricity missionaries with electricity contracts in hand to visit the homes of people planning to switch to gas, to persuade them of the virtues of electricity.[38]

But the campaign against gas in households failed. Beginning in the late 1960s, householders—even in the Ruhr, the stronghold of coal-based electricity—began converting enthusiastically to natural gas, especially for heating, and building companies started equipping new homes with gas heat. (One reason the campaign for electricity failed was reportedly that conservative German householders found it too upper crust. According to market research they mistrusted the all-electric home as fit only for "society women and playboys."[39]) There followed a spectacular move to gas. Every year between 1970 and 2000 some 220,000 households on average switched to gas.[40] From the late 1970s on, whenever German householders chose a heating fuel, they opted for natural gas over 60 percent of the time, despite the fact that natural gas was slightly more expensive than heating oil.[41] By 1980 the share of German households that used gas had reached nearly a quarter, and by 2000 it was almost half.[42]

In the power sector the rivalry played out differently. In West Germany,[43] the effect of the argument that gas was too valuable to waste on power can be clearly seen from the numbers: the share of gas in the power sector went from practically nothing in 1960 to a peak of 18 percent in 1975 and then fell back to 5 percent in 1985.[44] Coal's share was 80 percent in 1960; though much diminished, it nevertheless remained significant at about 50 percent throughout the period. In contrast, nuclear power's share grew rapidly, from a share of less than 1 percent in 1965 to 33 percent in 1990. Thus a consequence of the "too valuable" argument from the mid-1960s to the late 1980s is that the power sector was left to coal and nuclear power.[45] In Germany, the initially strong growth of nuclear power had fateful political consequences, setting the stage for the antinuclear movement in the 1980s, as we shall see in a moment. Meanwhile, the continuing role of coal has remained a source of political conflict down to the present day.

The Nuclear Response in France and Germany

We turn now to the story of nuclear power in France and Germany. The core question is why nuclear power became the main response of the French to the energy crisis, with a total commitment that remained intact down to very recent times, whereas in Germany the commitment was always uncertain and was effectively abandoned after a decade. It spawned instead a powerful antinuclear movement, which has transformed German politics. This question turns out to be central to our story, because the antinuclear movement in Germany leads straight to the *Energiewende,* while in France the environmental movement remains relatively weak and poorly organized.[46]

Nuclear power in Europe and Russia was the ultimate expression of the top-down statist approach to energy. Thus, it is not surprising that when the first oil shock broke, nuclear power was one of the main responses of both the European states as well as the Soviet bloc. Yet success proved uneven: the Soviet program encountered severe delays; the German program was held back, as environmental opposition grew into a mass political movement; and the British program, hampered by poor political leadership, was a relative failure, as was nuclear power in Italy and the Netherlands.[47] Nearly everywhere nuclear power, as an alternative to conventional fuels, turned out to be far more complex, time-consuming, and costly than anticipated. But most serious of all, it soon faced a collapse of public trust.

The one exception among major European economies was France. There nuclear power was a brilliant success. To the present day it supplies three-quarters of French electricity, and it has done so without a major accident or even serious incidents for nearly half a century. In the process nuclear energy has largely closed the door to natural gas in the power sector. It is only now, as the long-lasting French marriage to nuclear power shows signs of fraying, that the question of natural gas in power takes on new urgency.

So why was the French nuclear program such a success, while the German program was ultimately rejected by both society and the government and is now on the edge of a final shutdown? Both programs started at about the same time; both were promoted by the same interests; and both were equally ambitious in their initial scope: to build and operate 50 gigawatts of nuclear capacity by 1985. But the German program soon fell short, while the French program—under three presidents representing both the left and right—went from strength to strength, backed by a solid consensus in elite and public opinion. Why?

By the beginning of the 1970s, even well before the first oil shock, the French had many reasons to feel anxious about their energy supply. The national coal industry was stretched to the limit, and France was even importing coal, even though its share in meeting primary energy demand had plummeted from 85 percent in 1950 to 16 percent in the early 1970s. Cheap imported oil had rushed into the gap, and by the early 1970s oil supplied 70 percent of French primary energy demand. Natural gas was just beginning to be perceived as an option.[48] In short, as the French planners surveyed the scene, there seemed to be nowhere to turn—except to nuclear power and conservation.

Conservation in French minds was initially the equal of nuclear power in importance as a response to the oil shock. When Prime Minister Pierre Messmer appeared on national television in December 1973 with the leader's first address to the nation in the wake of the oil crisis, it was conservation he chose to talk about, not nuclear power. Some of the measures he announced may seem far-fetched today, and some may seem downright comical: in addition to higher energy prices and lower speed limits and thermostat settings, road races were banned, many air flights were canceled, public monuments were no longer lighted, and even the French television monopoly was told to go off the air at 11:00 pm. One of the more unexpected legacies of the French conservation program was the high-speed rail program, which was slipped into President Georges Pompidou's agenda as an energy-saving program, under the nose of the finance minister (and future president) Valéry Giscard d'Estaing, who was an opponent. "It's a good idea," said Pompidou, and the finance minister had to watch in silent frustration as the measure was approved.

But right next to conservation came nuclear power. It was virtually the last decision of a dying president. In the spring of 1974 Pompidou was near death from a rare form of lymphoma. He had suffered from it for years, but time was finally running out. Yet though weakened and in pain, he was able to preside over one last meeting of the Council of Ministers, on March 6, 1974. It proved to be a fateful session for French history: after a long day of discussion, the president and his ministers endorsed a massive acceleration of the French civilian nuclear program, leading off with thirteen new reactors, totaling 13 gigawatts of new nuclear capacity over the following two years—more than the entire thermal generating capacity of the country in 1972. Less than one month after the meeting, Pompidou was dead.

The nuclear program was presented to the French public as an emergency response to the oil crisis, but in reality the way had been prepared over the previous two decades.[49] Under General Charles de Gaulle's presidency and even earlier, the French had amassed vast experience in military uses of nuclear energy.[50] The civilian program benefited from the skills and experience acquired as a result. The entire program was overseen by a small cadre of high-level technocrats, most of them graduates of the prestigious Ecole Polytechnique and members of the elite Corps des Mines and Corps des Ponts et Chaussées—familiarly known in France as "X-Mines" and "X-Ponts."

Meanwhile, in Germany events took a very different course. We begin with the small town of Wyhl, nestled between the Vosges and the Black Forest in southwestern Germany.

Germany—the Shattered Consensus

The town of Wyhl lies in the heart of Baden-Württemberg, directly across the Rhine River from Alsace in France. It is a quiet rural community of orchards and vineyards, hardly the setting for a revolution. Yet it was here that in 1975 a popular protest broke out against a planned nuclear power plant. The local authorities, who wanted to turn the surrounding valley into an industrial complex, needed power for it. They were unaccustomed to having their plans challenged, and when faced with resistance they resorted first to subterfuge and then to force. The resulting violence caused a scandal throughout Germany, split the local political elite, and ultimately forced the cancellation of the project. The name "Wyhl" is a landmark in the history of the green movement in Germany: it marked the birth of the antinuclear movement and the end of the consensus that had driven the nuclear program forward up to that point.[51]

The German nuclear power program at its outset was as ambitious as that of the French, at least in rhetoric. Little attention was given to safety issues at first: among the nuclear scientists who advised the government in Bonn, there were no experts on risks (although the companies were presumably well staffed).[52] At the time of the first oil shock, West Germany already had eight nuclear units in operation, although with an as yet modest total capacity of 2.3 gigawatts In September 1973 the Social Democratic Party (SPD)–Liberal coalition headed by Brandt announced a dramatic expansion of the program. The new aim was to have fifty units in operation by 1985, with a capacity of 50 gigawatts—a program the same size as the Pompidou-Messmer one adopted in France a year later. The reasons were much the same in both countries: nuclear power was still thought of as the energy of the future, and it offered the alluring prospect of energy independence.

To anyone reading today the history of the French and German nuclear programs, what is most striking is the strength and continuity of the political support for nuclear power in France, compared to the more hesitant and fragmented commitment in Germany. As a result, the French program became stronger over time, while the German program weakened and was ultimately rejected by public opinion and abandoned by the government, leaving the German power sector more dependent on coal. What explains the difference?

Timing accounts for a great deal. By the time of the first oil shock, French nuclear planners had been working on a civilian nuclear program for two decades.[53] Pompidou's decision to approve the Messmer plan was in essence an acceleration of a program that was already well under way.[54] In contrast, the German program was at a much earlier stage in 1973 and was slower to get started. The initial protests then caused further delays: "Threatening state activities, alliances between protesters, bold tactics, elite support, and authorities' mistakes reinforced each other . . . contributing to a national stalemate on nuclear policy and the downsizing of the federal government's nuclear energy plans."[55] By the early 1980s, the political and economic setting had changed: the growth of the German economy had slowed; projections of power demand had been cut back; and opposition to nuclear power had been incorporated into the platforms of several political parties—notably the SPD, which turned from initial support to opposition, and above all the Greens, who owed much of their early success to the antinuclear movement.[56] In short, after a decade the German program ran out of political support. The last German nuclear power plant was commissioned in 1982. By that time, the program had achieved only 17 gigawatts of the target for 1985 of 45–50 gigawatts set in 1974.[57]

In contrast, the strong political commitment behind the French nuclear program can most clearly be seen at the two moments when it was most severely challenged—not by antinuclear opposition from below (which, though episodically violent, soon faded) but by management problems with construction, internal battles within the French technocracy, and electoral politics. On both occasions the challenge coincided with presidential successions. In 1975 Giscard d'Estaing, the newly elected successor to Pompidou, allowed himself two years of suspension before recommitting himself to the nuclear power program. Construction was running behind schedule and over budget, forecasts of French power demand were being cut back as economic growth slowed, economists and engineers in competing ministries fought over the nuclear targets, and Électricité de France (EDF) struggled with siting and contracting problems. But in the end, Giscard came down firmly on the side of the nuclear camp, and the program's ambitious targets were reaffirmed.[58] As Bernard Esambert, a close advisor to Pompidou, summed it up, "In the nuclear area, Giscard continued Pompidou's policy. He continued to accelerate the building of nuclear powerplants. We had seventeen under construction at the time of Pompidou's death. Giscard kept the pace going. There was a perfect continuity between the two presidents."[59]

In 1981 the priority of the nuclear program appeared to be threatened again, this time following the election of a socialist president, François Mitterrand. Mitterrand had campaigned on a program that echoed many of the themes of the environmentalists, who demanded a halt to the nuclear power program. "For six months," recalls a high executive of EDF, "we were very frightened." It appeared that Mitterrand was inclined to cancel several controversial projects, especially one in Brittany called Plogoff. A few months after Mitterrand's election, the chief of EDF, Marcel Boiteux, went to see the new president at the Elysée Palace. At the end of their meeting, at which Boiteux briefed the new president on the state of the power program, Mitterrand offered Boiteux a deal: "You leave me Plogoff and I'll save the others for you." Plogoff was duly canceled, but the rest of the program went ahead—to the disappointment of the environmentalists, who felt betrayed but were powerless to oppose the decision.[60]

What kept the French program driving forward? It was no longer the same sense of impending energy shortage that had motivated the emergency program in the early 1970s. Indeed, demand for power was slowing down, and by the early 1980s it was clear that nuclear power was overbuilt relative to the economy's needs. As Boiteux conceded in an interview, by that time "we were building too many nuclear megawatts." As much as anything, the explanation was technological pride and a fierce desire for national independence. Nuclear power, both military and civilian, had become became France's signature accomplishment of the 1970s and 1980s, the comforting symbol of France's role as a leading technology power.

Moreover, in the decade following the riots of 1968, the French had become less patient with the quasi-anarchistic violence that de Gaulle once described with typical pungency as *chienlit*.[61] The difference can be seen in the reaction of public opinion to the two most violent protests against nuclear power, at Creys-Malville in France in 1977[62] and Gorleben in Germany in 1979. On both occasions the protesters resorted to extreme tactics, the authorities overreacted, and there were casualties on both sides. But whereas the result in Germany was a wave of popular support for the demonstrators, followed by the cancellation of the project, in France public opinion swung sharply against the protesters, and the result was a consolidation of popular support for the nuclear power program.

The French political system was more centralized in those days than it has become since, and French public opinion was more trusting of technocracy. In the German federal system the *vertikal* of power (as Russians today would call it) is much weaker, and there are more access points at which a protest movement can delay or weaken plans made in the federal capital.[63] A revealing example is the role of administrative courts in the German battle against nuclear power. On repeated occasions the local administrative courts intervened on behalf of the protesters, and while they were frequently overruled by higher courts, the result was further delay to the German program, which brings us back to the fact of timing. Many of the younger judges in the lower courts had been influenced by the student protest movement of the 1960s, and they were inclined to be sympathetic to the antinuclear protesters' cause. Members of the antinuclear movement, in turn, soon learned that they could have more impact by turning to the courts than to the streets. As a former street fighter named Joschka Fischer—who went on to become environment minister in Hesse and then foreign minister under Gerhard Schröder—discovered, one of the sharpest weapons in his hands was the faithful enforcement of the laws by the courts.[64] "The best allies of environmentalism," comments an observer who was in the judicial system at the time, "were the administrative courts, which had meanwhile taken on the '68 student generation as judges."[65]

These events may seem distant today. The protesters of Wyhl, Brokdorf, and Gorleben are grandparents now, as are the judges who intervened on their behalf, the local Christian Democratic Union (CDU) and SPD politicians who split with the national leadership, and the pioneers of the Green Party who made the antinuclear movement their core issue. Yet they have become part of the fabric of the political establishment of Germany. The antinuclear movement is part of their DNA. As the historian Joachim Radkau observes, "The dispute over nuclear energy has been the largest public controversy so far in the history of the Bundesrepublik: its duration, from the action at Wyhl to the present, belies all claims about short-lived fashions in today's media world."[66] In addition, one must add a number of features specific to Germany, such as antimilitarism, the association with Hiroshima and Nagasaki (stirring shared feelings in a people who, as defeated victims of carpet bombing, knew what it must have felt like to have been at *Stunde Null*[67]), and the fear of a plutonium economy and the fast-breeder reactors that the nuclear engineers promised as the next generation. This too was central to what led to the distinctly different German and French reactions to nuclear power.

But we are still missing part of the puzzle. How did the anxiety of the age, catalyzed in particular by the oil shocks, translate into a broad environmental movement, especially in Germany? In the last part of this chapter we develop two themes. The first is that environmental anxiety arose as a by-product of the very prosperity that had preceded the oil shocks, as awareness of the ill effects of industrial growth began to spread among the public. Second, environmental anxiety was anxiety of a new sort, driven less by concern over threats to economic growth than by fear of the consequences of growth itself. Over the course of the decade and a half from 1970 to 1985, environmental anxiety evolved and broadened, from a concern over point sources of pollution as a health hazard to an increasingly global anxiety about threats to human ecology and the future of the planet.

The Image of Gas: Still Positive

In the 1960s and 1970s natural gas was still viewed as environmentally friendly, and as long as the focus of anxiety was an energy shortage and its consequences, natural gas was seen as part of the solution. If there were constraints on the rate at which gas penetrated various sectors of the economy, those had nothing to do with the greenhouse properties of methane—which in any case were still unknown.

We have said that the symbolic key word of this period was "anxiety." On one level, this was anxiety over slowing growth; on another level, it was anxiety over the side effects of growth itself. The latter was driven by the increasingly persuasive evidence of damage caused by various forms of pollution such as acid rain (for example, alarm over the *Waldstarb* [the death of trees] that lies, alongside opposition to nuclear power, at the origin of the Green movement in Germany), and partly also by the growing scientific understanding of the impact of increasing greenhouse gas emissions on the stability of the environment as a whole.[68] It was in the 1980s that these two sorts of anxiety came to be perceived by growing numbers of people, including some of the world's most influential leaders, as competing—and even, by the 1990s, as embodying irreconcilable values.

Public awareness of the connections between growth and environment damage evolved in two ways. The first was a geographic broadening. Until the 1960s, the environment was perceived as a local issue, mainly chemical leaks and car emissions, "killer smog," and the like. (Significantly, the very first agency in Germany to take up environmental issues was the Ministry of Health.[69]) Then in the 1980s, it came to be seen increasingly as an issue that crossed national boundaries, such as in the cases of acid rain and the diffusion of pesticides through the food chain. Finally, by the 1990s, the environment became a global issue, as the effects of human activity on climate began to be measured and modeled scientifically.

The second direction of evolution was on the level of political ideology. As the definition of "environment" broadened, the nature of "anxiety" began to change as well. In the first phase, pollution is visible, audible, and olfactable (if one may be allowed the word). Its health effects are dramatic and directly perceptible. But as one moves through the next two phases, the effects of environmental degradation become invisible, probabilistic, and hypothetical. Anxiety is increasingly the fear of the unknown.

As public anxiety over the environment grew, coal and nuclear power increasingly came to be seen as "bads." Natural gas, in contrast, was still perceived as the "virtuous fuel": cleaner than coal, safer than nuclear power, and more secure than imported oil, as well as being convenient and popular with customers. At this stage, solar power, wind power, and other "renewables" (as they were coming to be called) were limited to special situations, still at the edge of science fiction as far as the general public was concerned. It was only in the 1990s that doubts began to emerge, because of natural gas's contribution to CO2 emissions. Yet even then, gas continued to be viewed as the logical "bridge fuel" to the future.

Environmentalism in Germany

In a sense, there had always been an environmental movement in Germany. Since the nineteenth century there had been associations devoted to the preservation of nature. But the nature movement had been so thoroughly co-opted by the Nazi propaganda machine that it fell into disfavor after World War II in West Germany—although not in East Germany, where nature provided a welcome internal refuge from the regime.[70] In West Germany, the nature movement went into temporary dormancy, as people threw themselves into postwar reconstruction and were happy to buy into the messages of progress and prosperity. Until 1970, there was no such thing as environmental policy in Germany.[71]

Why did this begin to change in the early 1970s, not only in Germany but all over northern Europe? One reason is the growing impact of economic growth itself. Another is the model of the United States, which preceded Europe both in its public awareness of the problem and in its legislative response to it.[72] The latter was particularly important: the National Environmental Policy Act of 1970 and the creation of the Environmental Protection Agency in the same year caught the attention of the German policymakers and media. There was an explosion of public awareness during the years 1970–1971. According to a public opinion poll in September 1970, only 41 percent of those polled were familiar with the expression "environmental protection" *(Umweltschutz).* By November 1971 the figure had risen to 92 percent.[73] Along the way, the political salience and priority of the environment rose steadily. Health and environment ministers rose from junior status to key players—some ultimately to become prime ministers and chancellors, as in the case of Angela Merkel, who was minister for the environment and nuclear safety from 1994 to 1998.

This kind of thinking—the new face of anxiety—became especially popular in the mid-1980s in Germany, when an obscure industrial sociologist named Ulrich Beck won overnight fame with a best seller called *Risk Society: Towards a New Modernity,* the central point of which was that anxiety about the unknown effects of modern life and technology was changing the very nature of politics. Henceforth, according to Beck, the major question of politics would no longer be the distribution of goods but the allocation of bads.[74] In the case of the environment, this raised two questions: Who would bear the costs? And who would suffer the damage?

When seen in this context, the politics of nuclear power in Germany become more understandable, as part of the evolution of the problem of the environment and the changing face of anxiety. As the anxiety over energy scarcity began to fade in the 1980s, the main driver for nuclear power disappeared. In contrast, the anxiety over nuclear power grew as a classic "Beckian" phenomenon—as a fear of the unknown, the diffuse, and the remote. Once nuclear power was seen in this light by a broad public, it was clear that there was only one possible outcome: nuclear power has to go.[75]

In July 1974, local farmers and winegrowers held a mock funeral to protest the plans of the local *Land* government to license construction of the nuclear power plant at Wyhl. But it was not just the power plant that the local demonstrators objected to, it was the way the *Land* government was ramming it down their throats:

> Late in the afternoon, as government officials concluded the licensing hearing . . . a funeral cortege entered the meeting hall. From their seats on the dais, the officials looked on in disbelief as a group of citizens filed past bearing a black coffin marked "DEMOCRACY." Back on the streets, the mourners paraded solemnly to the village mayor's home, where they eulogized the deceased. . . . They had been moved to act when the officials excluded them from . . . the licensing hearing by suddenly switching off the audience microphones.[76]

The protests became a battle not just against nuclear power but a defense of grassroots democracy.[77] Enemies of nuclear power in Germany and Austria denounced the "atomic state" to "describe the connection between nuclear matters and a sort of creeping abuse of government authority, which limited democracy by asserting that citizens were not equipped to debate nuclear energy and thus unable to govern themselves in an age of high technology."[78]

The protest at Wyhl—at that time the largest public demonstration in West German postwar history—came as a surprise to German public opinion. Until that time there had been little overt opposition to nuclear power. Radkau, who has studied the origins of the antinuclear movement, concludes that there was no link to the student protests of the late 1960s. The first commercial nuclear power plant ordered in Germany, a 1967 project by BASF in Ludwigshafen, attracted no attention from the student movement. By the time of Wyhl, the student movement of the 1960s was in decline.[79] The origins of the opposition to nuclear power lay elsewhere, primarily in the region around Wyhl, and expanded from there. As Radkau observes, "The protest at Wyhl at first belonged more to the tradition of peasant revolts."[80]

Yet over the following nine years, what started as a local protest movement in a farming community, almost literally at the grass roots, evolved into a national political party that gained power in numerous local communities and key *Länder* (German states), culminating in the election of twenty-eight Green deputies to the Bundestag (the federal parliament) in 1983. As they entered the hallowed halls of the Bundestag on March 29, 1983, the deputies seemed to symbolize the successive generations of the protest movement and the broadening of the issues they represented: "There was Walter Schwenninger with a long hand-woven peasant sweater alongside Dieter Drabiniok and Gert Jaansen with their flowing locks and wild beards; Marieluise Beck appeared with a pine tree pockmarked with acid rain; Petra Kelly was also there, carrying a large bouquet of fresh flowers."[81]

The ultimate symbol of the Greens' coming of age was the swearing in, two years later, of Josckha Fischer as minister of environment for the *Land* of Hesse. Fischer took the oath of office as environmental minister dressed in jeans and white Nikes, while the president of Hesse, Holger Börner, administered the oath of office in the dark business suit that had been de rigueur for such occasions—until then.[82]

By 1998, thirteen years later, in a development that would have been thought utterly impossible only a decade before, the Greens entered the government in a coalition with the Social Democrats under Chancellor Gerhard Schröder, with Fischer as foreign minister. By this time, he had put aside the jeans and sneakers for a more traditional dark suit accessorized with a stylish necktie. While some of Fischer's former comrades on the left might sniff at such blatant *Gentrifizierung,* Fischer's sartorial evolution nevertheless stands as an apt symbol of the coming of age of the grassroots antinuclear protest movement and its institutionalization in an order that its members had previously rejected.[83]

Summing up the accomplishments of the antinuclear movement between 1975 and the entry of the Greens into the federal parliament, a German historian writes: "In the mid-1970s, grassroots activists changed the course of democracy's development in Western Europe. They forced open new debates, engaged new people in politics, and confronted elected officials with their inability to adequately address their concerns about nuclear energy within the liberal democratic order. . . . Over time, therefore, anti-nuclear activism itself changed the way its protagonists practiced democracy."[84] This, in turn, led directly to the *Energiewende* and prepared the way for the final suppression of nuclear power in Germany. We return to these topics in later chapters. But first, we turn briefly to the Soviet side of the environmental story.

Environmentalism Soviet Style

Environmentalism appeared as early as the 1960s in the Soviet Union, although there also it was initially confined to single issues rather than a broad condemnation of the environmental damage inflicted by the Soviet industrial model. A major controversy broke out over a plan to build a cellulose plant on the Selenga River, one of the sources of Lake Baikal—the largest body of fresh water in the world. The purpose of the project was to produce rayon, an early synthetic fiber based on wood pulp, for bomber tires for the military. Ultrapure water was needed, and Lake Baikal was the perfect source. But Lake Baikal occupies a unique place in the hearts of Russians everywhere, and the news of the military's plan set off a powerful public campaign against it. In the supposedly totalitarian Soviet Union, a coalition of scientists, journalists, and intellectuals banded together to attempt to defeat the project. Such was the hue and cry that the Politburo itself became involved, and a special commission of the USSR Academy of Sciences convened a public hearing to evaluate the project. In the end, the campaign was unsuccessful (the plant was built and continues to operate to this day) but from the early 1970s on the environment was increasingly present in the Soviets' minds, especially among young people.[85]

There was similar sentiment building elsewhere in the Soviet Union, but it would be an exaggeration to say that that the environment per se was a public cause. That would have been a challenge to the Soviet industrial model, and it would have been politically dangerous to be too outspoken in public. But environmental awareness and opposition were building nevertheless, if mostly behind closed doors and among the younger members of the elite.

It was the Chernobyl disaster in 1986 that marked the turning point, and Mikhail Gorbachev's subsequent efforts to use public outrage over Chernobyl to build support for his reforms by encouraging the public to agitate in favor of glasnost (openness), one aspect of which was the environment.[86] Thus, one of the first people to begin building a political following as an environmental defender was Boris Nemtsov, a researcher in radiophysics in the Volga city of Nizhnii Novgorod (a previously closed city then known as Gorky), who was among the first to campaign against the dangers of nuclear testing and contamination. In the wake of Chernobyl he succeeded in mobilizing public opinion against the construction of a nuclear power plant near Nizhnii Novgorod. Nemtsov was one of the young scientists who had already begun campaigning for the environment as early as the 1970s—but as yet inside the walls of the institutes where they studied.[87]

Thus the period from 1970 to the mid-1980s was transformative in many ways. It began with optimism; it ended with anxiety and pessimism. It began with a settled and in many ways traditional political landscape; it ended with new political forces and structures welling up from society and forcing conventional parties and politicians to scramble to adapt. It began with shared enthusiasm for economic growth whatever its price; it ended with the rise of environmentalism. It began with a strong and self-confident Soviet Union; it ended with a weakening one, which faced a stagnating economy. By the 1990s, the agenda of the world—what it perceived, what it worried about, what it debated—had changed radically.

All this was not entirely the product of the oil shocks—after all, the 1968 movement of peace, love, and rebellion preceded the first oil shock by five years—but energy played a key role throughout. This was particularly true in the diverse responses of various European countries to a commodity cycle that began with high prices and ended with low ones in a half turn of the wheel that left policymakers by the mid-1980s to deal with the consequences of their predecessors' decisions, as they entered what turned out to be a decade and a half of low energy prices.

In this chapter we saw how two countries, France and Germany, dealt with the energy shocks in two opposite ways, the first by embracing nuclear power and the second by rejecting it. Then we homed in on the German response, to see how the antinuclear movement arose from the grass roots to generate a new political party, the Greens. With their entrance into the established political structure of Germany, the way was open to the next phase, the rejection of nuclear power and the legislative antecedents of the German *Energiewende.* In energy policy as in other matters, Germany by the late 1980s was unrecognizable compared to the Germany of twenty years before.

Until well into the 1990s, the benefits of natural gas in the local environment—in particular, in terms of air quality—were still powerfully appreciated, especially in reducing acid rain. In Germany, natural gas came to play a leading role, especially in providing residential heat. But in the power sector, the penetration of coal was constrained, as we have seen, by administrative barriers, and by the strong positions occupied by nuclear and coal. But the rejection of nuclear power has left a hole in German energy policy, which for the time being is still filled by coal despite the remarkable progress of renewable sources. If in addition Germany exits from coal, as it is committed to do completely by 2038, and if renewables are not able to fill the gap, how will Germany fill the gap? What, then, will be the future role of natural gas, and where will it come from? We return to this question in the Conclusion.

As we shall see, the Soviet Union was quite different. After a detour to Norway and the North Sea in Chapter 4, we return to the Soviet Union and quite another sort of revolutionary change—the decline and collapse of the Soviet Union and the role played in them by oil and natural gas.

CHAPTER 4

Norway and the Rise of the North Sea

On the face of it, nothing could be more different from the experience of the Soviet Union than that of Norway.[1] The immense gas transportation system of the Soviet Union that was developed in the 1970s and 1980s was built as a single whole spanning the western third of the country, where most of the population and industry were concentrated, in conscious imitation of the massive electrification campaign of the 1930s. If the latter was the signature program of the early planned economy—in Lenin's famous formulation, "Communism equals Soviet power plus the electrification of the entire country"—the gas campaign of 1970–1985 was its last major success story (a theme to which we return in Chapter 5). It was the ultimate illustration of what a centrally planned economy could achieve, a vast infrastructure and a world-leading industry created at great cost in one of the most hostile environments on the planet. And it met the energy needs of millions of citizens and thousands of factories. To this day, Russian *gazoviki* celebrate in books and films the romance of the early explorers and engineers who developed the gas resources of West Siberia, bringing the blue flame of Russian methane to all of Europe.

But if the Soviet gas campaign was a classic example of communist planning, the discovery and development of North Sea gas were signature achievements for capitalism, with equally stirring stories of daring and heroism. The cold of Siberia was matched by the waves and storms of the North Sea, and the endurance and bravery of the divers and builders of the North Sea platforms and the undersea pipelines on its shelf were every bit the equal of those of the Soviet drillers and pipeline construction crews in the mosquito-ridden swamps and tundra of Siberia. Tales of industrial heroism are out of fashion nowadays, and few people today give a thought to what it took to build the systems that bring that convenient blue flame to their homes and kitchens. But one must pay due tribute: the gas campaigns on both sides of the Iron Curtain brought out the best in both systems and the people who distinguished themselves in them.

Yet beyond the parallel stories of romance and derring-do, it is the differences between Norway and Russia that are the more significant for our story. The "capitalism" that developed the Norwegian North Sea was a unique blend of Norwegian state socialism (itself a rather unique animal) and international private-sector entrepreneurship—in other words, it was capitalism Norwegian style. This has turned out to be significant in three respects. The first is the contrasting Norwegian and Soviet systems of ownership and control. The second is the opposing Norwegian and Soviet approaches to industrial and technology policy. The third is the two countries' different relationships to the European Union and the resulting business models under which gas was exported to Europe, especially the approach to contracts. Norway successfully adapted its gas policy to its situation as a small economy exposed to the world economy and the large European market next door. In contrast, the Soviet gas system, for all of its strengths, was the last hurrah of what proved to be a dying command economy largely closed to the global economy. Even today the Russian gas industry continues to labor under problems inherited from its Soviet past. In contrast, the Norwegian model, as it evolved in the 1980s and 1990s and down to the present day, remains a success. As a result, Russian gas, despite its abundance and promise, is a continuing policy challenge for Russia and Europe, while Norway is a positive model that the Russians watch closely and have in some respects begun to emulate.

Soviet gas was from the beginning more geopolitical than Norwegian gas. It was developed in the middle of the Cold War, and there was always the suspicion—especially in Washington—that if Soviet gas gained a dominant position in Western Europe, it would be used as a lever in Cold War politics. That fear turned out to be exaggerated: throughout the Cold War, the Soviets were punctilious in observing their export contracts to the West. Where the threat of leverage was real, however, was in the East European satellites. To maintain their empire, as we saw in Chapter 2, the Soviets delivered their gas to Eastern Europe on a barter basis at cut-rate prices in exchange for loyalty and cooperation. But it is an interesting question who got the better of the deal. The Soviets complained that the East Europeans constantly squeezed them, supplying shoddy merchandise and overpriced labor in return. The leverage was real enough, but the price, at least as seen from Moscow, was high.

The development of the Norwegian North Sea, in contrast, was run primarily on commercial principles, based on a close partnership between the state and private-sector capital and expertise, especially international oil companies. Indeed, the relationship with the international oil companies is the defining trademark of the Norwegian gas industry in contrast with the Soviet one. In both cases the gas network could not have been developed without the contribution of foreign companies. The only difference is that in the Soviet case the foreigners were confined to an arm's-length relationship as suppliers (except in the case of the Orenburg and Yamburg pipelines, where the East Europeans contributed labor as well), whereas in the Norwegian case the foreign companies were invited in from the first as licensees, equity shareholders, and operators.[2] In retrospect, one might argue that the Soviets would have done better to partner with foreign companies on a turnkey basis (as they did in the chemical industry at about the same time[3]), but in the case of the gas industry there were many obstacles to relying on foreign partners, of which not the least was the Soviet leaders' fear of geopolitical leverage from Washington.

In this chapter we take the Norwegian story through the early 1990s. At that point the Russian and Norwegian cases begin to diverge quite strongly, as the Soviet Union collapses while Norway comes under the growing influence of the European Union—indeed, becoming part of it for all practical energy purposes.

The Rise of the North Sea

The Western oil majors had not waited for the 1973 oil shock to begin reorienting their investments away from the members of the Organization of the Petroleum Exporting Countries (OPEC), and particularly toward offshore operations.[4] Shell and Exxon (the latter under the name of Esso) were early leaders. On the eve of the 1973 oil shock, Shell was well on its way to turning itself into a predominantly offshore company: in 1972, over half of its concession acreage was offshore, the largest share of the majors; and by 1974, half of its exploration and production (E&P) drilling and 70 percent of its E&P expenditure was going to offshore operations. By the time of the oil shock, Shell was ready to move quickly. But Shell was not alone: in 1973, 20 percent of the world's oil production and 5 percent of its natural gas already came from offshore wells (although to be sure most of them were shallow by today's standards), and the share was growing fast.[5]

Yet the worldwide trend to offshore operations initially bypassed the North Sea. As late as the second half of the 1950s there was widespread skepticism among mainstream geologists that there was much oil or gas to be found there. In 1958 the Norwegian Geological Survey reported to the Norwegian Ministry of Foreign Affairs that it was impossible that the Norwegian continental shelf could contain any oil or coal.[6] Danish geologists went the Norwegians one better: they were so pessimistic about the prospects of the Danish continental shelf that in 1963 the Danish state awarded all offshore rights for fifty years to a group led by a private company, A. P. Møller.[7]

Skepticism about the North Sea's oil resources was particularly strong inside BP, which, unlike Shell and Esso, had developed an internal view that the world's oil resources were approaching exhaustion. A succession of BP chief geologists believed that there were no major oil provinces left to be discovered in the "free world" outside of the Middle East. This was based on the seeming inexorability of the worldwide decline in the reserves-to-production ratio at that time. As Peter Kent, then BP's chief geologist, wrote in 1970 in an internal exploration review, "We have perhaps been searching for a mythical Holy Grail, and the evidence becomes stronger that it does not exist, that there is only one super-oil province [the Middle East] in the free world."[8] It was partly this state of mind, as well as chronic indebtedness and a shortage of capital, that accounts for BP's skeptical attitude toward offshore exploration for oil. As we shall see below, most of BP's subsequent investment in the North Sea was focused on the southern blocks of the British sector and concentrated largely on gas, with which it was ultimately successful.

The discovery of Groningen in 1959 encouraged geologists and companies throughout the oil world to take a second look. The Groningen deposit was unusual because it lay at a great depth. Further investigation showed that the gas most probably originated from a gigantic coal bed that extended far into the North Sea, nearly to the British coast.[9] Once this was realized, the hunt was on: "By the autumn of 1963 about twenty companies were actively exploring the North Sea. . . . The seismic surveys were done by helicopter, because the area was still not free from mines, eighteen years after the war."[10] (Helicopters were crucial to the development of the North Sea. As Shell's official history notes, "At the peak of the building period the flight movements of the helicopters were as numerous as the plane movements at Heathrow."[11])

Disaster, alas, played a part: in 1957 a storm destroyed a Shell platform off Qatar, and twenty-two men lost their lives. But the company was determined to go forward, and two years later a more storm-resistant platform, the *Seashell,* was completed in a Dutch shipyard and towed 6,400 miles to Qatar. Though built for Qatari conditions, "it was one of the forerunners of the great fixed platforms that later sprouted in the North Sea."[12] Consequently, when Shell decided to redirect its E&P investment away from OPEC, and it had become apparent that the North Sea was very promising, the necessary technology was ready.

But the North Sea posed challenges of an entirely different order. The half-dozen years following the strike at Slochteren had seen mounting excitement in the Northern European gas industry. Yet the North Sea was notoriously treacherous. As one veteran sea captain sighed, "There's nothing quite as vile as the North Sea when she's in a temper."[13] And she was in a temper nearly all the time: one study showed that "over the sea as a whole, there are fewer than 2% of calms throughout the year."[14] Waves up to one hundred feet high were not uncommon.[15] Indeed, the first major gas discovery in the North Sea, at West Sole in 1965, had also been marked by tragedy. Barely three weeks after the first successful test flare at West Sole, the drilling platform, the *Sea Gem,* collapsed in a storm and thirteen men were drowned.[16]

To deal with such challenges an alliance was a natural fit, to share technology and experience—and risk. In 1964 Shell and Esso formed a joint venture, known as Shell Expro, to explore for oil and gas in the North Sea. Before long Shell Expro had developed into the largest capital investment made by either of the partner companies anywhere in the world. "By 1976," notes Shell's official history, "more than 80 percent of Shell's capital expenditure on oil and gas production outside North America was in the North Sea."[17]

Yet, as Shell's history recounts, "At the end of 1965 Shell still did not expect that the North Sea would add much to its crude oil reserves, but by 1970 11 fields had been found."[18] These were almost entirely in Dutch and British waters. It was initially perceived that the blocks south of 56 degrees north were more promising than those to the north of it, and consequently there was more competition for the southern blocks than the northern ones. No one suspected the enormous fields that lay hidden in the north. Nevertheless, in 1964 Shell Expro bid for the northern blocks as a preemptive move and obtained all of the fifty blocks offered. In the second half of the 1960s, it became apparent that the northern blocks were far richer than the southern ones.

But who owned them? At this point fate intervened. As Petter Nore recounts in his excellent thesis on the history of Norwegian petroleum,

> The legal agreement establishing the median line between the UK and Norway portions of the shelf was signed in April 1965. Any attempt by the UK to challenge the Norwegian interpretation in an international court would have taken many years to settle, if the normal speed of such cases is anything to go by. And the UK was in a greater hurry to extract oil from the North Sea than was Norway. All of Norway's present oil and gas fields are today situated in what would have been disputed waters had Great Britain persevered against the Norwegian interpretation.[19]

The Norwegians were remarkably fortunate. As Fredrik Hagemann, the first director general of the Norwegian Petroleum Directorate, acknowledged in an interview years later, "Had the border been drawn some tens of miles farther east, Norway would have had no part in North Sea oil and gas resources at all."[20] The story of the negotiations with the Danes is even more fortuitous. According to an oft-repeated story (at least in Norway), the Norwegians were aware that the relevant Danish minister was well known for arriving in his office in the morning in a high state of inebriation. Norway's proposed median-line division between Norway and Denmark was accordingly put before him in the morning and was duly signed. Included in the Norwegian zone was the sector that subsequently turned out to include Ekofisk, the first major oil discovery in the North Sea.[21]

Persistence and Luck: The Discovery of Ekofisk

Sometimes it pays to take a vacation. In July 1962 Paul Endacott, vice-chairman of Oklahoma-based Phillips Petroleum, was touring the northern Netherlands with his family when he made a most interesting discovery: a drilling derrick in the neighborhood of the town of Groningen. At this time the news of Nederlandse Aardolie Maatschappij's discovery at Slochteren—which the company had kept carefully under wraps while it continued exploration—was still largely unknown to the wider public. Only small discoveries had been made elsewhere in northwestern Europe. And skepticism about the North Sea itself was still widespread among the major oil companies. The head geologist in one large company famously claimed that he "would drink whatever oil could be found on the Norwegian shelf."[22] Gas he did not even think worth mentioning.

But when Endacott (his curiosity aroused) got back to Phillips headquarters, he urged Phillips managers to investigate the possibilities of exploration in the North Sea. This set in motion a chain of events that led seven years later to the first supergiant discovery in the North Sea, the Ekofisk field in the Norwegian sector of the sea, and shortly afterward to the first exports of oil to Great Britain and gas to Germany.[23]

At the outset, for the reason already mentioned, Phillips had the Norwegian offshore operations to itself. Indeed, what initially drew Phillips to Norway was that as of the autumn of 1962, Norway was the only country that had not yet been contacted by other oil companies. But this did not last long. By the following spring Phillips learned that the Norwegian Ministry of Foreign Affairs had awarded a number of approvals for seismic exploration. For obvious reasons there were as yet no special agencies in Norway with expertise in oil and gas matters. Phillips's initial guide and go-between in its first contacts with the Norwegian government was Trygve Lie, former general secretary of the United Nations and Norway's most distinguished diplomat, who was assigned by the Ministry of Foreign Affairs to handle the initial contacts with the international oil companies.[24] Phillips hastily sent in an application of its own, but the Norwegian government was initially unprepared to receive it.

At this point, the head of the Foreign ministry's legal department, Jens Evensen, took up the challenge of developing the foundation for Norway's legal claims to the Norwegian continental shelf. At the end of the same month, Norway declared national sovereignty over its sector of the continental shelf—a huge piece of real estate bigger than the entire land mass of Norway—and issued a temporary law on exploration and exploitation of the subsea reserves. Evensen, one of Norway's most brilliant civil servants, continued to play the key role in Norway's offshore territorial negotiations, especially with Great Britain, for the next decade. A succession of Norwegian prime ministers have paid tribute to Evensen's contribution to securing Norway's rights to offshore resources, which in turn spawned the country's oil industry.[25] A recent biography calls him "the man who made Norway larger."[26]

Even once the sectoral division had been settled, it took several more years before the necessary approvals were obtained, equipment and manpower were assembled, and actual exploration began. But the first two years of drilling in the Norwegian sector yielded little more than a series of expensive dry holes. By late 1969, as the drilling season neared its end and the North Sea's famous winter storms began brewing, the picture was discouraging in the extreme. Thirty-two wells had been drilled on the Norwegian continental shelf without a single commercial discovery. The chief geologist on the Phillips project pleaded for one more well in a promising new zone, but he was turned down by company headquarters. The message from Oklahoma was clear: no more wells.[27]

But what to do with the drilling rig and the crew? The rig had another year to run on its lease, and daily charges accrued whether it was used or not. Efforts to sublease it found no takers, and Phillips was stuck with the bill. This time, when the project geologist made his case for another well, headquarters reluctantly agreed. As the weather worsened, the rig was towed to the site and work began. After two days of drilling, as the platform supervisor recalls, "Oil mixed with gas and drilling mud came up. . . . I found a bucket which I filled with oil." That night, he wrote excitedly in his daily report, "I can cover the North Sea from here to the North Pole with oil."[28] It had been a classic case of the last well, but it delivered, and soon the name of Ekofisk—a made-up name formed from the words "echo" and "fish"—shot around the oil world.

Thus, well before the first oil shock of 1973–1974, the international oil companies had already begun a far-reaching transition away from their previous dependence on the Persian Gulf and OPEC. Offshore exploration and production accounted for a steadily growing share of their activity, and in the years immediately preceding the oil shock one exciting discovery followed another, both in the Norwegian and British sectors—the Forties field in 1970, then the Frigg, Auk, Brent, and Cormorant fields in 1971.[29] Then, as the historian Daniel Yergin notes in *The Prize,* "The 1973 oil crisis turned the rush into a roar,"[30] starting with the discovery of Statfjord in 1974.

Norway and Great Britain

The response of Norway is often compared to that of Great Britain, the other great beneficiary of North Sea oil and gas. Some authors speak of a "North Sea model," of which the central trait is a mixture of state regulation and private control. We shall have more to say about the British approach later on in this book; for the moment, suffice it to observe that by the 1970s, Norway and Great Britain had adopted radically different economic policies. Confronted with a ravaged economy after World War II, Great Britain had opted for state ownership of the "commanding heights," including the entire network of town-gas companies.[31] Though the government of Norway also favored a strong state-guided welfare system and Keynesian macroeconomics, it went in a different direction, with a more balanced blend of state and private ownership.

There is first the matter of size, and second the role of the state. Great Britain, despite its troubles in the 1970s, was a mature industrial economy, in which energy, though important, was just part of a much larger and highly diversified industrial structure. Norway, by contrast, was a much smaller country, most of whose GDP in the 1960s still came from farming and fishing and the industries that supported them, such as shipping and hydropower (which produced fertilizer).[32] The British gas industry had grown up organically, from a host of local municipal companies that produced town gas from coke for local consumption; it was therefore effectively a by-product of the steel and coal sectors (and at the time a backward and unprofitable one at that).[33]

In Norway, gas and oil were new. Great Britain burned coal and imported oil, while Norway was largely self-sufficient, thanks to its waterfalls and hydropower.[34] Because of its small size, Norway could not afford to close itself off from the world economy and capitalism. In contrast to Great Britain, it rejected state ownership as an instrument of industrial policy. As late as the early 1970s, only twelve Norwegian industrial firms had a majority state share.[35] Yet it had an extremely tight regulatory regime, which was maintained through the end of the 1950s. People of a certain age remember their childhood as a time of austerity. "You had to apply far in advance to buy a car, even a small one," one Norwegian oilman told me. "Everything was rationed, especially imported things like cocoa and sugar. We could buy only one bar of chocolate every two weeks."

But in the early 1960s, as the Norwegian economy began to prosper again, the rationing was replaced by a more open and market-oriented economy with less regulation. A succession of predominantly center-left Labor governments, though committed to Keynesian macroeconomic policies and the Bretton Woods system, nevertheless adopted a less interventionist stance than the British government did and—to preserve Norway's access to the world economy—rejected the urge to nationalize. But at the same time the Norwegian government played a major role as an investor. It accumulated capital, which it then transferred to the private sector; however, it was the latter that ultimately controlled its allocation.[36]

As a result, when natural gas came to Norway and Great Britain, it did so in two very different ways. In Great Britain, a state-owned monopoly (initially the British Gas Council, which was succeeded by British Gas) controlled distribution and sales literally down to the burner tip—the Gas Council's monopoly extended to the sale of kitchen stoves.[37] It negotiated with the private-sector companies that produced offshore gas, while participating as both partner and competitor in the offshore operations upstream. However, as a state-owned company, it was under the constant (and highly unpredictable) regulatory authority of successive left and right governments, whose ideologies differed greatly. In contrast, the Norwegian response was to create a national champion that, as time went by, increasingly functioned as a private company.

The international oil companies found that they were more comfortable with the Norwegian approach to oil policy than with that of the British.[38] For example, "viewed from afar," writes Exxon's official historian, "the path chosen by the Norwegian government appeared more threatening to the oil companies than that taken by Great Britain, but a distinct difference in tone and process made the Norwegian approach easier for Exxon to accept. The rules were known, and the government generally had the power, the national support, and the skilled public officials to enforce them consistently." In Norway, the energy sector tended to draw the top graduates of Norwegian schools, whereas the UK Department of Energy did not, as a rule, get the highfliers. In Great Britain, moreover, there was greater vulnerability to shifting political winds. Thus, the Exxon history gives high marks to Norwegian energy policy, which it ascribes to "strong institutions somewhat insulated from the pressures of day-to-day politics." And there follows the ultimate accolade: "In this sense, Exxon faced a government whose commitment to efficiency and profits for the nation resembled, in spirit, Exxon's approach to business."[39]

The Norwegian Model: The "Ten Oil Commandments" and the Creation of Statoil

In 1972 the Storting (the Norwegian parliament) passed a measure that has come to be known as the "Ten Oil Commandments."[40] This document, which governed gas policy as well, occupies the same symbolic place in the Norwegian industry as the famous "nota de Pous" of a decade earlier in the Netherlands (see Chapter 1). It laid down two key principles in particular: (1) petroleum resources were to be developed for the benefit of the entire nation (although there was no mention yet of a petroleum fund) and to remain under national management and control; (2) one of the main goals of petroleum policy should be the development of an indigenous industrial base to support the exploration and development of petroleum. The Ten Oil Commandments remain the founding document of Norwegian energy policy to this day.[41]

One issue not addressed, however, was the relationship to the rest of Europe. The commandments contained no mention of gas exports (the first Norwegian gas exports did not begin until five years later, and even then they were much smaller than the country's oil exports).[42] Indeed, one of the commandments was that Norway's petroleum must be landed in Norway (the aim was to develop a petrochemical sector), although the document cut the suit large by allowing for "the exception of individual cases where national policy concerns call for a different solution."[43] That particular commandment was soon abandoned, along with the petrochemical ambitions. Today only a fraction of Norway's petroleum, whether oil or gas, is landed in Norway. The bulk is exported directly from offshore fields via undersea pipelines,[44] and the division of the gas between the export and domestic markets—unlike the Soviet case—has never been a major issue.[45]

Europe was very much on the minds of the Norwegian drafters in 1972, but for a different reason. Norway was at this time in the middle of a bitter national debate over whether to join the European Economic Community. In the end, after a contentious campaign, a slender majority of Norwegian voters opted to stay out.[46] Most of the Norwegian urban population and business interests were in favor of joining, but most of Norway's farmers and fishermen were opposed. More fundamentally, the "no" vote was an uprising by a large part of the Norwegian public against what was perceived as a loss of sovereignty, including control over natural resources. It was a grassroots movement, underestimated by the political elite. The conflict was nothing new: it dated from Norway's first application in the 1960s. Yet the defeat continues to rankle among former "yes" voters, especially in the energy industry. As Arve Johnsen, later the first CEO of Statoil, commented sardonically in his memoirs about his early experiences of Norway's Europhobes as a young Labor Party activist, "I quickly learned that for some there was no relationship between life and learning when it concerns international solidarity." This was a frequent theme with Johnsen over the years. He goes on to write acidly about "our country's huge inferiority complex, fear of the unknown, and of mutually binding economic and political cooperation with our European sisters and brothers, including our neighbors in the north."[47] In retrospect, the "relationship between life and learning" has turned out to be considerably more complicated than Johnsen anticipated. Indeed, many Norwegians today may congratulate themselves on having stayed out, both then and in the second referendum that followed in 1994, in which the "no" side won again. But that is another story.

As momentous as the 1972 decision was, its immediate consequences for energy policy were minor, for the simple reason that at this time a European energy policy hardly existed.[48] It was only over the course of the following twenty-five years, as the European Community evolved into the European Union and energy came to the forefront of EU policy, that Norway's gas relationship to Europe became a major matter. For the time being, Norwegian gas was sold to Europe in essentially the same way as Dutch gas (and, for that matter, Russian and Algerian gas), on the basis of long-term bilateral contracts with prices linked to oil. The issues that subsequently caused conflict over Norwegian gas—chiefly having to do with competition and antitrust law—did not emerge until the 1990s, as we shall see in Chapter 7.

Statoil, when it was created in 1972, was given a 50 percent share in the Statfjord field and Norpipe, which transported oil from Ekofisk to Teesside in Great Britain and gas from Ekofisk to Emden in Germany. Elsewhere in the world, nationalization was in the air, but instead the Norwegians built a unique model based on cooperation between a strong state and the international companies. Note again the differences between the Norwegian and the Soviet models—the Soviets did not invite international players to come in as partners. The explanation is straightforward: the Soviets were already highly experienced in petroleum matters. What they needed was the pipe, compressors, and money.

Arve Johnsen's Butterfly

The creation of Statoil and the Norwegian Petroleum Directorate in 1972 were above all the work of Labor Party governments. The Labor Party, which had strong socialist roots going back to the beginning of the twentieth century, had been the dominant political party in Norway since World War II. The minister of industry, Finn Lied, appointed Jens Christian Hauge, a former minister of defense who had been leader of the Norwegian resistance in World War II, chairman of the board of Statoil[49] and named Arve Johnsen, his state secretary and fellow Labor Party member, the first CEO of Statoil. (It was Johnsen who coined the name Statoil, recently changed to Equinor.) All these key officials were allies from the Labor Party, going back to the late 1960s, when Johnsen and several party colleagues had worked together on the Labor Party's Industry Committee to develop the party's oil policy.[50] Over the following years the Labor Party was the connector, especially in times when there was a center-right government. "Tight links between Statoil, the bureaucracy, and the government were established through the Labor Party," observes a historian.[51] These connections endured into the 1990s.

Johnsen was promoted to CEO of Statoil when he was thirty-eight. Pictures of Johnsen in the early 1970s show a still-boyish face, but with a self-confident smile and a strong jaw. If he was intimidated by his august position, he certainly did not show it. In fact, it was not yet very august. Johnsen later recalled that when he made his first trip to Stavanger to start the new company, he carried with him a cigar box to keep the company's money in, and initially he slept in a sleeping bag on the bare floor of his apartment.[52] Within a few years things had changed, and by the time of his departure fifteen years later, Johnsen was running a company with over 11,000 employees.[53]

Johnsen was renowned as a strong boss—British observers referred to him as the "Denis Rooke of the Norwegian oil and gas industry," referring to the redoubtable head of British Gas. But Johnsen had a poetic side that he showed to very few people. In an interview he recounts that among the bouquets given to him at his accession there was a butterfly, and though it soon died Johnsen preserved it carefully throughout the fifteen years that he ran the company. In 1987, as he resigned his office, he recalled his leave-taking: "I locked the door, put the key on the desk of my secretary Bente Godø, removed the nameplate and left Statoil. At home in Madla I picked up the butterfly, laid it in a small box and buried it in the wild. Its time had passed." It was a symbolic burial indeed. As Johnsen wrote, "Statoil was my fourth child."[54] Others agreed. As Hauge commented, "Statoil was Johnsen's life's work."

During the 1970s and early 1980s, Statoil was considered a great success and an example of what strong state support could achieve, but after the second oil shock of 1979–1980 and the vast increase of oil revenues that followed for the next few years, it was increasingly perceived in Norway that Statoil had grown too powerful. There was also growing resentment over the long-standing tie between Statoil and the Labor Party. Labor lost heavily in the general election of 1981, and a conservative government ruled Norway for the next four years (from late 1981 to early 1986) in varying coalitions. In 1984 it proceeded to "clip the wings"[55] of the company by transferring a large bloc of the shares in the licenses that Statoil owned to a newly established body, the State's Direct Financial Interest (SDFI), whose mission was to administer the state's interest as direct owner.

The wing clipping was accompanied by considerable animosity on both sides. Kaare Willoch, a former prime minister, said he preferred to call it a "clipping of the claws."[56] "The danger," Willoch said, "was that Statoil (if allowed to keep all the revenues from the oil) could become *alone* the biggest part of the Norwegian economy. There was a need to safeguard the income for the state. In Norway there is a long-standing tradition that energy resources belong to society rather than to private owners." The principal victim of the "claw clipping" was Johnsen. In 1987 he was brought down over the huge cost overruns caused by an overambitious refinery project at Mongstad. "Arve Johnsen was bewildered by the amount of money coming in from the oil industry," Willoch commented. "It seemed as though Arve Johnsen thought that he had to spend the money that came in, or it would disappear. He built a refinery, and for what?"

But time has softened passions, and today Statoil even finds something positive to say about the episode. "The 'wing clipping' saved Statoil," said one former senior vice president. "We were allowed to keep Statfjord, and the cash flow from that project helped us to keep going through the oil-price crash of 1986 and after. And since we had been stripped of several licenses, our investment requirements were much smaller. The state ended up shouldering the risk and taking the losses."[57] One way or another, the state's role remains dominant: today, Statoil and the SDFI (now called Petoro) between them own some 70 percent of the equity in Norwegian oil and gas assets, as well as the most important transmission systems and terminals for natural gas.[58] As one Norwegian observer concludes, "What the government accomplished was to arm itself with a powerful economic instrument to intervene in the petroleum sector, without necessarily having to interfere in Statoil's managerial decisions."[59]

Thus, over the years the relationship between Statoil and the state (essentially the Ministry of Petroleum and Energy) evolved. At the beginning, in the "infant days of the Norwegian oil industry," Statoil was clearly a "junior partner to government," and more specifically to the Labor Party.[60] Later on, under the influence of the European Union and more liberal ideologies, the state's role increasingly became that of regulator and equity owner. Finally, when Statoil was partly privatized in 2001, it became primarily a commercial company. Harald Norvik (chairman from 1987 to 1999) was the last leader to come from the Labor Party.[61] Norvik was symbolic of the change. He pushed for Statoil's privatization; he launched a global alliance with BP; and after his resignation from Statoil in 1999, he became the chairman of the board of the Oslo Stock Exchange. In that capacity (and with the broad smile one can imagine), he wished Statoil welcome to the "Norwegian center of capitalism" as it joined the exchange after its partial privatization in 2001.[62]

Selling the Gas: Norwegian Export Policy

In the late 1960s and early 1970s, marketing gas was something new for both the Soviets and the Norwegians. By what rules and principles would the Norwegian gas be exported, and how were these different from those for Soviet gas? What lingering consequences have these differences had for the subsequent politics of gas exports from both countries?

The Soviet approach to gas exports was shaped by the existing institutional and political structure for foreign trade, which was that of a state monopoly. A foreign trade monopoly staffed by foreign trade professionals (initially borrowed from the oil-export monopoly) negotiated the first gas contracts, subject to import-export targets that were themselves determined by the planners on the basis of the needs of the domestic economy. Since each gas contract was essentially a countertrade deal involving pipe, machinery, and finance as well as gas, there were numerous teams involved from the separate bureaucracies that handled gas, equipment, and money. Coordination was a constant problem, as we have seen, which the Soviets handled (with more or less success) by putting a senior government official in overall charge. Initially the gas ministry was only one player among many, although as time went on and the scale of the Soviet gas exports grew, the gas ministry gradually gained the upper hand and eventually absorbed the foreign trade side altogether (see Chapter 2). Foreign companies played no part in this system.

The Norwegians, in contrast, started from scratch. Norway had no previous experience in selling hydrocarbons, and in any case there was no state monopoly on foreign trade. The Norwegians' solution was to entrust the negotiations for financing and exports from the first two major gas fields to the foreign operators.[63] Negotiations over the gas contract for Ekofisk were led by Phillips on the Norwegian side and a consortium of European companies headed by Ruhrgas. (In those days, the competition rules prevailing in the European Community were much less strict than they have become since, and the gas was sold by monopolies on both sides—a buyers' group on the Continent, and a sellers' group led by Phillips in Norway.) The first contract was signed in January 1973.[64]

The model for gas exports in Europe at that time was the classic Groningen contract (described in Chapter 2), and that is what Phillips and the Norwegians followed in their first exports to the Continent. By this time the Groningen model had become the norm throughout Northern Europe—including, as we have seen, the Soviets' first contracts with Austria and Germany. Unlike the Esso-Shell team that had had to fight its way into Germany and France a decade earlier against indifferent or hostile entrenched interests[65] (which the Soviets to some extent experienced as well),[66] the Norwegians found the commercial ground already prepared.

Even so, it took four and a half more years to complete the Norpipe pipeline to Germany, which landed at a newly built terminal at Emden, a small port town between Groningen and Bremen. At 443 kilometers in length and 36 inches in diameter, it was the longest welded steel structure in the world at the time. Building it required surmounting challenges of all kinds, ranging from the weather to inexperienced work crews. The German navy spent months sweeping old mines left over from World War II, one of which was discovered along the pipeline route.[67] At the last minute, the Danish government demanded that the pipeline be buried in a trench along the Danish portion of the route. The Danes were adamant: no burial, no operation. The standoff was finally defused when Norway's minister of industry, observing that his Danish counterpart had left for vacation, flew to Copenhagen and negotiated a settlement with his temporary replacement. Yet when the grand inauguration of the pipeline took place in Emden in September 1977, no gas came through it. The gas was delayed by a further two weeks, as Danish soldiers worked desperately to fill sandbags to be laid over the pipeline to provide a temporary burial.[68] Finally, later in September, Norpipe began transporting the first Norwegian gas from Ekofisk to Emden (see Map 4.1).[69]

Learning to Sit in the Front Seat

Over time, as the Norwegians gained in experience and confidence as well as information, they gradually took over the lead roles in the operation and exports of their gas fields.[70] One way of visualizing the relationship between the Norwegian gas producers and their international partners is to follow the history field by field, as each one was discovered, negotiated, developed, and finally connected to the export system. The sequence from Ekofisk ran through progressively larger and more complex fields, each with evocative names: Statfjord and Gullfaks in the 1970s and early 1980s, Sleipner[71] and Troll in the later 1980s, and finally Snøhvit and Ormen Lange at the beginning of the new century. At Ekofisk, Phillips was almost entirely in charge, and at Statfjord, Mobil played the same lead role. But by the time of the next major field, Gullfaks, the Norwegians were ready to take on the roles of developer and operator. In 1978 the license was awarded to a wholly Norwegian group consisting of Statoil, Norsk Hydro, and Saga, and production began eight years later. Today, Statoil speaks of Gullfaks as "a final exam for Statoil."[72] Johnsen singled out the importance of these fields as a turning point in Norwegian policy:

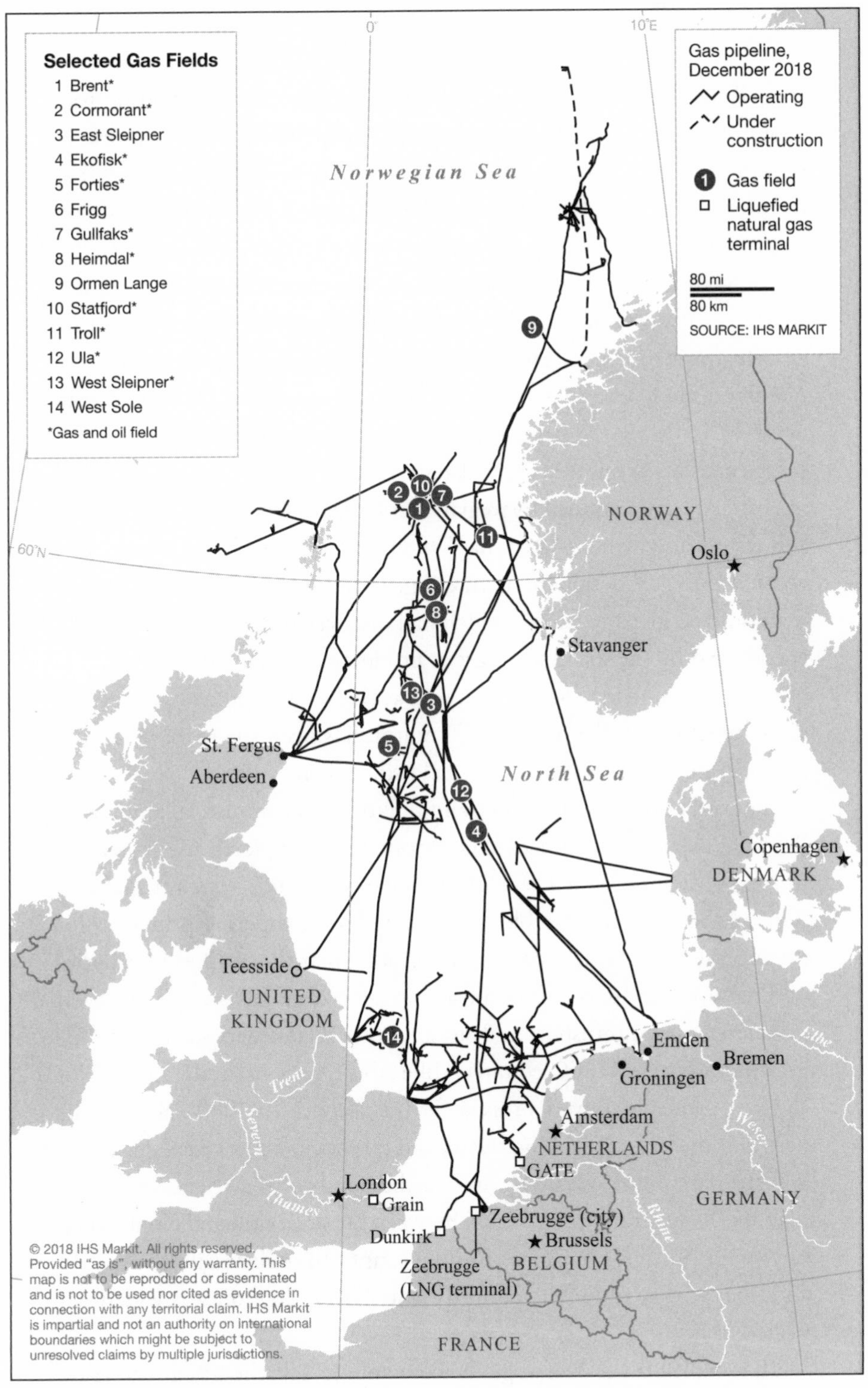

Map 4.1 Major gas industry infrastructure of the North Sea and Norwegian Sea.

"Of all the decisions made by oil companies and Norwegian authorities about oil and gas policy, none are comparable to the government resolution in June 1981 to explore Gullfaks, Statfjord, and Heimdal, as well as to transport the gas from Statfjord and the other fields. The resolution was historic."[73]

The giant Statfjord field was especially significant as a major step forward in Norway's ambitions and capabilities for producing and exporting gas. Beginning in 1985, gas from the field was shipped to the Norwegian mainland via a line called Statpipe to a gas-treatment plant at Kårstø north of Stavanger. As Rune Bjørnson, executive vice president of Statoil for the natural gas business area, relates, "Statpipe laid the basis for the whole gas infrastructure on the Norwegian continental shelf. . . . [C]onstructing this system also demonstrated that Statoil could handle the entire gas value chain from production to marketing."[74] Initially, the operatorship at Statfjord was given to Mobil, which had discovered the field. But in 1987, encouraged by the successful start of production at Gullfaks, Statoil took over the operatorship. As Johnsen said at the time, "You cannot learn to drive by sitting in the back seat."[75] Statoil has been the main developer and operator of every Norwegian field since then.

Statfjord also marked a high point in Norway's commercial ambitions. With gas export prices linked to crude oil, prices for North Sea gas tended to follow the prices for Brent. As Brent rose, so did Norway's expectations at the negotiating table. There were two competing principles for pricing gas at the time—parity with crude oil at the wellhead, or parity with competing fuels (which meant in practice a basket of refined oil products), minus transport and storage costs. Gas sellers such as the Algerians and other OPEC countries like Abu Dhabi naturally pushed for the first principle; buyers pressed for the second. In 1980–1981, at the height of the panic following the second oil shock, buyers were desperate, and the bargaining advantage was in the hands of the sellers. In particular, there was strong competition between Great Britain and Continental buyers for Norwegian gas, as Europe sought to replace oil with gas. In addition, at that time Dutch and British gas production was widely expected to decline.

Thus in 1980 the Norwegians were able to sell the gas at crude-oil parity from a small field called Ula to a small West German gas company,

Gelsenberg. Gelsenberg also agreed to purchase gas from Statfjord at crude-oil parity. But as the wave of panic passed, it was increasingly felt that the Norwegians had overreached themselves. James Alcock, at the time purchasing director for British Gas, disapproved of the high price demands from Norway, so much so that he asked sarcastically whether "gold dust parity" would be the next claim.[76] Buyers began to push back. In self-defense the continental buyers formed a consortium led by Ruhrgas and negotiated as a group. At first this did not change the terms. Gas from Statfjord was sold to the consortium at crude-oil parity for $5.50 per million Btu, or about $32 per barrel of oil equivalent. These were the highest prices in the world at the time and (in real prices) the highest ever reached by Norwegian gas.

At this point the Norwegians were landing their gas at only one location, Emden in Germany. They soon came to perceive this as a source of risk and sought to build a second outlet, this one taking gas to the Netherlands and Belgium. Diversity of pipelines has been a fundamental part of the gas industry's creed from the earliest days. But diversification can have all sorts of motives, from abstract geopolitics to the personal clash of strong egos. The story of Statoil's second pipeline was one of the latter. A Norwegian eyewitness tells the story:

> The negotiations with the Dutch were particularly tough. In one episode Arve Johnsen was negotiating with a senior figure from Gas-Unie named Antonius Grotens. Grotens wanted to take the gas at Emden but not pay for transportation in Holland. Johnsen rejected this out of hand. But the two men were so angry that you sensed they could easily have come to blows. Instead, they both took savage stabs at the peas on their plate, as if by "killing the peas" they would stop short of killing each other. Afterward, Johnsen gathered our team and said, "This has got to stop." And someone said, "Why don't we land the gas at Zeebrugge [in Belgium] instead? It's a perfect hub location." This proved obviously very popular with the Belgian gas company Distrigas, especially since up to that time Distrigas had been treated as the poor relation at the lower end of the table. The Dutch were helpless to do anything about it—and as a result they lost a fortune in transit revenues. Maybe they should have killed fewer peas.[77]

The Sleipner-Troll Dispute

But as the 1980s went on, the Norwegians' bargaining position weakened. This was part of a general pattern throughout Europe, as oil prices peaked and began to decline. As gas expert Jonathan Stern writes: "Even before the collapse of oil prices in 1986, the keen competition for available supplies experienced in the 1970s had evaporated. The 1980s saw the European gas market turn from importers competing for supplies to exporters competing for markets."[78] As oil prices stagnated and then began declining in the mid-1980s, the Norwegians were increasingly on the defensive.

But here commercial negotiations became intertwined with geopolitics. As the Norwegian gas exports had gained in scale, they had also grown in strategic importance in the eyes of Norway's allies, especially the United States. Washington saw Norway as a potentially vital counterweight to Soviet gas, while Norway saw the role of the international oil companies as part of its anchor in the Western alliance and hence its defense system. Until the early 1980s these multiple roles coexisted smoothly, especially as long as energy prices were rising. But in the mid-1980s, when oil prices began to fall, Washington's security concerns and Norway's ambitions as a gas exporter collided. The result was one of the few instances in which US president Reagan and British prime minister Margaret Thatcher came into open conflict. The two issues were the price of Norwegian gas and Norway's role as a strategic exporter, and the battlefield was a pair of fields called Sleipner and Troll.

The Sleipner gas field lies about 1,000 meters under the bed of the North Sea and about 250 kilometers west of Stavanger on Norway's southwest coast, in water that is 100 meters deep.[79] First discovered in 1974, it is still producing gas today, although its chief claim to fame now is that it is the world's first offshore carbon capture and storage facility.

But for all its impressive dimensions, Sleipner was not the largest gas field that had been found in Norwegian waters in the 1970s. That honor fell to the Troll field. The Norwegians knew that one day they would develop Troll and that it would become the jewel in the crown of their petroleum industry. But they did not want to do it just yet since the technical challenges were immense: the water in which the field stood was very deep and the seas wild. Sleipner had to be first.

But the US government also had its eye on Troll. The Americans recognized that the Troll field was of such a scale that it could compete with Soviet gas in the strategic provisioning of European energy markets. It lay under the seabed of a NATO ally—moreover, an ally whose northern land border with the Soviet Union made its leaders aware of and highly sensitive to Cold War threats and tensions. What could be done, US leaders wondered, to accelerate the interest of European gas companies and the Norwegian authorities in developing Troll? US interest in Sleipner and Troll grew even stronger after the failure of an attempt by the Reagan administration in 1982 to block the construction of a new Soviet export line to Europe by using an embargo. In the face of united opposition from its European allies, the United States abandoned the embargo but increased pressure on the Norwegians to move quickly toward developing the Troll field.[80]

The negotiations between British Gas and Statoil over the Sleipner field had begun in 1982, and by 1984 they had led to a draft agreement that both sides initially welcomed. British Gas was eager to close the deal. Looking at their contracted future supplies, the likely future demand, and their legal obligation always to have gas available for the customers, company officials knew—or thought they knew—that they needed the Sleipner gas. The Norwegians were equally eager, since they needed to get the deal done before they could turn their attention to Troll. The US government wanted to see this matter settled, too, again so that the Troll field (the really important prize) could get under way.

Yet at this point, to the surprise of all parties, Prime Minister Thatcher personally rejected the Sleipner deal. One might be tempted to suppose that her decision was driven by the shifting fundamentals of the oil market. International oil prices were already falling, and the price agreed in the Sleipner contract had begun to seem high. It may well have been that Thatcher wished to give priority to Great Britain's own offshore gas projects, combined with disagreements over the pricing of gas in the UK domestic market.[81]

This interpretation is strengthened by the publication of excerpts from a letter to Thatcher from one of her advisors at the time, John

Redwood, in which he wrote, "Owing to past follies over gas prices, we are faced with the ridiculous position that a country full of energy resources may well have to rely on imported gas to meet its needs in the early 1990s."[82] This was the perception of the Norwegian negotiators as well. As one former senior Statoil executive comments,

> We spent years at the negotiating table. Mrs. Thatcher's decision came as a surprise. Being in the middle of it, my understanding of it was a fear from UK operators, and in particular from Shell, that the Sleipner deal would make future gas projects on the UK shelf difficult. Also too high prices in the contract and agony over the self-ruled British Gas were elements. But they were minor compared with the industry's understanding of the UK continental shelf. Looking back, we have to admit that there never was a need for Sleipner gas, as negotiated. Thatcher saved billions on her current account and secured development of UK gas fields.[83]

Yet there was more than met in the eye in Thatcher's decision. A careful reconstruction reveals that Thatcher's decision was largely based on domestic, even parochial, concerns unrelated to geopolitics or economics. These included personal rivalries among the ministers involved, the insider power of technocrats who had Thatcher's ear, embryonic ideas in economic theory about creating a marketplace on the network for industries like gas and electricity, and intellectual debate around these ideas in academic and senior Conservative party circles. Thatcher's decision may even have been influenced by the pressure within the Conservative party at the grassroots level from small businessmen who hated British Gas's monopoly of selling gas stoves in British provincial high street shopping centers.

By this time the Reagan embargo had collapsed, and Washington's interest was moving on to other things. Yet the Norwegians were reluctant to admit that their moment had passed. Johnsen declared in an interview in 1983: "Norwegian gas will never be inexpensive gas. It will command a premium price, partly because of high costs, partly because deliveries are based on long-term agreements from a stable area."[84] In the end, the sale of gas from Sleipner and Troll was bundled into a single contract, which the European buyers' consortium signed in December 1985. It was not until the Labor Party returned to power under Prime Minister Gro Brundtland that Norway finally backed down and accepted that its bargaining power had disappeared.[85] When the second Brundtland government came to power in 1986, Norway adopted a form of market pricing, and it was able to sell the gas from Troll and Sleipner on that basis, although on less attractive terms than the Norwegians had obtained up to that point.[86] As Stern writes, "The importance of the Troll contract for overall European relationships was that, with the Norwegian acceptance of these principles, the fully commercial nature of the contracts—without any political or security premia—was established."[87] This was a model that the Russians did not accept for another generation, and then only very reluctantly.

Norwegian Industrial Policy: The Search for Autonomy

To develop a major oil and gas province in the North Sea required building an industrial support system in an industry in which the Norwegians had no experience. Hagemann recalled: "There is no doubt but that we were totally dependent on the foreign oil companies to get the ball rolling in the North Sea."[88] But the Norwegians had important assets: a vigorous, homegrown shipping industry; a well-educated and experienced workforce accustomed and willing to work with heavy equipment at sea; and advanced skills in design and construction. As Exxon's official history notes, in Norway "Exxon found a high-quality workforce that was 'very technically astute' and could be gotten up the curve extremely fast."[89]

As early as the mid-1960s, before any major oil or gas finds had been made in the Norwegian sector of the North Sea, Norwegian shipowners and shipyards led off by investing massively in the prospective new offshore market.[90] Barely a decade later, the Norwegians had broken the American monopoly on offshore drilling rigs and platforms. One of the early triumphs of the Norwegian platform industry was the Condeep, a gigantic concrete production structure that rested on the ocean bottom and rose over 150 meters. Towing the Condeeps from the fjords where they were assembled was, as Exxon's historians note, "among the most spectacular events in the oil industry, with spectators stunned by the sight of such giant structures moving out to sea."[91] Mention of the Condeeps brings back many memories among Norwegians in the oil sector. "I was a college student in the early 1970s and I worked in the Condeep yard during my summer holidays," one veteran oilman told me. "Our job was to pour the concrete. It had to be poured continuously to avoid cracks as it dried, so we worked in eight-hour shifts around the clock. It was a tough job."

The Condeep was a result of the joint efforts of a consortium of Norwegian shipbuilders and designers who pooled resources and captured the North Sea offshore market for a decade.[92] The initial success of the Norwegians was due to a combination of strong state support and private-sector entrepreneurship. As the Norwegian energy expert Øystein Noreng writes, "Norwegian shipowners have a tradition of gambling and taking risks. They are accustomed to risking large sums for the sake of quick profits, which has produced an acute sensitivity to new opportunities. The gambling attitude, however, led them to overinvest in rigs and tankers in the wake of the oil crisis in 1973."[93] In contrast, it took British industry more than a decade after the first oil and gas discoveries in the North Sea to begin mobilizing. This was the cause of considerable hand-wringing in the British media. The *Economist* once famously dubbed North Sea oil "God's last chance for Great Britain" and deplored "the lack of virility and adaptability of British business," which was causing it to trail far behind "tiny Norway."[94]

By the mid-1970s the Norwegians were the envy of Europe. An admirer wrote in the *Economist*: "It is ironic that in social-democratic Norway, businessmen are far more convincing standard-bearers of capitalism than their British counterparts. Norwegian private enterprise has moved into the offshore oil industry with breathtaking speed and mastered a new technology which it is exporting all over the world. It is a record to shame every British shipbuilder."

However, the early success of the Norwegian shipping and oil services industry was followed by hard times, owing initially to the first oil shock of 1973 and subsequently to the heightened competition for business that followed the second shock in the 1980s. As a result, Norwegian industry lost much of the early position it had staked out at first. With the 1973 Arab-Israeli War and the subsequent embargo, market rates for ships collapsed. By the mid-1970s more than one-fourth of the Norwegian fleet was idle. Even after market conditions improved, the Norwegian shipping industry faced mounting competition from other countries such as South Korea. Between 1975 and 1995 the contribution of shipping to the Norwegian economy fell from 9 percent to 2 percent.[95] The Norwegian shipping industry has never fully recovered.[96]

The impact on the Norwegian petroleum service industry was equally severe. Already in the mid-1970s there was a growing fear that the industry had developed too much capacity. By the early 1980s the era of the Condeep was nearing an end, and Norwegian industry had nothing available to replace it. It was a time of fierce international competition, coupled with rapid technological advances. Four different major offshore projects were being built at the same time with four different designs of platforms, including the first tension-leg platform. One consequence was a rapid globalization of the offshore support industry, which at this time began to include South Korean contractors. "Never before," write Joseph Pratt and William Hale, "had the industry seen such a burst of advances in platform design."[97]

But Exxon was never sold on the concrete platforms and did not use them outside the North Sea. In contrast, "Norway led the way in designing, constructing, and installing concrete platforms, which became the norm throughout the Norwegian sector of the North Sea."[98] In the nearby British sector, the Brent platforms were both steel and concrete. This difference reflected in part the oil companies' desire to please the Norwegian government by using Norwegian contractors.

The oil service industry follows the boom-and-bust pattern of oil prices, compounded by the ups and downs of the shipbuilding industry, and by the 1980s the Norwegian sector was in a deep crisis because of overcapacity, inadequate attention to marketing, and rivalries among many small companies. Larger American and British competitors, soon followed by the Asians, soon captured market share, and the Norwegians were unable to follow up on the brief triumph of the Condeep.

In short, the early success of Norwegian entrepreneurship, allied with government policy, brought a flowering that lasted just a decade and a half. Technological innovation in the offshore industry, combined with the increasing depths at which the industry operated around the world, presented the Norwegian companies with challenges they proved unable to meet at the time. The Norwegian companies were too small to compete with South Korean shipbuilders. A generation later, there was a renaissance of Norwegian entrepreneurship, based on new high-tech specialties and start-ups—but that is the subject of another book.[99]

Human capital was just as important to the Norwegians as a key to industrial development. Norway used the licensing and bidding process to build in requirements. One Exxon executive remembers the reality of bidding in Norway: "They wanted to know, What can you give us training-wise? And so every time we bid on a block, one heavy, heavy part . . . was sending Norwegians to our research lab and even sending our Norwegians to MIT or Princeton or some school where they wanted that kind of technology."[100] Similarly, in another bidding round in the late 1970s, the Norwegians required that bids include "a list of potential research and development projects totaling $3.5 million to be undertaken in Norway."[101] Exxon noted that it found this policy easier to respond to than the strong-arm entreaties of the British government to build refineries and so on.

Thus, the availability of skilled manpower and infrastructure from the shipping industry enabled the Norwegian government to pursue an aggressive policy of developing and nurturing a homegrown supply sector to support the booming oil sector. Despite its subsequent difficulties in the 1980s and after, Norway's early experience has left a vital legacy in the form of a major international and gas oil company (Statoil) and a strong high-tech sector. In a remarkably short time, with the support of successive Labor governments, Norway's private sector built one of the world's leading industries for offshore oil and gas technology and equipment—which the USSR and its Russian successor did not.

Comparisons with the USSR

Comparing the Norwegian and Soviet or Russian experience in founding and developing their gas industries might seem at first to compare the incomparable. Despite the strong socialist and statist traditions of the Labor party in Norway, plus the fact that in Norway as in the Soviet Union the resources in the ground belonged to the state, the two systems were as different as night and day. When the Norwegians spoke of planning, they meant something more like indicative guidance within what remained overall a market economy—quite unlike the all-encompassing physical input-output matrix and the total elimination of the market that formed the intellectual model for the Soviet command economy. And as far as the world economy was concerned, the fundamental difference can be summed up in two words: the Norwegian system was open; the Soviet system, despite a partial opening after Joseph Stalin's death in 1953, remained fundamentally closed down to its final days in 1991.

Nevertheless, the differences between the two gas industries at the creation are instructive, in that they help to explain some of the important contrasts between the roles of Norway and Russia in Europe's gas economy today. We come back now to the three topics at the core of this chapter: the role of the state, export policy and the pricing of gas, and industrial and technology policy.

First, on the role of the state: One might have expected that the existence of two poles of power in Norwegian society—the government and the private sector—would produce unstable, conflict-ridden politics, while one's image of a command economy leads one to anticipate a smooth and orderly process of goal-setting and execution. The reality was the opposite. The Soviet state was so hidebound, imprisoned by its annual targets and the politics that produced them, that the only way it was able to innovate was by a process of survival of the fittest, which engendered strong bureaucratic entrepreneurs whose talent lay in overcoming the inertia of the system through guile and grit, and making it move. The basis of the Soviet economy, in practice, was less scientific planning than hard-nosed negotiation.

The Norwegian system too produced powerful leaders, and their function, not unlike that that of their Soviet counterparts, was to manage the interface between industry and the state. Prime examples were Johnsen, the first CEO of Statoil (1972–1987), whose main mission was to manage relations with the Labor Party that made energy policy, and Hagemann, the even more perennial head of the National Petroleum Directorate (1972–1997),[102] whose job was to look after issues such as health and environment and advise the government on policy issues—in particular, on depletion policy and licensee awards. Evensen provided the talent in international law that secured a doubling of Norway's territory, half of it offshore. In contrast, the Soviet system produced empire builders such as Aleksey Kortunov and similar figures in the space and defense industries, whose genius lay in overcoming the dysfunctional inertia of the command system. The empire Kortunov created—Gazprom—endures as his monument to the present day. The Norwegian entrepreneurs prospered within the system, their Soviet counterparts in battle with it.

In Norway, a gas field was developed only when there was a buyer lined up. One had to build a pipe, which could cost billions of dollars, and be sure that there was someone at the end of the pipe who was prepared to buy the gas from the moment it arrived and, crucially, to continue to buy it until the money invested in the field and the pipe had been recovered. So the decision to develop this or that particular Norwegian gas field depended on—and had to follow—the sale of the gas to one or more European buyers. The result was a piecemeal pattern of development, in contrast to the all-encompassing Soviet planning system with its ambitious concept for a countrywide integrated system, although in practice the realization of the Soviet projects in practice often diverged considerably from the grand plans.

Second, on industrial policy and the drive for autonomy: The fundamental difference between Norway and the Soviet Union is that in Norway the relationship between new and old industry was mutually supportive and hence highly productive, whereas in the Soviet Union it was competitive and debilitating. Specifically, old industry in Norway consisted of the shipbuilding industry, which launched itself enthusiastically into the development of offshore engineering and technology. In the Soviet Union, the development of turbines, compressors, pipe, and other related equipment was held back by the monopoly control of the military-industrial complex. The ironic result was that Norway was able in short order to produce a homegrown industrial base for its gas industry, whereas the Soviet *gazoviki* were forced to depend on imports. Thus, a system that prized autarky above all else ended up depending willy-nilly on capitalist partners and indeed, in no small measure, on its socialist satellites as well.

This difference echoes down to the present day. Until recently, Russia had no need for an offshore oil and gas industry and consequently has never developed one. Until the collapse of the Soviet Union, its naval shipyards and technological centers were largely military. (The Ministry for Shipbuilding, in particular, was one of the nine ministries that made up the the Soviet military-industrial complex.) Consequently, today's Russia faces the challenge of building a domestic offshore industry virtually from scratch, while Norway enjoys a forty-year head start. And the Russian military sector remains an obstacle even today. For example, it took Gazprom twenty years to build its first offshore platform (at Prirazlomnoye in the Pechora Sea), largely because of the inability of the local military shipyards to adapt to the requirements of a civilian project.[103]

Another unique feature of the Soviet structure was the tension between domestic needs for gas and the requirements of gas-export contracts. Both in Soviet times and since, gas has been sold into the domestic market at much lower prices than exports. As a consequence, Gazprom has had a consistent preference for exporting gas and guards its export monopoly fiercely (although it has lost monopoly control of LNG, as we shall see). The traditional tension between domestic and export markets has now changed form, as a result of the rise of independent gas producers—chiefly oil companies—which have now captured nearly half of Gazprom's share of the domestic market. Again, there is no counterpart to this in Norwegian experience.

Norway is, in many respects, the significant other in the European gas business. Whereas most of the gas from the UK sector of the North Sea goes to supply the British domestic gas system, Norway—plentifully supplied with hydropower—exports most of its gas. Norway, with much larger reserves than Great Britain, remains along with Russia the chief supplier of pipeline gas to the Continent, with Russia accounting for a third and Norway for a fifth. In 2012 Norway's gas production passed its oil production for the first time, and the gap will continue to widen since its oil production is declining steadily.[104] According to official estimates, Norwegian gas production will plateau at about 100 billion cubic meters per year in the mid-2020s and then gradually decline to about 90 billion cubic meters per year by the mid-2030s.[105] By that time Norway will have produced and sold over 4 trillion cubic meters of gas, accounting for some two-thirds of its presently known gas resources.

Although these numbers are large, it is clear that as a long-term supplier Norway is not in the same league as Russia, and while further discoveries will surely take place over the next generation, Norway's significance in Europe's gas security equation will gradually decline. Yet its importance for the European gas story lies less in its share of production and exports than in the political structure and business model it represents in contrast to Russia. Those differences, rooted in the earliest days of their respective gas sectors, remain embedded in the very different structure and commercial practices of the Norwegian and the Russian gas industries today. The result today is a peaceful and constructive relationship with Europe in the case of the Norwegians (though it was not always so), and a contentious one in the case of the Russians.

This chapter has argued that state capitalism Norwegian style has been, on the whole, more successful as a business model than state capitalism Russian style. It does not follow that the Norwegian model will continue to be as successful as it has been in the past. Norway faces three long-term challenges as a competitor for the European market.

First, the next generation of Norwegian gas lies in the far north, in the Barents Sea. This may turn out to be too distant from the European market to enable the Norwegians to reach it economically with pipelines. Thus, unlike the Russians, who have used the legacy rents from Soviet gas to build a network of Ukraine bypasses to bring gas from the far north (the Yamal Peninsula) to Europe, the Norwegians have no such infrastructure in place. The alternative for Norway is to develop the Barents as liquefied natural gas (LNG), but here Norway faces the problem that it has not developed a major capability to produce and export LNG (with the exception of LNG from Snøhvit), and it is a late starter in a crowded field. Norway faces the challenge of learning a new business.

That in itself would not necessarily be a major obstacle, were it not for a second problem: Norwegian society is increasingly divided on the subject of exploration and development of the Barents and offshore northern islands such as Lofoten. While Norway will hardly stop exporting gas, the practical effect of the increasingly emotional debate over the environment in the far north may be to slow the investment that Norway will need to compete in tomorrow's European market.

Finally, where the Norwegians may turn out to be weak, the Russians may turn out to be strong. Their gas in the far north is already developed, the pipelines are built, and the Russian export model has increasingly adapted to the requirements of doing business in Europe. Tomorrow's gas in Europe may be more likely to be Russian than Norwegian. But for that to happen, Russia will need to meet challenges of its own, which we turn to in Chapter 10 and the Conclusion.

However, here we have raced a half-century ahead of our story. We return now to the 1980s and Leonid Brezhnev's last hurrah.

CHAPTER 5

Soviet Gas

The Last Hurrah

The Soviet Union in the early 1970s, as we noted in Chapter 1, was at the height of its power, prestige, and self-confidence. The economy was still growing strongly; consumer standards were rising; and the Soviet Union enjoyed respect throughout the world as one of two great powers, the only rival to the United States. In the second half of the 1960s, Soviet GDP had grown faster than it had in the previous three decades. There was, to be sure, the embarrassing fact that the Americans had landed a man on the moon ahead of the Soviet Union, causing the Soviets to abandon the race for a manned landing. And inside the institutes and the planning apparatus, specialists were warning that the growth rate of the economy, alarmingly, had suddenly slowed. Yet the Soviet empire was quiet, domestic politics were stable, and East-West détente made the Cold War seem less immediately threatening. The Soviet Union in those years seemed serene, almost complacent.[1]

Within a few short years the serenity was shattered. By the second half of the 1970s the slowing of the economy had become apparent to all, and by the beginning of the 1980s GDP had begun to shrink. Repeated crop failures caused food to run short in the shops, and the problem was worsened by a system that left nearly two-thirds of the harvests to rot in the fields.[2] A vast program to promote irrigation had resulted mainly in a waste of capital, accompanied by more severe environmental problems.[3] Only the windfall brought by the dramatic increase in oil prices after 1973 enabled the Soviets to muddle through. But it was only a palliative. By the time of Leonid Brezhnev's death in 1982, as oil prices peaked and then began to fall, the resulting financial crisis set the stage for the desperate reforms undertaken by Mikhail Gorbachev in the second half of the 1980s. Few people inside or outside the Soviet Union imagined that the crisis would prove to be terminal.

At the center of the turmoil was the energy sector. In this chapter we recount the gathering problems with oil, nuclear power, and coal; the impact of the two oil shocks on the Soviet economy; and the role of natural gas as it evolved from its troubled adolescence to become the final success story of the dying Soviet system—Brezhnev's last hurrah from beyond the grave.

The Odd Couple: Brezhnev and Kosygin

Even his once-famous bushy eyebrows are now largely forgotten. Not many Russians nowadays spare a thought for Leonid Brezhnev, who ruled the Soviet Union as General Secretary of the Communist Party from the mid-1960s to his death in 1982. Even in his lifetime "Old Lyonya" was the butt of popular jokes that mocked his peasant speech, his supposedly low level of literacy, and of course the eyebrows. In the later 1980s his successor, Mikhail Gorbachev, made Brezhnev the symbol of *zastoi* (stagnation), the paralysis from which Gorbachev meant to deliver the Soviet system.

But Brezhnev was no fool. As the Party boss of Dnepropetrovsk during the postwar famine in Ukraine and of Kazakhstan at the beginning of the Virgin Lands campaign, he had experienced firsthand the successive disasters of Soviet agriculture. As Nikita Khrushchev's protégé, he had helped to implement Khrushchev's schemes to rescue Soviet agriculture by dividing the Party apparatus into agricultural and industrial halves until the day came in 1964 when Brezhnev led the plot to overthrow his mentor. The memory of Khrushchev and his fate remained with Brezhnev for the rest of his long career, and his policies during his eighteen-year tenure as General Secretary could be summed up as a reaction against the improvisation and administrative chaos that had characterized Khrushchev's rule. The watchwords under Brezhnev were consensus and stability, especially his formula, "stability of cadres." The so-called Generation of 1939, of which Brezhnev was the ultimate example, ruled largely undisturbed up to his death in 1982.

Brezhnev's closest partner during most of this long reign was his prime minister, Aleksey Kosygin. The two men could not have been more different, beginning with their appearance and public style. Brezhnev exuded a rough-hewn bonhomie and enjoyed a drink (or two), a good joke, pretty women, and his collection of fast cars. Kosygin was dour, gloomy, saturnine, and abstemious. They were also far apart in their political styles. Brezhnev was the ultimate hands-off manager. Kosygin was all business, the classic policy wonk.

Nikolai Baibakov, who was chairman of Gosplan throughout the Brezhnev era, described the scene in the mid-1970s when he and Kosygin went to Brezhnev's state dacha at Zavidovo to brief the General Secretary on the upcoming annual plan. Baibakov and Kosygin, as usual, came armed with armloads of dossiers and facts and figures. After listening to Baibakov with increasing weariness for the better part of two days, Brezhnev finally exploded: "Nikolai, go to the devil! You've stuffed our heads with your numbers and I can't make sense of anything anymore. Leave the numbers to Aleksey Nikolaevich [Kosygin], and just brief us on the concrete decisions." At the next meeting of the Politburo a few days later, according to Baibakov, Brezhnev grumbled, "I listened to Baibakov for two days, and now I can't sleep."[4]

Whereas Brezhnev preferred life at Zavidovo or his vacation dacha on the Black Sea, Kosygin was constantly on the road, visiting construction sites and oil fields, peering at everything, and asking questions. Whereas Brezhnev wanted "just . . . the concrete decisions," Kosygin never met a number he didn't like. For a decade they made a good couple. Brezhnev was the shrewd political tactician, while Kosygin had the strategic vision and daring. In this respect Kosygin was Khrushchev's true heir: like Khrushchev, he believed ardently that the Soviet system could still be made to work, and he was not afraid to undertake radical experiments on a nationwide scale. The so-called Kosygin reforms aimed to make factory managers and collective-farm chairmen more efficient by turning profit into the yardstick of performance and the basis for bonuses. Kosygin persuaded Brezhnev to go along, and during the second half of the

1960s the wager seemed to be paying off, as consumer standards rose and Soviet GDP leaped forward.

From Complacency to Panic

The euphoria did not last. Indeed, even in the 1960s the Soviet Union had experienced occasional crop failures, and the leadership was aware that food was a growing problem. The first major imports of grain were paid for in gold—372 tons in 1963, over one-third of the USSR's gold reserves, and nearly the same amount in 1965. But what was episodic in the 1960s became chronic in the 1970s and catastrophic by the 1980s. Soviet imports of grain, still a relatively modest 2.2 million tons in 1970, grew to 29.4 million tons in 1982 and reached a peak of 46 million tons in 1984.[5] Former prime minister Yegor Gaidar writes: "By the mid-1980s, every third ton of baked goods was made from imported grain. Cattle production was based on grain imports."[6]

The Soviets had never maintained large hard-currency reserves. Instead, they exported increasing quantities of gold to offset poor harvests, which came more and more frequently—in 1973, 1976, 1978, and 1981. Yet here they were helped by events in far-off Washington. In 1971 the administration of President Richard Nixon decided to leave the gold standard and brought the Soviet Union its first windfall when the world price of gold bounded up from $38.90 an ounce in 1970 to $183.77 four years later.[7] Yet even with higher gold prices, the Soviet Union was forced to borrow money to pay for grain, and by 1974–1975 it had become an international debtor.

Then, thanks to the first oil shock of 1973–1974, oil came to the rescue. As Gaidar writes, "The hard currency from oil exports stopped the growing food supply crisis, increased the import of equipment and consumer goods, ensured a financial base for the arms race and the achievement of nuclear parity with the United States, and permitted the realization of such risky foreign policy actions as the war in Afghanistan."[8] The global jump in oil prices brought a second welcome windfall after gold, yet the decision to rely on oil to pay for imports of food and consumer goods also caused a growing split within the Soviet leadership. For Brezhnev and the heads of the military-industrial ministries, the oil bonanza was the way to keep the traditional priorities and practices unchanged without risking a popular backlash. But the economic wing of the leadership was deeply unhappy. Baibakov felt betrayed. The oil and gas resources of the country were a gift of nature that should be used to support the industrial development of the country. Based on interviews with Baibakov many years later, his biographer attempted to capture his state of mind: "And now? Foreign currency—for what? Consumer goods? A rationale for making no changes in the economy? Even in his worst nightmares he couldn't imagine such a fate for the oil and gas resources produced at such sacrifice." For Baibakov, it was a "stab in the back."[9]

It also brought the end of the Kosygin reforms. Baibakov's biographer continues, "Why continue with the difficult—and moreover dubious from an ideological standpoint—reforms, when there were such financial revenues available?"[10] For Kosygin, the end of the economic reforms was a defeat that marked the fraying of his partnership with Brezhnev and the beginning of the end of his career. Not only did Kosygin deplore the growing reliance on oil exports, but he was also increasingly uneasy about the way Soviet foreign policy was evolving. By the second half of the 1970s, détente was fading and the atmosphere in the Kremlin was becoming more hawkish. The tensions between Kosygin and Brezhnev came to a head over the Soviet invasion of Afghanistan in 1979. In September 1980 Kosygin wrote a bitter letter to Brezhnev and the Politburo, in which he condemned the entire economic and foreign-policy course pursued since the mid-1970s, and especially the growing reliance on exports of energy and raw materials to finance an aggressively ideological foreign policy. It was "ill thought through," Kosygin wrote, "adventuristic, lacking any reliable economic or political basis, and therefore discredits the USSR and deepens confrontation with China." In conclusion, he demanded to be relieved as prime minister. One month later, Kosygin was out of a job, and one month after that, he was dead.[11]

There were parallel tensions in energy policy, but these took longer to come to the surface—partly, no doubt, because of the greater urgency of the agricultural problems and partly also because of the increase in oil-export revenues that followed the Arab oil embargo of 1973–1974. Revenues from Soviet exports of crude oil and refined products to the countries of the Organisation for Economic Co-operation and Development (OECD), which had been less than $3 billion in 1972, bounded to nearly $10 billion in 1974 and continued rising rapidly thereafter, reaching $13 billion in 1976.[12]

The flood of cash initially lulled Brezhnev and his colleagues into a false sense of security. As one reads the leaders' speeches from the first half of the 1970s, it is hard to find any sign of high-level concern over energy. Insofar as the subject was mentioned at all, it was mainly in terms of the long-range future—which the leaders, along with most of the scientific establishment at the time, saw in terms of coal and nuclear power.[13]

Yet simultaneously there was another, quite different stream of messages coming up through the ranks from the Siberian oil and gas fraternity. These warned that the oil fields were being pushed too hard, exploration was being neglected, gas pipeline construction was in trouble, and production costs were rising fast. It was as though two contrary narratives were being offered to the leadership—the first one portraying the energy situation as essentially benign and calling for a leisurely repositioning toward coal and nuclear power in the next century, and the second one an increasingly urgent signal of trouble immediately ahead in the oil fields. But until about 1977 the leadership heard only the first message and mostly ignored the second one.[14] The best indicator of this is investment trends: from 1970 through 1977 the share of energy in industrial investment stagnated, reaching 28.1 percent in 1977 compared to 28.7 percent in 1970.[15]

But in 1977 complacency abruptly gave way to panic, as the Kremlin's dashboard suddenly lit up with flashing red lights across the entire energy sector.[16] Coal production in the critical Donbas began to fall. Gas production from Shebelinka, Ukraine's largest field, peaked and headed downward.[17] The nuclear program fell behind schedule. But the worst news came from the West Siberian oil fields, which by then had become the chief source of energy growth. Here all the key indicators suddenly turned negative: the number of new fields identified; the flow rates of new wells; and, most important of all, the overall growth rate of Siberian oil output. By late 1977 it began to look as though oil production might peak by the end of the decade.[18] The only exception to this grim picture, as we shall see in a moment, was West Siberian gas, to which we return below.

In the spring of 1978, over the protests of Kosygin, Baibakov, and the planning establishment, Brezhnev ripped up the energy goals of the 10th Five-Year Plan (1976–1980) and embarked on a crash program to save the five-year oil output target. Suddenly Brezhnev was everywhere, touring West Siberia, giving keynote speeches at Party functions, and arguing for the urgency of salvaging the plan. Money began to pour into the energy sector—especially oil, where annual investment grew by over one-third in four years, from 4.1 billion rubles in 1976 to 6.6 billion rubles in 1980.[19] It was an extraordinary performance for a man renowned for his hands-off style in policymaking, and one who moreover was already in declining health. According to the Moscow rumor mill, the Siberians had gotten to Brehznev. But in retrospect it appears that Brezhnev had been more influenced by something else. In 1977 the CIA published three reports on the performance of the Soviet oil sector, predicting that Soviet oil output would peak in 1980 and decline sharply thereafter, which would turn the Soviet Union into a net oil importer.[20] The reports received enormous attention worldwide, but it appears that their most devoted reader was in the Kremlin. It is ironic, of course, that the core of the CIA's analysis was based largely on Soviet publications and merely repeated what Soviet oil experts had been writing for years.[21]

Brezhnev's emergency response made it possible to meet the 1980 oil output target, but at the price of increased waste and inefficiency. The costs of oil development and production rose sharply. The amount of capital going into the oil fields grew by nearly two-thirds over four years. The planners warned that to keep the share of energy production attributable to oil at its 1980 level of 44 percent throughout the first half of the 1980s, oil investment would have to increase by nearly 3.8 times—more than half again as fast as the investment increase for all of Soviet industry. Clearly, some other course was required.

By the fall of 1979, Brezhnev had hit upon the answer: big gas from West Siberia.[22] Yet it took about a year for the shift to the big gas policy to work its way through the bureaucracy, and Brezhnev did not get his way without a fight. There was still a widespread perception in the energy establishment (as there was in Europe as well) that gas was too scarce to waste on ordinary heat and power. This explains the continued stress on coal and nuclear power by people such as Baibakov, Kosygin, and

Anatolii Aleksandrov, the president of the USSR Academy of Sciences. At the same time, more broadly, Brezhnev's policies and authority were under growing attack from several quarters. Though Kosygin's health was failing, he joined Brezhnev's critics, and public polemics broke out between them. Brezhnev responded with new initiatives. The Afghan expedition was one, and the energy campaign was another. These factors help account for the suddenness and sweep of the Brezhnev gas campaign: the responses were vigorous because the point was precisely to show vigor.

In February 1981, announcing the targets for the coming 11th Five-Year Plan (1981–1985), Brezhnev stressed the top priority of gas:

> I consider it necessary to single out the rapid development of Siberian gas output as a task of first-rank economic and political importance. The deposits of the West Siberian region are unique. The largest of them—Urengoy—has such gigantic reserves that it can meet for many years both the internal needs of the country and its export needs, including exports to the capitalist countries.[23]

Brezhnev proposed to increase natural gas output by nearly 50 percent in five years (from 435 billion cubic meters in 1980 to 630–645 billion cubic meters in 1985), the bulk of the growth to come from West Siberia. Natural gas would provide 75 percent of the net addition to the fuel balance. To reach these targets, the gas sector would spend as much capital over the coming five years as it had in the previous fifteen.[24]

Putting West Siberia at the center of gas policy implied, of course, an equally massive increase in pipeline construction. The targets were phenomenal: More than two-thirds of the total gas investment would be devoted to building six huge trunk lines employing 20,000 kilometers of 1,420-millimeter pipe between West Siberia and the European USSR—a diameter larger than any used elsewhere at that time. In November 1981 Brezhnev hammered the point home, calling the six planned Siberian trunk lines "without a doubt the central construction projects of the five-year plan" and adding, "they must be finished on time without fail."[25]

What made Brezhnev think that the gas sector could possibly play the key role that he was scripting for it? The short answer is that by 1980 the gas sector had become the star performer of the energy sector, and indeed of all Soviet industry. It was the only energy source whose output in 1980 actually reached the targets set for it.

But therein lies a mystery. A decade before, the gas sector had been in deep trouble, as we saw in Chapter 2. What had happened during the decade of the 1970s to transform the gas industry from a troubled laggard to a leader? We turn back the clock to the early 1970s to pick up the story.

Continuing Battles on the Soviet Side

The deadline was 1973. That was the year the first gas exports to Italy and East and West Germany were contracted to be scaled up sharply—from under 5 billion cubic meters per year in the early 1970s to 7.4 billion cubic meters in 1973, and then faster and faster, from 17.2 billion cubic meters in 1974 and 24.7 billion cubic meters in 1975—driven by booming demand in Europe and supported by a proliferation of agreements with both Eastern and Western Europe. New giant fields in West Siberia were being discovered steadily, and reserves appeared limitless. Between 1966 and 1969 Siberian proven reserves grew from 714 billion cubic meters to 4.4 trillion cubic meters, equaling nearly half of the reserves of the rest of the world put together. The sheer scale of the Soviet discoveries dwarfed the expected needs of Europeans: the West German government, for example, was still planning for a gas demand of only 80 billion cubic meters per year by 1980.[26]

There was only one problem: the Soviet gas industry at the beginning of the 1970s was in trouble. To meet its growing delivery targets and its export commitments meant producing gas from wells as yet undrilled and shipping it thousands of miles through pipelines that did not yet exist. Aleksey Kortunov's vision from the beginning had been that this vast network would not only support exports to Western Europe but would also supply Eastern Europe and major Soviet regions, notably the Baltic Republics and the Urals. As soon as the Soviet government approved the concept of developing West Siberian gas in 1965, Kortunov and his team were ready with their grand plan—but only on paper.

The first countertrade deals with Western Europe were thus of crucial importance, since they provided for the import of the first 1,420-millimeter pipe.[27] The countertrade and the pace of the West Siberian program were thus intimately connected: as soon as the first deal was signed in 1969, the Ministry of Gas submitted a plan to the Central Committee of the Communist Party of the Soviet Union for the accelerated development of gas in Tiumen' Province, and the plan was rapidly approved. With the prospect of further contracts ahead, ambitions grew. By the time of the 1969 version of the plan, the target numbers had jumped to 280 billion cubic meters, to be realized by 1980. In a 1970 memo to the Council of Ministers, Kortunov spoke of reaching 1,000 billion cubic meters by 1985.[28]

The first German pipes arrived on schedule in June 1970, but the main problem was bringing them to their location and laying them. In summer the whole area turned to swamp, and pipe had to be brought in by barge; in winter it lay deep in snow, and delivering the pipe required special tractors that pulled sections of pipe over roads made of ice. There were two options: either follow an existing old railroad route, which was easier but longer, or cut straight through virgin forest, which would save on pipe. Kortunov chose the latter. But the ministry's lack of experience in laying pipe in such demanding conditions caused delays. The pipe had to be laid above the ground because of the hard permafrost underneath—a mixture of ice and sand that remains frozen throughout the year but turns to slush when any weight is placed on it. An added problem was that the gas came up hot from the wells, so it had to be refrigerated along the route to prevent it from melting the permafrost, which would cause the pipe to slide and buckle. All these problems took time to resolve.

By this time it was becoming clear that Kortunov had underestimated the difficulty of achieving what he had promised. He complained constantly about the poor support he was getting from other ministries, which were responsible for supplying all the necessary equipment—such as refrigeration, pipe layers, trucks, and above all compressors. The 1,420-millimeter pipes were designed to operate at 75 atmospheres of pressure, which required one 25-megawatt compressor station every 100 kilometers or so to push the gas through the pipe. But Soviet industry did not yet produce such large compressors or the gas turbines that powered them, and the gas industry was forced to rely on outmoded 10- and 16-megawatt models produced for the aviation industry. The workhorse of the Soviet pipeline system for the next twenty years was retired engines adapted from the Ilyushin-16 civilian airliner, hitched together five to each compressor station.[29]

Because of these problems, by 1971 the gas program was far behind schedule, and the Soviets were in danger of missing the crucial deadlines for the first deliveries of gas to Germany and Italy. At the 24th Party Congress in March 1971, progress in the gas program was hailed by Brezhnev and Kosygin, but in reality the planners were already aware that the 1975 targets would have to be scaled down, and the first Siberian gas would only be available in 1976 instead of 1973.[30] Kortunov vehemently resisted lower targets in his speech to the Congress. He vowed to complete the first Siberian pipeline by the end of 1974, and he lengthened the pipeline route to avoid as much permafrost as possible. By this time more German pipe had become available, and a longer route was now feasible. But by 1972 it was apparent that even the revised targets would not be met.

The only available answer was Ukraine. The pace of construction of an east-to-west Ukrainian pipeline was stepped up, bringing gas from Shebelinka to western Ukraine. Central Asian gas was also to be brought to Ukraine to supplement the available gas from Shebelinka.

But there were constant delays, and these caused bitter arguments between republic leaders and Moscow. The Baltic Republics lobbied hard against any postponements in the Siberian program. The Latvians in particular did not want to be at the far end of a pipeline stretching all the way from eastern Ukraine. In the end their complaints prevailed, and a direct pipeline from Valdai to Riga, promised and then postponed, was put back on the target list. But even so, the problems in supplying Siberian gas prevented the Balts from getting their full supplies—and in addition they found themselves competing with contracted exports to Finland.[31]

As delays accumulated, whispers began to run through the Kremlin and Central Committee headquarters on Old Square that the old man of the gas industry had lost it and that it was time for a change. Kortunov was under extraordinary pressure. He was effectively in charge of construction in two major ministries (oil and gas) simultaneously. This kind of arrangement—placing a high-priority cluster under a single tsar—was not uncommon in Soviet times (the combination of Siberian hydropower and aluminum was another famous example), but by the beginning of the 1970s the load on Kortunov had become untenable. First, the West Siberian oil and gas industries were diverging, as it became apparent that most of the gas was located in the northern part of Tyumen Province, far from the established oil infrastructure. The logistical and engineering challenges of laying pipe through virgin taiga—not to mention the politics of fighting other ministries—took up more and more of Kortunov's time and energy. Meanwhile, because of Kortunov's longstanding gentleman's agreement with Baibakov, the focus of the gas ministry became so lopsidedly focused on West Siberian oil that it was holding back the orderly growth of the gas sector there, as well as in the rest of the country. Lastly, and most serious in the context of the top-heavy Soviet way of doing things, Kortunov's health had begun to fail under the immense double load he was carrying.

Matters came to a head in 1972. Confronted with the increasingly bad news from the gas program, especially in the laying of the pipelines, the Party leadership decided to deal with the situation by splitting the gas sector in two. The Ministry of Gas would retain its name but would be stripped of most of its responsibility for construction, particularly for the trunk lines out of West Siberia. Instead these would be handled by a new ministry specializing in construction for both the oil and gas industries. There was less change here than appeared at first. Creating a separate ministry for what amounted to West Siberian oil and gas construction was in large part simply a formalization of what had long been the practice. The only remaining question was what to do with Kortunov.

Late in September 1972 Kortunov was summoned to the office of Andrei Kirilenko, the Politburo member in charge of heavy industry and a close associate of Brezhnev from Dnepropetrovsk days. Ushered in with him was Baibakov. There followed a painful scene, which had clearly been scripted in advance. Kirilenko asked Kortunov point-blank, "Aleksey Kirillovich, whom would you like to see as minister of gas?" Kortunov stared at the floor. It was suddenly clear to him that his brainchild was about to be taken away from him. A heavy silence hung over the office for several minutes, broken only by the entrance of an assistant with a pot of tea. While she poured, Baibakov discreetly handed Kortunov a note on which he had scrawled, "Aleksey—support Sabit." "Sabit" was

Sabit Ataevich Orudzhev, at the time the first deputy minister of oil and a lifelong friend and ally of Baibakov from Baku. Kortunov understood what was expected of him. He raised his gaze to Kirilenko and replied, "I would recommend Sabit Ataevich." Kirilenko, visibly relieved, said: "The Central Committee feels the same way. The Central Committee will recommend to the next session of the USSR Supreme Soviet your appointment as minister of oil and gas construction."[32]

Given the mounting delays in the pipeline construction campaign, the leadership's decision was perhaps inevitable, but Kortunov perceived the split as a humiliating defeat—it was, after all, the very opposite of the single integrated gas organization he had advocated for years. It was not a peaceful or pleasant transfer of power. Orudzhev began his tenure in early October 1972 with a meeting of the ministry's leadership. The announced topic promised a difficult session: "Enormous shortcomings in the work of the gas transportation network." There were "sharp accusations." This caused such bitterness that many Kortunov loyalists decided to leave along with him and transfer to the new ministry.[33] There were painful separations, as people chose whether or not to follow Kortunov. Mikhail Sidorenko, the first deputy minister, and Stepan Derezhov, the head of international relations, stayed with the gas ministry, but most of Kortunov's people left with him, including his longtime "European ambassador," Anatolii Sorokin. Most of the West Siberians, such as Iurii Batalin, also went with Kortunov. As Orudzhev named his deputy ministers, they formed a patchwork of transfers from other places—such related agencies as the Ministry of Oil, the oil and gas department of the Central Committee of the Party, Gossnab (the State Supply Committee), and the Ministry of Geology. Most of them were oilmen rather than gas specialists. Many of Orudzhev's department heads stayed on through the 1980s, and some of their relatives are still in Gazprom today.

Suddenly, as a result of these changes, there was an administrative barrier between the gas ministry and the new construction ministry. Throughout late 1972 and early 1973, even as his health failed and his strength ebbed away, Kortunov continued to bombard senior officials in Moscow with bitter memoranda, detailing Mingazprom's failure to supply his new ministry with the pipe and compressors needed to meet his construction targets. This was a typical Soviet scenario, and the results were equally typical: the senior officials would respond with stern notes, threatening dire punishments if the situation were not corrected immediately. But within weeks the next jeremiad would arrive from Kortunov, complaining that nothing had changed. This too was typical, as was the physical and psychological strain under which senior Soviet officials constantly worked. Fourteen months after the split, Kortunov was dead.[34]

Kortunov's importance as the founder of the Soviet gas industry and the originator of the gas bridge with Europe cannot be overstated. Without his vision and drive, organizational talent, and political skill, the development of West Siberian oil and gas might have been delayed by as much as a generation. Gas exports to Europe would have remained modest, for lack of sufficient ready reserves and a pipeline system through which to ship them. Above all, the rapid displacement of coal and oil by gas in the Soviet primary fuel balance—one of the last successes of the Soviet planned economy—would have taken much longer. By the beginning of the 1990s, when the Soviet Union fell apart and the Soviet oil industry with it, it was the gas industry, by then Russia's most important source of primary fuel, that kept the Soviet cities heated and lighted, while oil was exported for desperately needed dollars. That was Kortunov's legacy to the country he so ardently believed in.

Yet in retrospect it is his successor, Orudzhev, who put a permanent stamp on the gas ministry, creating a structure that is recognizably still that of Gazprom today. It was Orudzhev who was the true creator of the Gazprom model. First, he oversaw the creation of specialized departments *(upravleniia),* which were led by people, chiefly oilmen, who came from other ministries and brought in their own protégés. Consequently, from Orudzhev's time on there were several competing patron-client networks inside Gazprom, which remains true today. Second, Orudzhev distanced himself from the West Siberians, who were Kortunov loyalists. This may account for the strong priority given to a competing pipeline—the Orenburg pipeline (discussed below)—and subsequently to the "Orenburg mafia" that led Gazprom from the 1980s on.[35] Third, however, despite the presence of Derezhov (who parted ways with the Kortunov crowd and stayed with Mingazprom), it was the Ministry of Foreign Trade that kept control of Gazeksport and played the dominant role until Victor Chernomyrdin seized it in 1990.

Orudzhev, Kortunov's successor at the gas ministry, was quite different from the austere and demanding Kortunov. Orudzhev had had a distinguished career in the oil industry, but his experience lay largely in production, and it took him some time to realize that in the gas business the essential challenge is transmission. But he was lucky in that he inherited the positive results of the emergency measures taken by Kortunov, Baibakov, and Kosygin to meet the gas crisis of the first half of the 1970s. In 1975 the gas industry met its targets with room to spare, and suddenly it was the star performer in an economy that was sputtering. In gas as in war, Kortunov might have reflected bitterly, timing is everything. If it was Kortunov who first cracked open the window to the export market, it was Orudzhev who opened wide the door, as Soviet gas exports went, in a stunning increase, from 3.3 billion cubic meters in 1970 to 54.2 billion cubic meters in 1980.[36] But Orudzhev too paid a price in ruined health: he died prematurely in April 1981.

The Soviets Deliver—Barely

On October 1, 1973, in a ceremony attended by Orudzhev and Hans Friderichs, West Germany's federal economic minister (under Willy Brandt until 1974, and then under Helmut Schmidt), Friderichs spoke of the deliveries as a "further important cornerstone." Orudzhev, more picturesquely, spoke of the Soviet gas crossing the border "without passport and visa . . . burning like a torch at Waidhaus."[37]

But it was Ukraine that paid the price for the on-time deliveries to West Germany. As early as September 1973 it became clear that Mingazprom would not be able to fulfil delivery targets for power plants and municipal buildings in the major Ukrainian cities, and the delivery targets to Ukraine were cut back by one-third. Other republics were affected as well. By October gas deliveries to power plants in the western USSR were already falling short, and the plants were forced to burn reserve fuels (oil or coal) instead, while supplies to thermal plants that heated dwellings in Kiev, L'vov, and other cities had to be cancelled altogether. Party and government officials from the affected republics complained bitterly, as did the ministers of the industrial sectors affected—power and electrification, the chemical industry, and ferrous metallurgy.

Ordinary people suffered more acutely. A letter to Brezhnev dated October 1973 from the citizens of Drogobych, a medium-sized city in the Lvov region of western Ukraine, complained of the lack of gas heat for apartments and schools:

> For four years already we have endured a disastrous situation during the autumn-winter period. . . . The amount of gas delivered is insufficient for supplying dwellings, childcare institutions, and medical and administrative facilities. Houses are very cold, and since apartments are not designed to be heated with firewood or coal it is impossible to cook. . . . Grown-ups, not to mention children, often fall ill.[38]

In January 1974 the weather was extremely cold, and the prime minister of Ukraine appealed directly to Kosygin:

> Technological processes in metallurgical and chemical plants have been disrupted, factories have stopped working, and heating plants in residential areas are on the verge of breakdown. The situation is aggravated by the fact that many enterprises lack reserve fuel, making it impossible, on particularly cold days, to transfer gas to municipal needs.[39]

In a complaint that became familiar in later decades, after Ukraine and Russia had become separate countries, the head of Gossnab, Veniamin Dymshits, noted that "gas users along the trans-Ukrainian route consumed more gas than they were entitled to," with the result that gas from Shebelinka was not reaching western Ukraine.[40]

The West Europeans, who were receiving their contracted gas with only occasional technical interruptions, were largely unaware of the disarray on the other side of the border. The fact that export deliveries were kept on schedule in 1973 and 1974 was taken in the West as reassuring evidence that the Soviets would be reliable exporters. Yet in reality, as the gas historian Per Högselius writes, the "emerging East-West gas system was a shaky construct, built in haste and based on a capricious blend of inferior Soviet methods and technologies and Western pipes and equipment.

Keeping the new export system operational became as challenging a task as constructing it in the first place. Emergency events were reported on a more or less continuous basis."[41] Ironically, precisely because the Westerners had been "so skeptical about the Soviets' trustworthiness . . . importers had taken a variety of measures to protect themselves against supply disruptions. . . . As a result, end-users were never affected more than marginally by the problems that did occur." Moreover, south Germany was increasingly integrated with the rest of the German gas system. When gas from the North Sea began to be delivered in the north, it could be used interchangeably with the Soviet gas, since its composition was essentially the same.[42]

Thus, the Soviets made their export deadline to Western Europe—barely, and chiefly at Ukraine's expense. But largely invisible to the West Europeans were other conflicts going on behind the Iron Curtain, between the Soviets and their East European satellites. We turn to those now.

Conflicts over East European Gas Dependence

The Russians had long complained that the East Europeans were getting a free ride at their expense in the form of underpriced Russian exports. Just as Ulbricht and Khrushchev had quarreled over this in the 1960s (as we saw in Chapter 2), so the arguments continued in the 1970s. Prime Minister Kosygin was particularly incensed by what he saw as a systematic pattern of reverse discrimination: East European imports to the Soviet Union were shoddy and overpriced, he firmly believed, while Soviet exports were underpriced to practically symbolic levels. The result was a widening gap between the living standards of ordinary Soviets and those of their comrades in the satellites. Throughout the second half of the 1970s Kosygin bombarded the Politburo with memoranda and policy proposals, all aimed at reforming the trade system within the Council for Mutual Economic Assistance[43] and making the East Europeans pay their share. He called for the East European regimes to invest more in heavy industry and less in the consumer sector. The East Europeans fought back vigorously, warning that any change in the terms of trade would inevitably weaken the position of the communist regimes in Europe.[44]

Kosygin was overruled. The Brezhnev camp—which included much of the foreign-policy and military-security elite (as opposed to the civilian industrial sector that Kosygin represented)—was more realistically aware than the domestic planners were of how fragile the legitimacy of the the East European satellites was, and of the ever-present danger of destabilization, particularly following the suppression of the Prague Spring in 1968.

Much of this debate was about energy. The East European regimes had long supplied energy to their populations at heavily subsidized prices. In the German Democratic Republic (GDR), for example, consumers paid 17 pfennigs per cubic meter for town gas that cost 33 pfennigs to produce, and the state made up the difference with millions of marks in subsidies from the budget.[45] When the era of natural gas dawned and the Soviets began planning exports to Europe, the main question was naturally how much of the exported gas to deliver to Western Europe and how much to the East European satellites? Exports to Western Europe would yield pipe and finance. What would exports to Eastern Europe yield, other than political risk insurance—an uncertain and increasingly expensive commodity? It is no surprise that the Brezhnev "political" camp leaned one way and the Kosygin-Baibakov "economic" camp the other. Since gas exports were first discussed, the Soviet side was divided.

The key to a compromise within the Soviet leadership was East European labor. In early 1974, Nikolai Patolichev informed the East Germans that the export level of Soviet oil and gas to the Comecon countries would be cut by 9 percent in 1974. Henceforth, Soviet energy exports would be contingent on active participation by the satellites in building a new export pipeline, to be called Union *(Soyuz),* from the Orenburg field near the Russian-Kazakhstan border to Eastern Europe.[46] The core idea behind the partnership was that in exchange for labor and equipment, the East European allies would receive gas at reduced prices.

The explanation of the timing is straightforward. After the first oil shock the Soviets could suddenly make a lot more money exporting oil to the world market, and they needed the hard currency to buy food in the wake of a major crop failure in 1973. All these things came together at about the same time: the long-standing Soviet irritation over the low prices charged for oil to Eastern Europe (a sticking point for Kosygin in particular), the need for hard currency and the opportunity to sell oil elsewhere for higher prices, the difficulties with West Siberian gas, and the changing of the guard at Mingazprom.

The Orenburg Pipeline

The general agreement for the joint pipeline from Orenburg was signed on April 21, 1974. In the GDR the project was designated as a Central Youth Project, and the Free German Youth organization was given the job of recruiting and training young workers for it. Even though the official Soviet name of the pipeline was Union, the East German portion was dubbed Friendship *(Druzhba)*—evidently the Free German Youth wanted to piggyback on the associations with the earlier oil export project by the same name. Ultimately over 10,000 East Germans worked on the segment assigned to East Germany, which was located in Ukraine between Kremenchug on the Dnieper River and Bar in western Ukraine.

There was apparently no difficulty in finding recruits, not least because the pay was very good. In addition to a regular salary (which was credited to them back in the GDR in East German marks), they also received a day wage *(Tagegelder)* in rubles, up to half of which could be deposited in a special account at an attractive exchange rate They were also given privilege cards that could be used inside the USSR to buy items not available to ordinary shoppers. The GDR government maintained a regular airlift between Berlin and Kiev and kept the East German workers supplied with German books, movies, and television programs. There was even a German-language newspaper, the *Trassenecho.* Similar arrangements were made for the workers from other East European countries participating in the project.

Yet conditions were difficult. The *trassniki* worked in temperatures as low as −30° F, many fell sick, and there were numerous cases of dysentery. They worked a sixty-hour week, in exchange for which they received a month's leave every three months. Construction began in September 1975 and was completed by the end of 1978; the last East German workers returned home in 1979. As payment, East Germany received 30 billion cubic meters of gas (2.8 billion cubic meters per year) through 1990, and after that the balance was paid in money. The contract ran to 1998.[47]

The project ended up costing the GDR almost twice as much as originally budgeted in the official cost estimate contained in the agreement, although some of the difference came from different conventions on exchange rates and accounting. It included a hard-currency allowance of $402 million for purchase of Western pipe and machinery, of which the GDR ended up spending about $355 million. Yet although the project was more expensive than originally planned, the GDR considered it a good investment, since the arrival of natural gas enabled a modernization of portions of the economy.[48] On the Soviet side, the Orenburg project was successful as a placeholder that gave the Soviet gas industry a breathing space in which to iron out the problems in West Siberia.

The US Gas Pipeline Embargo

Up to this time the United States had played no active part in the politics of Soviet gas exports.[49] Washington had made no attempt to block the first gas pipeline a decade before, nor did it interfere with the Soyuz pipeline. But a few years later, as the Soviets began planning a new export pipeline from West Siberia, conservatives in the administration of President Ronald Reagan, led by CIA director William Casey, urged the president to stop it by declaring an embargo on American-licensed technology, especially compressors. Reagan was initially reluctant, but the imposition of martial law in Poland in 1981 swung the debate in favor of the hard-liners. In 1982 he issued an executive order imposing sanctions on exports of energy-related technology. The confrontation that followed bore an eerie resemblance to the debate over the Nord Stream 2 pipeline today, but with some significant differences. Once again, it was a time of mounting East-West tensions over Soviet actions in Eastern Europe, although at that time Poland and its East European neighbors were in the Warsaw Pact instead of NATO and, unlike today, supported the pipeline instead of opposing it. Then, as now, Washington was determined to punish the Soviet Union for its aggressive foreign policies, but the defense of Ukrainian interests, for obvious reasons, was not on Washington's mind at the time. Then as now, West Europeans denounced the embargo as an American attempt to interfere with business and insisted that interdependence in gas with the Soviet Union was beneficial to their economies. In the end, despite Washington's objections, the pipeline went forward.

However, the significance of the episode does not end there. The Soviet leadership, urged on by the dying Brezhnev, reacted to the American embargo with an all-out push to equip the pipeline with Soviet-made compressors. In the end, about half of the compressor power on the pipeline was domestic. Today the Russians make their own large-diameter pipe and compressors. Brezhnev would have been pleased.

Indeed, although Brezhnev died in 1982, shortly after the gas campaign got under way, he would have taken considerable satisfaction from its success. The vision and ambitions of Kortunov and his team of West Siberians were fully vindicated. What they could not have foreseen, however, was that within a decade the gas industry would become the mainstay of a failed state.

CHAPTER 6

Crossing the Channel

The Neoliberal Tide Reaches Brussels

In the 1980s and 1990s, an intellectual and political revolution in Western Europe overturned the rules and regulations, and ultimately much of the structure, of the European energy industry. It happened that it was in these same years that the Soviet Union collapsed and the Iron Curtain disappeared. The result in the energy sector was that at the beginning of the 1990s, the statist-collectivist culture of the East, as it emerged from the isolation of the Soviet Union, came face to face in Europe with a newly reenergized capitalism from the West, together with a new regulatory regime emanating from Brussels. Two worlds struggled to understand one another, in politics, economics, and business. In the gas industry and the East-West gas trade the confrontation was particularly sharp. In this chapter we focus on the West European side of the story, while developments in Russia—particularly the Russians' response to the new business environment they encountered in the West—will be treated in following chapters. But to foreshadow the major theme ahead: the Russians entered the European gas market as players at the beginning of the 1990s under one set of rules and understandings, a set that was much more attuned to their own early experience of gas exports and could be summed up under the term "Groningen rules"—that is, long-term contracts with monopoly buyers, based on oil-price references. By 2010, they found themselves faced with a world transformed, one that they had not anticipated, did not understand, and for a long time did not accept. But that part of the story must wait until we have set the stage.

A Tide of Radical Thought

The term "revolution" is of course much abused. The neoliberal revolution was a broad tide of thought that originated mainly in the Anglo-American world and flowed to the Continent over the course of two decades. It has been called Chicago economics, Reaganomics, Thatcherism, the neomarket revival, the anti-Keynes reaction, the Hayek restoration, and many other names. I prefer the term "neoliberal," which emphasizes its capitalist foundations as well as its ultimately European roots and its retranslation into a European setting, notably by British officials and advisors. Its essence was a wholesale challenge to the statist, top-down, plan-oriented mode of thought that had dominated European thinking since World War II. That mode of thought was replaced by a cluster of ideas centered on the market: liberalization, privatization, globalization, and above all competition across an open market space and a level playing field, all backed by a new regulatory regime from a reenergized European Commission.

The neoliberal revolution in Western Europe was in large part a reaction to the events described in Chapter 3 on the age of anxiety. As a result of the two oil shocks, economic growth had come to a halt, inflation was rampant, and unemployment had reached levels not seen since the Great Depression. In Great Britain the situation was extreme. Widespread strikes gripped the country, and successive governments of both major parties were unable to deal with the crisis. During the "Winter of Discontent" of 1978, garbage rotted in the streets, and the dead lay unburied. The misery of the 1970s created the political momentum for neoliberalism in Great Britain.[1] In the United States and Great Britain, economists in a key handful of universities such as those of Chicago and Birmingham pinned the blame for these troubles on excessive state intervention and regulation and the inefficiencies of top-down state investment.[2] After decades in the intellectual wilderness, the ideas of Friedrich von Hayek and the Austrian school—which stressed the virtues of free-market capitalism and supply-side economics—began to attract interest again, while the prevailing statist, demand-side doctrines associated with the name of John Maynard Keynes were forced into retreat. Before long, a tidal wave of renewed faith in markets and private enterprise began to build in Europe and the United States.

The role of the United States, and particularly that of the economics department at the University of Chicago, as the primary source of the neoliberal tide, is central. By the late 1970s a broad revolt against the prevailing Keynesian doctrines and the excesses of government regulation had caused neoliberal ideas to migrate to many economics departments across the United States and Great Britain. Particularly significant was US thinking on competition. In Europe, this was a real innovation. Both before and after World War II, cartels and interfirm cooperation were regarded as the norm on the Continent. Such cartels protected jobs, fostered innovation, and prevented wasteful competition—at least, these were their virtues in the eyes of most Europeans. But in the decades after the war, US officials pressed their European counterparts to accept competition as a fundamental goal of economic policy, with considerable success at the regional and local levels. As we shall see in Chapter 7, competition law turned out to be the chief weapon in the hands of the European Commission in the late 1990s and 2000s as it sought to reform the European gas sector.

A competing source of intellectual influence, especially on the Continent and in particular in Brussels, was German ordoliberalism. For the German ordoliberals, cartels and monopolies were the chief enemy, but the best instrument for combatting them was the state. "Market forces and a competitive economy were the standard for the Ordoliberals," write Daniel Yergin and Joseph Stanislaw, but "they believed in a strong state and a strong social morality."[3] The leading figure of ordoliberalism in the postwar years was Wilhelm Röpke, who taught in Geneva, but whose economic philosophy resonated throughout Germany and became the basis for German economic policy under Economics Minister Ludwig Erhard.[4] For Röpke, competition in a free market was paramount, but only a strong state could make sure that markets performed this function. His oft-quoted words virtually define the difference between Anglo-American neoliberalism and German ordoliberal thinking, but they also define much of the social-democratic economic philosophy of Jacques Delors, who as president of the European Commission from 1985 to 1995 led the fight for a single European market:

> A pure free-market economy . . . cannot float freely in a social, political, and moral vacuum, but must be maintained and protected by a strong social, political, and moral framework. Justice, the state, traditions and morals, firm standards and values, are part of this framework, as are the economic, social, and fiscal policies which, outside the market sphere, balance interests, protect the weak, restrain the immoderate, cut down excesses, limit power, set the rules of the game, and guard their observance.[5]

The result in Brussels, though neoliberal in its outline and essence, was much changed from its American version and was blended with other traditions. Yet arguably the British influence was paramount, as a generation of Commission officials, many of them British (and some of them Irish), arrived in Brussels after the 1980s. They had absorbed the neoliberal message via their education, their postdoctoral stays abroad (especially in the United States), and their early work experience, whether as regulators in Great Britain or executives in private business. What they brought was not drawn solely from Chicago or the United States but from what was by the 1980s and 1990s a broad neoliberal consensus in the Anglo-American world[6] that focused not just on liberalization and the internal market, but more broadly on globalization and privatization. To appreciate their role, one must look at the level of career civil servants rather than political appointees—that is, heads of directorates rather than commissioners or vice-chairmen. There is frequently a continuity, as one leading "Anglo-Saxon" official replaced another over the space of a generation.

Meanwhile, in the Soviet Union, the neoliberal revolution was more than a decade away. Chicago, to Soviet ears, meant organized labor, or perhaps organized crime, but not economics. Ronald Reagan and Margaret Thatcher were geostrategic threats, not market models. Hayek was anathema, mainly because of his book *The Road to Serfdom,* which could be read only in smuggled copies or in the safety of a *spetskhran*—a classified library collection.[7] Insofar as there was a model for economic reform in Moscow, it came from Budapest or Belgrade. Thus, when neoliberal economics finally reached Russian ears in the late 1980s, it came in a rush as part of Mikhail Gorbachev's glasnost and perestroika, and while it was eagerly absorbed by radical young economists, mainly in Saint Petersburg, the understanding of these exotic ideas by the rest of the population boiled down to slogans—mainly privatization and decontrol of prices—without much awareness of (or, for that matter, interest in) the theoretical and philosophical underpinnings.[8] As a lasting consequence, even the most sophisticated Russians came into the 1990s with little awareness of the ideological changes that had taken place in London, Brussels, and the other capitals of Western Europe and their potential consequences for Russian business, including gas trade, as it moved into Europe in the early 1990s.

The story of the neoliberal revolution in Western Europe provides the essential background for understanding a further upheaval that followed directly after it: the spread of European Union law and competition doctrine throughout the member-states. But in the energy sector established interests—traditional monopoly utilities, particularly in the power and gas sectors—fought back, as statist elements in France and corporatist structures in Germany dug in to defend their traditional ways of doing business, safely shielded behind national borders. This was, at least at first, a struggle of Europeans against Europeans, as the Russians watched from a distance.

The irony of this story is that the primary source of competition doctrine and the market-based innovations that have transformed the gas and power sectors on the Continent stemmed above all from the country that was always the most reluctant member of the European Union and the most ambivalent about the European project—Great Britain. Yet it was British statesmen and officials who played the key roles in developing the Single European Market road map in Brussels and who, as the first competition commissioners in the EU Commission, were the pioneers in implementing it. Again, energy was a crucial battleground. The ideas that ultimately led to the controversial EU Gas and Power Directives of the 1990s and 2000s originated in ideas and reforms pioneered in Great Britain. The hub-based systems of trading used on the Continent today are modeled on ones first used in Great Britain. The revolutionary upheavals that eventually took place in the corporate structures and energy policies of the Continent were the culmination of political, economic, and intellectual developments that originated in Great Britain a generation earlier.[9]

Neoliberalism was absorbed in multiple places in Great Britain and blended to varying degrees with older liberal teachings. The London School of Economics (LSE) was a major source: Harry Johnson, one of the leading neoliberal stars at LSE, was originally at Chicago. David Newbery, who became one of the major neoliberal faculty members at Cambridge University and ultimately led the revolt against the prevailing Keynesianism in the economics department there, had been influenced by his experiences on the faculties of Stanford and Princeton Universities and at the World Bank. Stephen Littlechild, a major figure in the history of regulation of British utilities, originally came from the University of Birmingham (as did many others),[10] but he earned a doctorate at the University of Texas, and during his years in the United States he focused on deregulation in monopolies such as the airlines and telecommunications. There were many others whose careers thus spanned the Atlantic.

The translation of neoliberal ideas into EU policy in Brussels was, at the highest level, above all the work of two people, an odd couple indeed: Margaret Thatcher, prime minister of Great Britain from 1979 to 1990 and a convinced neoliberal, and Jacques Delors, a social-democrat. Their single greatest victory—despite their deep disagreement over nearly everything else—was the design and adoption of the Single European Market project, which then supplied the legal underpinning for the expansion of competition policy. The man who made this possible was Lord Cockfield,[11] whom Thatcher dispatched to Brussels as one of her close allies—"one of us," to use Thatcher's famous term—but with whom she subsequently broke. Yet the breach between the two proved fortuitous. "Much as I regret having to say so," Lord Cockfield wrote in his memoirs, "what emerged over the years was that the most powerful support I enjoyed in the Community was the Prime Minister's hostility."[12] Indeed, he added, "Her contribution consisted of trying to knock a hole in the bottom of the boat."[13]

The Revival of the European Idea

At the end of the 1970s the European project appeared becalmed. Commentators, historians, and statesmen agree that at the end of the 1970s the European Economic Community (EEC) was in the doldrums. A decade and a half of argument over British membership, the common agricultural policy, the multiple challenges of dealing with the two oil shocks, and other divisive matters had left the EEC exhausted. As Lord Cockfield recalled the scene in his memoirs, these difficulties "had brought the Community virtually to a halt."[14]

What sparked the revival of the EEC, and especially the European Commission, as a force for policy innovation?[15] After the conflicts of the 1960s and the exhausted passivity of the 1970s, there was a remarkable rebirth of optimism and enthusiasm on many sides. Several causes came together in the first half of the 1980s and drove Europe forward. The world recession caused by the two oil shocks began to fade, bringing new life to the EEC, and the anxiety over energy security that had followed the oil shocks began to pass, giving way to a greater focus on efficiency. "The flame was rekindled," writes one observer, "and the phrase 'relaunching the Community' came into popular use."[16]

Some authors stress the bottom-up movement that originated among a group of politicians who strove to revive the federalist project to create a political union. The group was led by the Italian statesman Altiero Spinelli and worked through the European Parliament, whose members in 1979 had become popularly elected.[17] Others point to the coincidence of three strong national leaders coming to power at about the same time (François Mitterrand, Helmut Kohl, and Margaret Thatcher), all of whom had their own reasons for backing a stronger Europe. There was also the accident of personality—in particular, the arrival in 1985 of Jacques Delors as the new president of the European Commission. Delors proved to be the most effective and visionary Commission president in the history of the European Union.[18]

For French authors, in particular, a major cause of the restart of the EEC was the pro-European convictions of Mitterrand, who began his presidency of the EEC with a pilgrimage to the various European capitals to rekindle enthusiasm. He found a kindred spirit in Kohl, and it was above all the partnership of these two leaders that gave new impetus to the EEC, beginning with the summit at Fontainebleau in June 1984. Delors paid tribute to Mitterrand as the chief catalyst: "He was the sole architect of the Fontainebleau Accord, without which the restart of 1985 would not have been possible."[19]

Yet another contributing cause, less noticed at the time but now considered to be highly significant, was the 1979 decision of the European Court of Justice (ECJ) in the *Cassis de Dijon* case.[20] It was a major step forward in the reinforcement of competition law as the indispensable basis for the Single European Market, and it affirmed the role of the ECJ as the principal ally of the European Commission in policing the implementation of the Single European Market project.

The *Cassis de Dijon* case was a landmark in opening the way for the single European market. The case involved cassis, a blackcurrant liqueur made in France. Mixed with white Burgundy wine, it makes a popular appetizer drink known as kir, after a certain clergyman, Canon Kir, who popularized it. A German importer had sought to market the product in Germany but was blocked by a national regulation that a "fruit liqueur" had to have an alcohol content of 25 percent or higher. The problem was that cassis had only 15 percent. The importer sued the German regulator, and the ECJ ruled in the importer's favor, thereby establishing a legal principle that was crucial for the single market. The court laid down the doctrine that any product that was lawfully produced and sold in one member nation could be sold throughout the EEC, and "any measure that now impedes, directly or indirectly, actually or potentially, free exchange within the Community" was ruled illegal. While the court recognized the possibility of exceptions for special circumstances, it placed the burden of proof squarely on the regulator.

"Nearly everything the European Union touches turns into law," trenchantly observes a scholar of European competition law.[21] This is one of the most remarkable features of the European Union. On most occasions, when the Commission prepares a directive, the ECJ backs up the Commission with its own verdicts. The decisions of the ECJ have generally been accepted by the national courts of the member-states and incorporated into national laws. As such, they are binding on the national governments. The process is lengthy—it generally takes several years for a member-state's complaint to reach the ECJ and produce a decision—but it has been decisive. The role of the court has been indispensable to establishing and maintaining the power of the Commission in promoting the single market. This has subsequently become a major factor in strengthening the single market doctrine in energy policy.

But not least as a cause of the revival of the European Union was the growing influence of neoliberal ideas, especially in the area of competition policy. This can be seen clearly in the genesis of the Single European Market project, and the role of British officials in launching it from Brussels.

Why the Single Market Program?

Granted that there was renewed interest in Europe in completing the unfinished agenda of the 1957 Treaty of Rome, why did it take the form of a high-priority push to complete the single market? There were other priorities. Indeed, in the "Solemn Declaration on European Union" signed at Stuttgart by the heads of government in June 1983, the single market came well down on the list of priorities, behind a long list of other policies such as monetary union and external strategies. And the declaration of the Fontainebleau meeting of the European Council in June 1984 did not even mention the single market, although it did focus on the free movement of individuals.

The single market may have been a subject of relative indifference on the Continent, but that was certainly not the case in Great Britain. The center of British interest was services, especially financial services. "You can see a lorry carrying goods stopped at the frontier," wrote Cockfield, who as Thatcher's envoy oversaw the creation of the Single Market Program. "But you cannot see a banking service or insurance stopped at the frontier."[22] Yet the Continental nations, especially France and Germany, "had a very illiberal approach and wanted to preserve it."[23] Hence Thatcher's choice of a successful businessman and senior civil servant to take the job. Thus began the historic partnership of Cockfield and the Commission president, Jacques Delors.

Delors and the Single Market

Delors was no neoliberal, but a Catholic socialist. Charles Grant, in his account *Delors—Inside the House that Jacques Built,* describes Delors as a rarity, "a successful socialist in an era of resurgent neoliberalism."[24] He was a European federalist through and through (hence his fervent belief in a common currency) and a staunch supporter of a stronger Commission. As he said more than once, "No move to integration can succeed without genuine institutional dynamism."[25] In short, Delors stood for everything that Thatcher disliked or distrusted about the EEC.

Yet they had more in common than might have appeared on the surface. They had both risen from modest beginnings, and thus on a subconscious level they understood one another. They shared a common idea of progress as something driven by industry and technology—hence the importance in their eyes of industrial policy. In the end, however, these beliefs made Delors more of an ordoliberal than a neoliberal. His attitude toward the single market policy was the prime example. Grant writes:

> Utilities such as Gaz de France and Electricité de France have always found a sympathetic ear in Delors's cabinet. The president never championed the Commission's plans to liberalize the gas and electricity markets, which by 1992 had stalled in the Council of Ministers. [*NB:* This jibed with Delors's general belief that Europe needed large companies if it was to hold its own in world markets.] Whatever some may say, I say 'Long live Euro-champions.'"[26]

Thus the paradox of the Delors decade, Grant concludes, was that "despite ten years of a French, socialist, and personalist president, the EC has become—across the whole range of its economic policies—more of a force for than against economic liberalism."[27]

So how did it happen that the single market became the flagship program of the Commission during Delors's presidency? It was not initially at the top of Delors's priorities. In 1985 Delors put economic and monetary union in first place, followed by the strengthening of the European Parliament.[28] But the Single Internal Market came to be the centerpiece of the Delors program and is now widely regarded as the main achievement of the decade in which Delors was president of the Commission. How was it promoted to pole position in the Single European Act?

Arguably, it was British neoliberal thinking—the single most important part of that broad church known as Thatcherism—that provided the most important contribution to the development of Europe. But the individual who played the leading role was Arthur Cockfield, who, although he was an early member of the Thatcher team, would not have been called a neoliberal. And although he was sent to Brussels to implement a key neoliberal idea (the elimination of regulatory barriers to the free flow of goods, capital, money, and labor), Cockfield soon interpreted his mission in a far more expansive way. In the process, he broke with Thatcher, who became his active opponent and did not renew his appointment as commissioner after his first term.

Arthur Cockfield (later Lord Cockfield) rose from a modest background to take degrees in economics and law from the London School of Economics and then went on to distinguished careers in four different fields. He was a high official in the Treasury, a managing director of Boots (a leading chain of drugstores in Great Britain), and a member of Thatcher's cabinet for treasury and trade. But his fourth career, his crowning achievement, began in 1984, when he was nominated by Thatcher to be Great Britain's senior commissioner and vice president. In that capacity he became the close ally and collaborator of Delors, the new president of the Commission.

Cockfield's decisive contribution was largely forgotten over the decades, until the recent events surrounding Brexit caused his name to be revived and his contribution to be remembered. A frequent columnist in the *Financial Times,* David Allen Green, called Cockfield "the second most significant UK Tory politician of the 1980s."[29] Great Britain's problem, wrote Green bitterly in 2017, is that it "had a Cockfield to put the single market in place, but it certainly does not have one to take the UK out of the EU."

Gifted with near-perfect recall of even the most obscure document and supremely self-assured in his command of facts and figures, Cockfield was both admired and feared by his colleagues, and he was one of the few who dared to stand up to Thatcher. Cockfield's slow and deliberate delivery was easily parodied—it was said of him that he "spoke like a White

Paper."[30] Sir Roy Denman, the longtime EU ambassador to the United States and a lifelong admirer, wrote in Cockfield's obituary in the *Guardian:* "His conclusions would be remorseless in their logic; their presentation, in a slow, grave, deliberate voice, had the force of a glacier on the move. . . . He spoke with the authority of the Recording Angel."[31]

Why did Thatcher send such a man to Brussels as her chief negotiator on the single market? As Denman writes, "Cockfield was far from being regarded as a Euro-enthusiast. The prevalent view in Whitehall was that Thatcher had sent a dour and elderly Eurosceptic to clip the wings of an overweening bureaucracy."[32] Thatcher was initially enthusiastic about her choice. When Delors met with Thatcher and Cockfield at Downing Street in October 1984 to discuss Cockfield's assignment over dinner, it was Thatcher who made the case for Cockfield to take the internal market portfolio, and Delors was easily persuaded.[33] As Cockfield reflected later, "My success in the portfolio negotiations marked the high-water mark of my relations with the Prime Minister."[34]

But whatever their calculations may have been, Cockfield confounded everyone shortly after he arrived in Brussels. He became a whole-hearted convert to the European cause. For the next four years, Cockfield was Delors's most important ally and collaborator. It is widely acknowledged that the success of the Single European Market initiative was due above all to this highly productive alliance. To it Cockfield brought the same qualities—"the glacier on the move"—that had made him successful in his previous careers. He was efficient and persuasive, and above all, he knew what he wanted. As Cockfield wrote, with typical self-assurance, "I was not simply or primarily a politician. I had run a major British company and I had run it successfully; this was that sort of job and that was the way I wanted to run it."[35]

But there was more to Cockfield's conversion than met the eye. Denman hints at possibly deeper reasons in Cockfield's obituary: "Had Cockfield been born in France, with its respect for intellectual brilliance and the tradition of an authoritarian technocracy, he would have become one of the central pillars of the governing class. So his move to Brussels turned out to be an inspired choice, though at the time it did not seem like it."[36] It is hard to resist the thought that one reason behind Cockfield's "going native"[37]—as Thatcher famously put it in her *Autobiography*—was that in Brussels he found himself in a congenial environment, in which his personality and gifts were respected and admired.

The Battle for the Single Market

"The term 'Internal Market' does not appear anywhere in the Treaties."[38] Cockfield explains that the original objective had been to create a single market for external matters—that is, a customs union. This was achieved in 1967. It had been assumed that once this was done, goods would circulate freely within the common market. But that did not happen. "Once the tariff barriers were removed it became evident that nontariff barriers remained as a very formidable obstacle. Moreover, despite the very categorical provisions of the Treaty, little or no progress had been made in important areas such as services, transport, and fiscal harmonization."[39] The European Commission had long been aware of this, yet in the atmosphere of the late 1960s and 1970s no progress was possible. "But in the early 1980s as the Community and the world as a whole emerged from the recession, the flame was rekindled."[40]

Cockfield did three things at the outset that created the conditions for his success in creating and putting through his program, which he called the Single Market White Paper. First, he obtained from Delors control over a wide range of sectors belonging to several different directorates. These included financial institutions and company taxation, the customs union, and indirect taxation. To Cockfield's pleased surprise, Delors not only agreed to this but offered him the entire industry portfolio as well. Cockfield carefully sidestepped "being lumbered with coal and steel" (as he put it), but he readily accepted Delors's offer, since the industry portfolio included all research and development and new technologies. Consequently, from the beginning of his tenure Cockfield had command of all the relevant branches that had to do with the internal market.[41]

The second innovation was to redefine the internal market as consisting not only of goods but also of services. This was a particularly good approach, because it guaranteed that Cockfield would have strong support from his home government and from British business: "There was enormous UK interest here because of our major stake in the service industries. Once one could demonstrate that there was no essential difference between goods and services one broke down the psychological barriers that had led to progress on 'goods' and stagnation on 'services.'"[42]

Cockfield's third innovation was to redefine the problem, by finding a single theme that would cover the literally hundreds of areas that constituted the European market, yet avoid vague phrases and empty platitudes. Cockfield's solution was to define his program around the elimination of obstacles. As he put it, "One ought not to be looking at freedom of movement *as such,* and certainly not at goods, services, people, and capital *separately,* but at the obstacles or barriers which prevented that freedom of movement." These consisted partly of physical frontiers: "As long as the frontiers are there they will attract controls: each control will be the excuse for some other control."[43] But in reality the greater problem was the impenetrable thicket of national standards and qualifications that defined in each member-state who could practice law or sell insurance or compete in a tender—or for that matter work as a plumber or an electrician. It was this jungle of restrictions on services that Cockfield now proposed to hack down.

Once he had created the framework for the White Paper, Cockfield went to work. "Delors left me to get on with it," recalled Cockfield.[44] Cockfield knew that if the 1992 deadline was to be met, most of the proposals needed to be approved by the Commission during Delors's first term (by the end of 1988). In the end, Cockfield, with Delors's steady support, very nearly met the target: of the three hundred or so proposals that made up the White Paper, the Commission reviewed and approved 90 percent. The Council of Ministers, which has the ultimate authority over the Commission's recommendations, lagged somewhat behind, yet it still managed to approve 50 percent of the White Paper's proposals by the end of 1988.[45]

Why was this approach so successful? Cockfield comments on the enthusiasm with which the draft White Paper was greeted throughout the Commission and then by the national leaders of the member-states. The single market was an idea whose time had come—or rather, come back. But skillful tactics were an important part of the story. Above all, Cockfield emphasized speed. He was aware that Delors had other priorities. In his inaugural speech to the European Parliament in March 1985, Delors had outlined four major issues that he proposed to concentrate on. Enlargement came first. The second was monetary union. The internal market was only third. But in the spirit of the old adage that "he controls the agenda who writes the first memo," Cockfield and his team developed the proposals contained in the White Paper on the internal market so quickly that by the time Delors needed to put forward concrete proposals for the next Council meeting (scheduled for June 1985 in Milan), Cockfield's program was ready and the others were not. In this way, Cockfield relates, "the internal market became the Flagship of the Enterprise."[46]

Cockfield also understood the value of publicity, and he made sure that the media followed his progress. The result was a steady flow of newspaper articles describing "the Cockfield document" as a "prodigious program" (in the *Times*), and Cockfield himself as "Great Britain's most popular export to Europe for years" (in the *Guardian*).[47]

But in the torrent of favorable media coverage there lay a danger. One can well imagine how the following statement in the *Financial Times,* when read back in Downing Street, would raise hackles: "The White Paper deserves support [because] the Commission has rightly exploded the fallacy that Europe can somehow enjoy the benefits of a market of 320 million people without substantial concessions of national sovereignty."[48] Yet it was precisely such "concessions of national sovereignty" that Thatcher was determined to avoid. She wrote in her *Autobiography:* "Unfortunately, [Cockfield] tended to disregard the larger questions of politics—constitutional sovereignty, national sentiment, and the prompting of liberty. He was the prisoner as well as the master of his subject. It was all too easy for him, therefore, to go native . . . and to move from deregulating the market to re-regulating it under the rubric of harmonization."[49]

In pointing to the contradiction between reregulating the market to deregulate it—a particularly important issue in the reform of the gas market that followed—Thatcher had put her finger on a fundamental tension in the Single Market Program. The potential conflict came into the open over the issue of majority voting versus unanimity. For years the issue had been finessed under the so-called Luxembourg compromise, a gentleman's agreement that majority voting would be the rule unless one of the members felt that some essential interest was at stake—in which case it was the informal understanding that the member concerned would enjoy a de facto veto. But Cockfield had come to feel that the Luxembourg compromise was no longer viable: "Majority voting on Internal Market proposals was an absolute necessity if the program was to be completed and completed on time."[50] Delors agreed. For him, the gentleman's agreement was being abused: "Out of 100 decisions gathering dust on the Commission's shelves, only 45 required unanimity. Fifty-five were held up because of the invidious effect of the Luxembourg compromise."[51] As a consequence, both men had come to believe that "'majority voting' required a treaty amendment; and the treaty amendment required an Intergovernmental Conference."[52]

But majority voting was only the beginning. Cockfield had also come to believe that dismantling the obstacles to the single market required a strong state—or at least a strong Commission. This, as far as Thatcher was concerned, was the camel's nose of European federalism, and with a Brit in the role of the camel. For Thatcher, who was opposed to all of this, and above all to a new European treaty, it was too much.

The issue came to a head in Milan in June 1985, barely half a year into the joint tenure of Delors and Cockfield on the Commission. As the heads of state and their delegations converged on the city for the meeting of the Council of Ministers, the tension was palpable. In a Council meeting, the pressure is directly on the heads of state. In the nice phrase of Stephen Wall, who was part of the British delegation and later served as the British ambassador to the European Union (or to use the official term, as the permanent representative): "The European Council is like a boxing match: the seconds are out of the room and the combatants slug it out, being cooled down, fired up, congratulated, or cajoled by the men and women who hold the proverbial towel and sponge in between rounds."[53]

So it was at Milan. The British came to Milan believing they had support for a gentleman's agreement that would not require a new European treaty. But Kohl and Mitterrand had prepared a secret position paper calling for a new treaty, which in turn would require an Intergovernmental Conference to discuss a draft. The Italians, who had secretly conspired with the French and the Germans but did not show their hand until the last minute, came out in support of the Kohl-Mitterrand proposal. The British still thought that they could block the plan by invoking the gentleman's agreement, but at that moment Bettino Craxi, the Italian prime minister and president of the Council, sprang a surprise. Breaking with precedent, he called a vote. This had never been done before in a Council meeting, but Craxi justified his innovation by calling it a procedural question. The Kohl-Mitterrand proposal passed handily by a wide majority, supported by all except the British, the Danes, and the Greeks. Suddenly, the British realized that they were isolated, their strategy in shreds. Swallowing her rage—at least in public—Thatcher joined the majority and voted for the proposal.[54] "I realized we must make the best of it," Thatcher wrote in her *Autobiography.* "I saw no merit in the alternative strategy . . . of the so-called 'empty chair.'"[55]

But in private, her fury was "Krakatoan."[56] Her rage and humiliation were shared by the rest of the British side. According to Wall, "British officials had not foreseen the ambush. It came as a slap in the face for Great Britain . . . the Italians had blindsided us."[57] The consequences of Milan were far-reaching. "Now Pandora's Box had been opened," writes Charles Moore in his magisterial biography of Thatcher. "Milan helped confirm Mrs. Thatcher in her Eurosceptic instincts."[58]

The coup in Milan demonstrated the tensions not only between Thatcher and the other Europeans, but also among the British. Sentiment among British officials in the mid-1980s, even within the Thatcher camp, was very far from the Euroscepticism that it later became. According to Wall, the rest of Whitehall shared Thatcher's enthusiasm for the single market, but not necessarily her suspicion of the European project as a whole. In a document called "Europe—the Future," written for the meeting of the European Council at Fontainebleau in June 1984, the larger global mission of the single market was clearly laid out:

> We must create the genuine common market in goods and services which is envisaged in the Treaty of Rome and will be crucial to our ability to meet the US and Japanese technological challenge. [The aim was] to harmonize standards and prevent their deliberate use as barriers to intra-Community trade . . . and liberalizing trade in services, including banking, insurance, and transportation of goods and people.[59]

This was clearly neoliberal language, injected directly from Whitehall into the European process. It brings us back to the larger questions of this chapter: What was the impact of British neoliberal thinking on the implementation of the Single Market Program, and by whom was it conveyed? What was the direct role of British players, and where did they come from? To what extent did their various contributions mirror actual British experience with energy reform?

The Brits in Brussels: A Thin Red Line

So far we have looked mainly at the direct connections between the Anglo-American sources of the neoliberal rebirth and their impact on the European Union via Thatcher and Cockfield and the Single Market Program. But to fully appraise the British contribution, one must look at the indirect influences as well.

Cockfield in his memoirs was bitterly critical of the attitudes of his own countrymen toward the EEC. "They recall little of its history," he wrote, "[they] know nothing of its philosophy, and even more striking is the virtually complete absence of hard factual knowledge."[60] He was speaking mainly of the politicians but also of officials in the Foreign Office. But with all due respect this was clearly an exaggerated view, marked no doubt by Cockfield's own disenchantment with Thatcher. Beginning with the United Kingdom's accession to the EEC in 1973, British officials began to migrate to Brussels. Some were Whitehall officials; others simply took jobs in the European Commission by a process of self-selection that reflected a diversity of motives, and stayed on to make careers there. One of those was Philip Lowe, who joined the Commission shortly after graduating from Oxford and went on over four decades to become director-general for competition (2002–2010) and energy (2010–2013), the key period when the liberalization of Europe's gas markets clashed head on with Russian expectations. An accomplished linguist who was equally at home in English, French, and German, Lowe was initially attracted to Brussels by his love of foreign places and cultures and by the appeal of the European idea. Another example from the same generation is Jonathan Faull, a lawyer who joined the Commission staff in 1978 after earning a master's degree in European studies at Bruges. Faull, a specialist in competition law, initially served in Cockfield's cabinet and then rose to become deputy director-general for competition in the 1990s, before occupying a host of other leadership positions.[61] In the 2000s, it was said in Brussels that Lowe and Faull were two of the most talented and influential directors in the Commission. Commenting on the extensive British and American influence, especially in competition law and policy, Faull said:

> Yes, there was strong British influence, and yes, because of key individuals, such as myself and Philip Lowe, not to mention the early commissioners. But more than that. Many people working in the Commission and specifically DG-Comp [Directorate-General for Competition] had been educated in England. In the early days, all the literature on competition law came from England—and from the United States. Conferences on competition law tended to be held in the U.K. or the U.S. The whole area of competition law was a very English-language environment. That is no longer the case today—the continentals have caught up; every university in Europe has courses in competition law today.[62]

There were never many people such as Lowe and Faull in the European Commission—in any case, there were informal quotas that limited the number of citizens from any one member-state—but they made up an increasingly influential part of the professional staffs of the commissioners and directors-general. The flow of such people dwindled after the 1980s and had practically ceased by the 1990s and 2000s. Lowe, looking back on his own career, commented: "The number of officials of U.K. nationality in the Commission at senior levels was much more important in the 1980's and 90's than subsequently. There were several reasons: lukewarm support for the EU institutions from successive UK governments, Eurosceptic tabloids, higher salaries in London than Brussels, and a decline in foreign language learning in England as English became the dominant world language."[63]

Visualize instead a more diffuse and subtle process, as two successive generations of Brits come to Brussels. Strikingly, there are few economists among them, and no energy specialists. What they bring is not drawn directly from Chicago or the United States, but from an emerging neoliberal consensus in the Anglo-American world, focused not just on liberalization and the internal market but more broadly on globalization and privatization.

Lastly, one must look beyond the top political appointees to the career civil servants—that is, the heads of directorates rather than commissioners or vice-chairmen. An important example is Adrian Fortescue, who had been one of the first British officials to arrive in Brussels after Great Britain's accession in 1972.[64] A diplomat who had studied classics at Cambridge, Fortescue was also well versed in economics, having earned a degree in business administration at the London School of Economics. After serving under Lord Soames, the first UK commissioner, Fortescue returned to Brussels as Cockfield's *chef de cabinet.* It was Fortescue who was responsible for coordinating the many components of the White Paper on the Internal Market and gaining the cooperation of the other *chefs de cabinet.* Cockfield paid tribute to Fortescue in his book on the European Union: "Fortescue was able to secure virtually total agreement among the *chefs de cabinet* both to the White Paper and the 300 proposals it contained before the matter was referred to the Commission again." The proposal, Cockfield added, "could have taken many months, if not years," had it not been for Fortescue's diplomatic skill and his knowledge of the inner workings of the Commission.[65]

At this point we change gears: from the uploading of the internal market program, with all its antecedents, we turn to the downloading that followed—that is, the process by which the proposals contained in the White Paper were transposed into national law and then enforced by a combination of actions by the Commission and the ECJ. This second phase, which was particularly important in the translation of neoliberalism into the gas sector, proved to be lengthier than the first and more difficult. If one dates the beginning of the first phase of the European revival from the symbolic *Cassis de Dijon* decision of 1979, the ideas contained in the internal market program took thirteen years before they were fully approved by the Commission and the Council in 1992. But the phase of translation into national law and regulations, particularly in the area of electricity and gas, took another fifteen years, through to the adoption of the Third Gas and Power Directive in 2007. And whereas the progress of the White Paper was propelled by favorable winds of all sorts, the second phase faced strong headwinds in the form of determined opposition by national governments and established interests.

By the end of 1988 Cockfield was gone. His legacy was the Single European Market project. But at this stage it had not yet been translated into policy, nor had it yet been applied to energy. At this point two things happened. First, competition policy became the leading Community (as opposed to national) policy. Second, competition policy tools were brought to bear on energy matters by the Commission—precisely because this is where the Commission (as opposed to national governments) had power.

The main roles fell to Peter Sutherland, the Irish commissioner who oversaw the competition directorate from 1985 to 1989 and whose term thus overlapped with Cockfield's; and Leon Brittan, who succeeded Cockfield as vice-chairman, taking over from Sutherland as commissioner for competition affairs. Both Sutherland and Brittan demonstrate the visible continuity of the English-speaking neoliberal influence in Brussels in the second half of the 1980s and early 1990s. But whereas Cockfield's main objective was the creation of the concept of a single European space, Sutherland and Brittan turned the concept into a powerful weapon, using competition law as their instrument.

The two men were very different in personality and in their relationship to Delors, yet both were highly effective.[66] Both were trained as lawyers. Sutherland had been attorney general of Ireland before he was sent to Brussels; he went on to become chairman of the General Agreement on Tariffs and Trade (subsequently renamed the World Trade Organization) and to have a distinguished career that included the chairmanship of BP and Goldman Sachs. "Sutherland asked Delors for the competition job," Grant relates, "because a close reading of the Treaty of Rome had shown him that, in theory, the Commission had more power in this field than in any other." Sutherland saw the ECJ as the most important European institution. "I saw the law," Sutherland used to say, "as a way of promoting federalism and the Court of Justice as the most important EC institution." Grant adds: "During Delors's 10 years as president, the Commission increased its powers over competition policy more than any other area. . . . By the end of his first term as president of the Commission, Delors had grown as close to Sutherland as to any of the commissioners."[67]

Brittan had a very different profile, illustrating once again the diverse paths by which British (and, in Sutherland's case, Irish) players entered the Commission structure.[68] He had been named to the British Cabinet as chief secretary of the Treasury in 1981 and served as a key aide to Geoffrey Howe.[69] Although he had the reputation of being anti-European, he proved a doughty champion of European interests. Born in the first month of World War II, the younger son of Lithuanian Jewish immigrants, "he was emotionally as well as intellectually European in his perspective."[70]

Brittan proved just as liberal and forceful in Delors's second term (1989–1992) as Sutherland had been in Delors's first. Surrounded by such colleagues, writes Charles Grant in his book on Delors's Commission, "Delors sometimes had to accept more liberal policies than he would have liked." Grant adds, "Delors had more respect for Brittan than any other member of his second and third commissions and describes him as "one of the most brilliant men I have met."[71] Brittan came to Brussels as a close ally of Thatcher, but "he too went native, in that he came to believe in a tighter union."[72]

More than any other figure at the key juncture of 1989–1992, Brittan played the lead role in the Delors Commission in turning the Cockfield White Paper into an explicit energy policy based on competition and modeled explicitly on the British example. In 1991, Brittan gave a key speech that acknowledged the direct influence of the British example of privatization as it had developed in the 1980s:

> Of course, I was not operating in a vacuum. The privatised but regulated gas and electricity industries in the UK offered one example of the introduction of a more competitive regime. Indeed, the Commission carried out a comprehensive enquiry into the competition policy implications of the new British arrangements and we asked for and obtained several significant changes in what was proposed.[73]

Both men were widely recognized for their intelligence and ability. Member-states were often criticized in those years for not sending their best and brightest to Brussels. In this group Sutherland and Brittan stood out. According to Grant, "The only stars of the first Commission were Cockfield and Sutherland. In the second, only Brittan could match Delors intellectually."[74] Both were perceived as forceful men, and powerful voices for liberal policies: "Peter Sutherland . . . was one of the few commissioners who stood up to Delors. . . . Brittan proved just as liberal and forceful."[75]

Armed with the Single Market Program as a road map, the two men made vigorous use of the antimonopoly provisions implicit in the Treaty of Rome, pushing into many areas beyond state aid. Dawn raids now struck companies suspected of price-fixing: in 1986 fifteen polypropylene manufacturers were fined 60 million ecus (40 million pounds). In 1988 Sutherland went after the telecommunications industry, breaking up several public monopolies. In the same year he attacked mergers in the airlines industry, notably forcing British Airways to give up part of the routes it had absorbed when it bought out a competitor, British Caledonian.[76] In none of these cases did the Treaty of Rome actually give the Commission these wide powers, but Sutherland and Brittan seized upon Article 90, which allows the Commission to bypass the Council of Ministers in certain cases and to act by decree; Article 85, which bans anticompetitive agreements; and Article 86, which bans the abuse of dominant market positions.[77] Not for nothing did Delors nickname Sutherland *le petit shériff.* As for Brittan, "the Englishman's qualities, reinforced by an impressive *cabinet,* made him the most influential commissioner other than Delors."[78]

Where the big utilities were concerned, however, particularly in gas and power, Delors was less than wholehearted in his support for the reformers. As a result, the two competition commissioners gave more attention to other monopolies, particularly telecommunications, than to energy and did not push hard for action on gas and electricity. If they had, Delors might have taken stronger action to oppose them. Thus, the Germans were able to turn back the European Commission's early regulation on gas and power. One reason was the close friendship between Delors and Kohl. One of Delors's French aides told Grant, "On many big issues we've worked to defend German interests." Grant goes on: "Thus when Leon Brittan, then the competition commissioner, tried to cut Germany's subsidies to its coal industry, Delors blunted his attack. Similarly Delors helped to exempt the Treuhand Anstalt (the body responsible for privatizing East German industry) from the full force of

EC rules on competition."[79] Thus it took another decade before the rules and regulations developed in Brussels were applied successfully to the German utilities, as we shall see in the next chapter.

This chapter has recounted the origins of the migration of the neoliberal revolution of promarket ideas that traveled from Britain in the 1980s to Brussels and the rest of the EU. The centerpiece of the movement was the Single European Market project, catalyzed by Delors and a team of British and German EU officials in the European Commission, with the strong support of the heads of state of France, Germany, and (with reservations) Great Britain. Together these people restarted the process of creating the EU through a combination of increased regulatory authority in the Commission, backed by the European Court of Justice and the national courts. The 1980s were a rare time of alignment of national priorities and personal leadership in both Brussels and the member-state capitals, and the result was a flowering of liberalization measures, particularly in the area of competition law.

The influence of the neoliberal renaissance was conveyed to Brussels by a team of individuals who came to Brussels by many different paths, but whose background, both direct and indirect, was predominantly Anglo-American and who worked in Brussels as part of the Delors-Cockfield team. Thus a set of ideas and concepts that had formed the core of Thatcher's economic policies was transplanted to the Continent from Thatcherite Britain. The result was a wholesale challenge to the traditional statist orientation of the European nation-states and its displacement by a set of ideas emphasizing openness, competition, and marketization. These ideas were at the core of the Single European Market doctrine, which was the greatest achievement of the Delors Commission.

However, the translation of the White Paper into energy policy was delayed for a decade and a half by the staunch resistance of the member-states and the established national utilities. Indeed, the battle was not truly joined until the late 1990s. That is the subject of Chapter 7.

CHAPTER 7

Brussels

Marching to Market

Two battles broke out on the gas front as the result of the increasingly active policies pursued by the European Commission. The first was the conflict that took place in Western Europe, chiefly in Germany, between the mid-1990s and the end of the 2000s, culminating in the defeat of the established utilities and the victory of market forces. The second confrontation was between the Commission and Gazprom, which began in 2011. In both of these battles the lead was played by competition law and its application to energy. In this chapter we look at the process by which the Commission gained the power to play this role, as a result of which the old order of the European gas industry was overturned.

As we saw in Chapter 6, in 1985 Jacques Delors and Lord Cockfield launched the Single European Market project, the centerpiece of which was a White Paper with three hundred specific proposals aimed at overcoming the obstacles in the way of the single market. However, the White Paper contained no mention of energy. How, then, was the White Paper on the Single European Market turned into an energy program and then into the Gas and Power Packages, despite the opposition of established companies and several member-states? Why was the industry so opposed to reform? What caused the Commission to persist? What effect did the Commission have in driving the changes that have taken place in the European gas industry, as opposed to other objective forces (chiefly economic and technological) that were even then altering the traditional gas market? Finally, how were these events perceived in Moscow?

The Old Gas Order in Europe

Through the mid-1990s, the gas business on the Continent of Europe had stoutly and successfully resisted the trend toward greater liberalization that had taken place in other developed gas markets such as those in Australia, North America, and Great Britain, where reforms had brought liberalized pipeline access, greater stress on market signals (especially new techniques for pricing gas), and the beginnings of gas-on-gas competition. The Continental gas system remained wedded to the old order of managed markets, long-term contracts, and interfuel pricing—essentially the same system that had been in place since the late 1960s, when the cross-border gas trade first developed.[1]

Yet four fundamental trends were already bringing change to the Continental gas trade. The first was the market's growing maturity and complexity, particularly the expansion of the international pipeline network to include both new producers and consumers at the edges of Europe. In addition, electricity generators were emerging as purchasers, as technological innovations in the power sector (chiefly combined-cycle generation) made gas increasingly attractive for electricity generation. Second, there was a growing trend toward market integration, as lower fuel prices and weaker netbacks caused gas suppliers to look to gas transmission—the midstream of the value chain—as a growing source of rents. Third, there was an increasing supply surplus available to Europe, especially with the end of the Soviet regime and the emergence of Gazprom as a newly entrepreneurial force on the European gas market. Finally, there was the growing impact of information technology. With the rise of sophisticated systems of such technology, and ultimately the internet, it became possible to design systems for trading on networks in real time. These four forces for change brought to the Continental market for the first time the prospect of substantial volumes of uncontracted gas, potentially weakening the traditional model of contracts.[2]

Meanwhile, the significance of gas in the European energy economy was growing steadily. A symbolic date is 1994, when for the first time the contribution of gas to the energy mix exceeded that of coal (although both remained well behind oil).[3]

While gas professionals were well aware of these deep trends taking place in the structure of their industry, they disagreed among themselves over their implications. The European gas world was divided between those who believed that the old system must yield to a more open and diversified market and those who defended the merits of the traditional order. As gas expert Jonathan Stern wrote in 1992, "Commentators are either enthusiasts—who can see no great difficulties inherent in fundamentally changing the commercial rules . . . or opponents, who regard third-party access as equivalent to the end of civilization in the gas business."[4] He added, "They may not have gone as far as to claim that its outcome would leave consumers freezing and starving in their homes, but this was the impression they left on their audiences."[5] The debate took twenty years to resolve, but the answer is no longer in doubt: the old gas world is gone, and the market has prevailed. Where elements of the old remain, as in eastern and southeastern Europe, they are on their way out. This was due in no small part to the steady pressure of the Commission, beginning in the late 1980s.

This pressure was strongly resented and resisted by the gas industry. Their sector, from their point of view, was a signal success story and an important one. It was not a simple matter, in other words, of monopoly or quasi-monopoly structures seeking to line their pockets. Ever since the 1960s, the gas industry had built a pipeline network and distribution system that had come to span the entire continent of Europe, and this was a matter of considerable pride. It benefited millions of consumers and supported hundreds of thousands of employees. It had hugely improved the quality of the air in cities all over Western Europe,

The industry could claim, with some justice, that it was more European than the European Commission itself. After all, the early Dutch gas missionaries had seen German, Belgian, French, and Italian customers as a natural part of their universe in the early 1960s. And in the 1980s, the construction of the Middle European Gas Pipeline (known as MEGAL, from its German acronym) to ship Soviet gas to Europe required European cooperation, as neither Ruhrgas nor Gaz de France (GDF) had the financial muscle to do it alone. The industry could point to the Europe-wide security of supply that was assured by the capability of switching gas flows from one pipeline to another whenever there was a problem, always with the Groningen swing supply in the background and serving nearly anywhere in Europe. During the Russian gas interruption of 2009, it was EON. Ruhrgas and GDF Suez that had made sure that Belgrade and Zagreb were supplied with gas, at no cost, because of their European mission. It was a story the gas industry believed in deeply.

Moreover, the gas industry could argue that its traditional structure had served Europe well in other ways, too, including the environment. For example, the cooperation of adjacent monopolies (through demarcation agreements in Germany and de facto national responsibilities for the national companies of France, Italy, and the Benelux) ensured pricing of gas and investment in network expansion that enabled gas to attack the dominant heating oil markets in households and industry (and later, the coal market in power generation). As a result, a great deal of interfuel substitution was achieved under the corporate architecture of the time.

Lastly, because in many states the gas industry was partly or wholly owned by governments, both national and regional, it supported a variety of good causes. Gas rents accrued to a wide range of industries in the form of subsidized prices and other forms of state aid. A famous example, noted in Chapter 2, was the creation of hothouse agriculture in the Netherlands, in which low-priced gas supported an entire industry that grew cucumbers, gherkins, flowers, and tomatoes. Flowers were a special case in point: the Netherlands was exporting 84 percent of the cut flowers it produced, two-thirds of which were being imported by Germany.

Into this happy picture had come the Commission with its strange neoliberal language of price transparency, level playing fields, and opposition to state aid. The very notion of a level playing field seemed odd. When Gasunie, the Dutch gas company, found itself embroiled in an early case with the Commission's competition authority over subsidized gas tariffs for "glasshouse growers" (a case that it lost),[6] a senior Gasunie manager exclaimed with indignation (genuine if good-humored), "Why shouldn't we let our tomato growers have cheaper gas than the other EEC countries? God gave Italy the sun, and He gave the Netherlands the gas—so that's a level playing field!" (The Russians, a quarter-century later, would put it no differently.)

The argument of this chapter is that the role of the Commission was crucial in setting the agenda for the liberalization of the gas industry in continental Europe. While the Commission proved unable to reform the industry single-handedly and ultimately required the help of the courts and the national regulators, as well as some key converts in the gas industry itself, it nevertheless succeeded in creating the framework for reform and translating it into enforceable legal rules.[7] In the process the style of the regulatory relationship with industry also changed, becoming more adversarial—more American, some observers would claim—than the negotiated style more characteristic of European tradition. This in turn accelerated the broad changes that were already transforming the industry. One major result of this process was that by the time the Russians came into the European midstream in the 1990s, they encountered a world that was on the verge of becoming unrecognizable, compared to the one their Soviet predecessors had grown up with since the 1960s.

There is irony, one should note in passing, in that the liberalization championed by the Commission took the form of increased regulation—something that fit rather more with the German ordoliberal vision than the Anglo-American neoliberal one. In that sense, as we noted in Chapter 6, the neoliberal revolution as practiced in Brussels was very much a hybrid, and as time went on German players in the Commission came to play some of the lead roles in the development of its enforcement powers.

The Commission's First Moves

The debate over the future of the gas industry overlapped with the evolution of the Single European Market project and the development of a new legal and regulatory framework to enforce it. At first glance this might seem coincidental. As we saw in Chapter 6, the revival of the single market project was strongly driven by forces that on the surface had nothing to do with gas—neoliberal ideology, relations among member-states, and individual personalities.[8] At first, the single market agenda was not applied to energy at all, and then it was applied only to electricity, not gas.[9] However, notes gas expert Michael Stoppard, "the lack of harmonization and the diversity of energy prices were becoming increasingly anomalous within the wider context of the general thrust toward a single market."[10] By the late 1980s gas had moved to the center of the Commission's attention. The result was the Internal Energy Market document of 1988, which represented the Commission's first attempt to translate the Single Internal Market White Paper into energy policy.[11]

Reform was pushed forward by two high-profile commissioners under Delors, one for energy and the other for competition. They had very different styles. Antonio Cardoso e Cunha was commissioner at the head of the Directorate-General for Energy (DG-Energy, known in those days as DG-XVII) from 1989 to 1993, in the second half of the Delors presidency. He was a politician from Portugal's Social-Democratic Party, having had a previous career as an engineer in the chemical industry. He was the first commissioner from Portugal and was presumably eager to set good precedents. He was "very committed to liberalization . . . it was almost a personal crusade."[12] Yet his approach was pragmatic: he preferred to find negotiated solutions rather than invoke the law, which occasionally led colleagues to underestimate him. Faced with the task of overseeing a neocorporatist directorate general for energy whose staff lacked the will to challenge sectoral interests, Cardoso went around it, establishing an independent task force headed by a senior official from the competition directorate, which had experience of liberalization in other sectors, and he formed alliances with his fellow commissioners from competition and industry.[13]

Thanks to this partnership, there followed a burst of increased activity. The first draft energy directive was drawn up by the Energy directorate in 1989. On transparency of prices, it was adopted without major opposition the following year. A second directive on common carriage and third-party access (according to which a shipper that does not own a pipeline may nevertheless gain access to it to ship his gas to a third party without undue obstacles), adopted in 1991, was initially limited to transit between utilities. But this was as far as Cardoso was able to go, because member-states and established energy interests began to mobilize against the draft directive. Cardoso's search for political solutions proved unsuccessful, and in the end he failed to get his proposals for the energy sector past the Council of Ministers. In retrospect, there is a feeling among his colleagues that Cardoso was perhaps not tough enough. His competition colleague, Leon Brittan, once paid mixed tribute to Cardoso when he said Cardoso "has made considerable efforts to get things moving in the Council of Ministers"—suggesting that he had not been very successful, which was in fact the case.

Brittan, who was a commissioner at the same time as Cardoso (1989–1993), was a very different sort. We noted in Chapter 6 his close relationship with Delors. His approach from the first was more combative: unlike Cardoso, he had enforcement powers and was determined to use them. DG-COMP increasingly turned its attention to energy. Something of Brittan's resolve can be sensed from a speech he gave in the spring of 1991:

> When I arrived in Brussels in 1989 I did not see very much competition in the energy field. . . . Now my instincts tell me to be suspicious about any industry which claims to be a special case, to deserve special treatment and particularly when these arguments are presented in an attempt to justify monopolies. . . . Our objectives can be clearly stated: Consumers should have a choice of suppliers, prices should be set by competition, suppliers should be able to deliver their production through someone else's network, state subsidies should be transparent and limited and there should be no barriers to trade between the Community's Member States.[14]

Despite these strong words, by the end of his tenure as competition commissioner, Brittan had not achieved much more in the energy field than Cardoso had, and Brittan had done still less in the area of gas. It was broadly perceived by most observers that the Commission had been on the losing end in the energy field. Writing in 1992, the historian Stephen Padgett concluded that "There is general agreement that energy policy must be ranked as one of the Community's major failures."[15] Indeed, Padgett and many others believed that the situation was unlikely to change: "The commercial and technological structure of energy markets is inherently anti-competitive." In addition, Padgett wrote, there were several characteristics that made the gas and electricity industries "national," so national governments wanted to control the industries. Chief among these characteristics was the dependence of electricity and gas on physical networks that tended to be based on national boundaries. It was not until 1998 that the Commission finally succeeded in getting a gas directive through the Council of Ministers. Jonathan Faull, who went from an early career in DG-COMP in the 1990s to become by the time of his retirement the most senior British civil servant in the Commission, summed up the reasons for the long delay in making headway in energy regulation in those days with the simple words, "It was so damned hard!"[16]

But in retrospect the verdict of failure was premature. Despite the apparent blockage of the draft energy directive, two things had begun to happen. First, largely thanks to the Commission's efforts, decision makers at the level of the member-states were beginning to consider what had previously been unthinkable concepts for so-called natural monopoly industries, including electricity and gas. Gas expert Simon Blakey noted at the time, "The Commission's impact on national-level decision-makers . . . can genuinely be said to have changed the playing field on which the gas industry conducts its business." Third-party access, transparency, and unbundling "were ideas whose time had come."[17]

A prime example of the changing playing field in the wake of the First Gas Directive was the inauguration of the UK-Belgium Interconnector pipeline, which suddenly brought the possibility of gas flows from the liberalized UK gas market to continental Europe. Two years after it had begun operation in 1998, the Interconnector—in which Gazprom took a 10 percent stake—was already serving as a major vehicle for arbitrage between British spot gas and Continental contract prices. Daniel Yergin commented on this historic development in the spring of 2000: "A yawning arbitrage gap is opening up in Europe between the price of long-term contract gas and spot gas. A new basis for pricing and trading gas on the continent—the march to market—has begun."[18] The Interconnnector brought the reality of lower British spot prices to the Continent, and European consumers took notice.

The second development, initially modest but soon the subject of major headlines, was the growing legal muscle of the competition directorate, especially in terms of its power to conduct raids and impose fines. The directorate had been granted these extraordinary powers as far back as 1957, in the founding document of the Common Market, the Treaty of

Rome. The famous Regulation 17 of the Treaty, in the words of the competition scholar Daniel Kelemen, "gave the Commission unprecedented powers":

> In other policy areas, the Commission could enforce EU law only vis-à-vis national governments. In competition policy, the Commission could deal with firms directly. . . . Commission competition officials could conduct dawn raids on company premises and impose substantial fines on violators. . . . [They] had the power to preempt investigations by national authorities. . . . Also Regulation 17 granted the Commission a monopoly on the power to grant exemptions.[19]

Commission veterans agree that Regulation 17 was both crucial and unique. As Faull commented, "The most important thing was the adoption of Regulation 17 in 1962. It would be extremely difficult to enact such legislation today."[20]

Yet though these powers had been available to DG-COMP since 1962, they were not widely used until the 1980s. Prior to the arrival of Peter Sutherland and Brittan, the competition directorate had imposed fines for violations of the Commission's anticartel regulations, but they had been modest and fewer than a half-dozen in number. With Sutherland and Brittan, however, a more aggressive phase in the directorate's use of its enforcement powers began. Though DG-COMP's weapons were not initially applied to the energy field (much of the initial attention of the competition directorate was focused on sectors other than energy, particularly telecommunications) the weapons were being sharpened.

"Muscles in Brussels"

Since the 1980s DG-COMP has gradually become the most powerful of the directorates in the European Commission.[21] In the early 2000s, an international survey of experts rated it as "the most trusted and admired" among competition watchdogs at all levels of the European Union. Its civil service positions were among the most prestigious in Brussels, and the competition portfolio was among the highly prized.[22] By the first decade of the 2000s DG-COMP had reached the height of its power. "With its 'dawn raids,' epic legal battles, and multibillion euro fines," writes Kelemen, "competition policy has long been one of the few sources of drama in EU regulation, demonstrating . . . that at least in the field of competition policy there were indeed muscles in Brussels."[23] That is even more true today under the present commissioner, Margrethe Vestager, a former Danish deputy prime minister.[24]

But it was not always the case. DG-COMP started out in the years after the Treaty of Rome in 1957 as a "sleepy, ineffectual backwater" of European Community administration. It had only a "handful of A-grade officials," and working there carried "little prestige."[25] The directorate was perennially understaffed and underresourced (as indeed it remained over the next three decades). Its rise in the 1980s and 1990s was striking but also unanticipated. How did it happen?

Potentially strong powers in the area of competition had been granted to the Commission from the beginning. Jean Monnet, the spiritual father of the European Community, had considered competition one of the most important features of the Treaty of Rome, and he saw it "as a mandate to dissolve cartels, ban restrictive practices, and prevent any concentration of economic power."[26] Yet in its early years DG-COMP (at that time known as DG-IV) acted with great restraint, even timidity. There had been strong opposition from several member-states to including any mention of competition at all in the Treaty of Rome, and opposition only grew stronger as time passed.

The first cause of the growing power and assertiveness of DG-COMP was the strong support of the European Court of Justice (ECJ). In the landmark *Grundig* case in 1966, involving an action against the German electronics manufacturer of that name for restraint of trade, DG-COMP for the first time solicited the ECJ's opinion. The high court not only upheld much of the Commission's decision, but it also affirmed the Commission's authority against the governments of Germany and Italy, which had joined the case in support of the defendant. Over subsequent decades, the consistent support of the ECJ has been a key factor in the rise of DG-COMP.[27]

The second major driver was a succession of strong commissioners, beginning as we have seen with Sutherland and Brittan from the mid-1980s to the early 1990s, followed by Karel van Miert, Mario Monti, and Neelie Kroes in the later 1990s and the early 2000s, down to Margrethe Vestager to 2019. The presence of strong personalities from the mid-1980s on was not a coincidence. Thanks to the strong support of Delors, DG-COMP became an increasing attractive post. But in addition, as the 1980s turned into the 1990s, competition was becoming the consensus doctrine, part of the increasingly dominant neoliberal *air du temps,* and this too attracted strong personalities.

DG-COMP is also marked by a unique culture. One longtime Commission official remarks: "DG-COMP stands somewhat on the sidelines of the Commission, not in the main line of business. It doesn't legislate, and it doesn't spend money—two matters that take up much of the time of colleagues in other directorates. It has the reputation of being so technical people think they'll never master the complexities. They stand in awe of it, and are even somewhat reluctant to be assigned there."[28]

DG-COMP is also an interesting example of the ways in which British—or, more broadly, Anglo-American—neoliberal ideas, migrated to Brussels and influenced the exercise of competition law, as discussed in Chapter 6. However, one should also note the strong presence of German officials in DG-COMP. Philip Lowe, a former DG-COMP director-general, observes:

> German competition law dominated thinking for many years. Earlier on, the Federal Cartel Office regarded DG COMP as very much its junior. And before I became Director general in 2002, all the previous Directors general had been German. Some commentators at the time said that the German antitrust community and the German government only tolerated my appointment because I spoke fluent German. . . . It is not possible to explain the reputation and influence of DG COMP without referring to the positive and fruitful way in which German and UK Competition Policy traditions interacted.[29]

In the 1980s and 1990s, the number and range of cases brought by DG-COMP grew to include abuse of dominant position, illegal market-sharing agreements, price-fixing, restrictive distribution agreements, and many other actions. DG-COMP searched the premises of suspected violators (the famous "dawn raids"), and levied increasingly hefty fines. Total Commission fines imposed in cartel cases, for example, ballooned from 540 million euros in 1990–1994 to 3.46 billion in 2000–2004 and 9.76 billion in 2005–2009, before subsiding slightly in 2010–2014.[30] As its activity expanded, DG-COMP was consistently supported by the ECJ, which affirmed its supranational authority and enforcement powers.

Yet as the power of DG-COMP grew, so did the opposition to it—particularly in the area of regulated networks, such as electricity and especially gas—and DG-COMP did not always have things its own way.[31] The same was even truer of DG-Energy. For nearly a decade, as we have seen, DG-Energy was unable to gain approval for a gas directive, and when it finally succeeded in 1998, the resulting document was so weak that it was regarded as a failure. From that point on, DG-COMP increasingly took the lead. Yet it took another decade before the Commission, even with the two directorates-general working in tandem, prevailed.

During that time, however, DG-COMP's focus on the energy sector steadily sharpened. The path of growing DG-COMP involvement in the energy sector can be traced through the successive directives that the Commission published on gas and power, beginning in 1998 and culminating in 2007, the year of the passage of the Third Gas and Power Directive, the so-called Third Package. Though these were written primarily by DG-Energy, they were developed in close alignment with DG-COMP.

In the published literature on DG-COMP and the role of the Commission as a force for reform in the gas industry, there are two schools of thought. For the first, the most important driving force for change in the gas industry was not the Commission but underlying economic and technological trends that were shaking up the gas industry, especially starting in the 1980s and increasing in the 1990s. For the second, the role of the Commission was paramount. In addition, while DG-COMP in the 1980s and 1990s was not strong enough to prevail on its own, the steady support of the ECJ made the Commission progressively stronger as the years went by.

The answer depends mainly on the period one looks at. There were broadly three phases.

Phase One: Competition Policy and the Single Market, 1985–1992

Delors's drive to relaunch European integration with the European Single Market project put the emphasis on competition as the centerpiece. As the White Paper of 1985 stated, "As the Community moves to complete the Internal Market, it will be necessary to ensure that anti-competitive practices do not engender new forms of local protectionism which would only lead to a repartitioning of the market."[32] DG-COMP took up the challenge. Under commissioners Sutherland and Brittan, the competition authority intensified its enforcement. As we have seen, it began raiding the premises of companies under investigation, carrying away paper records and, for the first time, computers. It began imposing heavy fines (although still modest by comparison with current levels, as discussed below).[33] Above all, it forced changes in corporate practices and structure.

One by-product of the single market program was an increase in cross-national mergers. Businesses began to complain to the Commission about the anticompetitive impact of mergers by their competitors. DG-COMP initially hesitated to advance into this field, but in 1987, a decision by the ECJ in a case against Philip Morris increased the extent of DG-COMP's power to review mergers. As DG-COMP expanded its activity, businesses began to take notice, asking DG-COMP for advance clearance before undertaking mergers. A new merger control regulation in 1989 increased the Commission's powers further and established that DG-COMP would be the dominant player in the merger control field.[34] Philip Lowe notes about the late 1980s:

> The first EU Merger Regulation, which set up the world-renowned European one-stop shop for mergers, was passed in 1989 in the very early months of Leon Brittan's mandate after 17 years of negotiations between the Member states. It established for the first time legal deadlines for the approval of mergers and was the model for many of the subsequent wider reforms of DG COMP's organisation and working methods.[35]

The story of DG-COMP's expanded powers in merger control illustrates an important point about the drivers behind the increase in DG-COMP's powers. It was not solely the result of the ambitions of commissioners or the desire of the directorate to expand its turf, but at least as much a product of the economic forces unleashed by the European Single Market project and the resulting surge in merger activity throughout the European Union. The result was pressure from merger candidates themselves to have a one-stop shop that could provide advice on acceptability of proposed mergers. Today, in the words of Faull, "As a result of the merger control regulation there is now an automatic turnover threshold, above which the EU gets automatic notification. Consequently, there is no discretion over whether to bring a case or not. It's automatic."[36]

These expanded powers in competition law soon began to have an impact on the Commission's policies toward the gas industry. By the end of the 1980s the traditional structure of the gas industry was coming to be seen, not only in Brussels but among a growing number of national decision makers in the member-states, as a source of obstacles to trade. The prices paid by industrial consumers were neither well-known nor accessible.[37] Lack of transparency prevented competition in a market that was supposed to operate under the competition rules of the Treaty of Rome. Instead of one European market, there was a series of national markets existing side by side. The few companies that were the major players maintained extensive cooperative agreements that amounted to restraint of trade. Among these were pricing principles that prevented competition among different sources of supply. At the beginning of the 1990s three issues had emerged as uppermost in the battle for a single market in gas: (1) exclusive rights, (2) third-party access, and (3) unbundling (that is, the separation of supply and transit into two separate companies). All three were part of the same battlefield—gas transit.[38] It was over transit that the battle was first joined.

The first real test of the Commission's powers in implementing the European Single Market project in the gas industry came in 1990, with a Directive on Gas Transit. This gave the Commission its first taste of opposition from member-states and established interests in the gas sphere. Vigorous lobbying by Germany and the Netherlands had delayed the measure by over a year, and by the spring of 1990 it appeared to have been derailed. But on this occasion, to general surprise, the Commission prevailed, thanks to the use of the qualified majority voting rule that had been set up by the Single European Act of 1986. This was a first: "Never before had a major industrial policy measure been voted into effect by a qualified majority under the Single European Act rules."[39] Even then, it was a near thing—only a last-minute decision by Spain to change its vote to yes enabled the measure to pass the Council of Ministers and be sent on to the European Parliament. Yet this measure set an important precedent. For the first time, the Commission, backed by a majority of member-states, succeeded in overruling what formerly would have been a national veto.

Yet as far as its immediate consequences for the liberalization of the gas market were concerned, the passage of the transit directive was more symbolic than real. It was a relatively weak measure. It required only that gas transmission companies in the member-states transport gas belonging to other gas companies across their pipeline systems for delivery within the Community. Thus Gaz de France would be required to transport gas from Belgium to Spain, if the respective gas companies of the two countries concluded a sale and purchase agreement. This was still very far from third-party access, under which gas companies would be required to make space available in their pipelines for third-party producers such as petrochemical companies or power utilities. Nor did the directive specify how a transporter should respond if space was not available in its network. However, as weak as the directive was, the gas industry perceived it as the thin edge of the Commission's wedge, which it ultimately proved to be. But more than a decade went by before the principle of third-party access was fully established in EU law.

Opposition from Germany was especially strong. Of the forty-three companies covered by the Transit Directive, twenty-nine were German.[40] (Indeed, more than twenty of them were either customers, suppliers, or joint-venture subsidiaries of Ruhrgas.) This was no coincidence. In Germany the gas industry was organized by region. It was largely unregulated, and it was mainly a private-sector business. Given this structure, the German gas industry and the government argued that the directive in effect imposed an early form of open access on gas in Germany, and they saw it as the camel's nose of third-party access. A statement by Burckhard Bergmann, then CEO of Ruhrgas, testifies to the strong feelings the directive aroused in the incumbent industry:

> The German gas industry is willing to adopt a constructive approach to the transit function. In this respect the European Community Transit Directive was superfluous. It has always been possible to come to commercial arrangements on all transit issues. Whenever projects failed in the past, ostensibly due to insoluble transit problems, the argument does not stand up to scrutiny. In fact, in such cases, an excuse was often sought to explain why light-hearted promises did not materialize for projects which had never been viable.[41]

Another innovative feature of the Transit Directive was its sweep. In principle, it applied to the entire "grid of origin or final destination"—that is, to any gas flowing to, from, or through any of the European Community countries and, for that matter, the European Free Trade Association as well. Thus, again in principle, it covered gas imported from Norway or the Soviet Union. Although this feature passed largely unnoticed at the time, it was the first time that an office of the Commission had asserted, if only by implication, that Russian gas could be subject to extraterritorial rules by the European Commission—a theme with a long future ahead of it, as we shall see. But since the Transit Directive lacked teeth, it had no practical impact at the time.

Phase Two: The Courts and the Member-States Push Back, 1992–1999

The European Court of Justice (ECJ) could give, but it could also take back. As DG-COMP expanded its activity and as private actors began coming to the ECJ to challenge DG-COMP's decisions, the court began to rein it in. In addition, a new court, the Court of First Instance, was created in 1989 to take over much of the ECJ's ordinary caseload. From that point on, DG-COMP found that while the courts were its best support, they were also demanding friends. The Court of First Instance, in particular, began to review DG-COMP's substantive fact-finding and economic analysis. (One by-product was that DG-COMP began to hire economists as well as lawyers, and today the skill set of DG-COMP is a blend of both.)

The pushback against the application of competition law to the energy sector was part of a larger backlash against DG-COMP in the 1990s. The directorate was perceived to have become both overambitious and ineffective.[42] DG-COMP was seen as lacking transparency; the grounds for its decisions were criticized as opaque, lacking sound analysis, and unpredictable. The Commission responded that DG-COMP was a victim of its own success: it faced an ever-expanding stream of complaints and requests for clearance, as well as a mounting backlog of cases. To deal with the workload, DG-COMP began taking informal shortcuts, issuing advisory opinions, unofficial comfort letters, and so on. The combination of these labor-saving measures, plus the growing scrutiny from the ECJ and the Court of First Instance, led to a decline in the cases brought and fines imposed during van Miert's tenure as commissioner, in 1993–1999. These were years of relative quiescence in DG-COMP as a whole, during which the competition directorate was on the defensive.[43]

However, at the level of the member-states there were changes afoot. The same three issues noted above dominated the discussions among the national-level regulators and governments: exclusive rights, third-party access, and unbundling. Of these the most controversial was third-party access. This had already been one of the main thrusts of the 1988 Energy White Paper and had been pushed strongly by Cardoso, but the Commission had been unable to obtain the national governments' approval for this measure in the European Council. However, at the national level cracks were already appearing in the traditional structure. One of these was the so-called Weissenborn case, in which the German regulator (the Bundeskartelamt) broke new legal ground by affirming that Verbundnetzgas, the East German gas utility, had failed to provide access to its grid for gas moving from Wintershall to Weissenborn, a paper manufacturer located on the Czech border.[44] The German regulator's decision was subsequently upheld by the ECJ.

Phase Three: "An Audacious Coup" (1999–2007)

The third phase of the growing power of the competition authority came under the vigorous chairmanship of Mario Monti (1999–2004), who served under Commission president Romano Prodi. Monti, who subsequently returned to Italy and served as prime minister (2011–2013), presided over a far-reaching modernization of DG-COMP, accompanied by a strengthening of its powers to investigate and impose fines.

Monti had already established a reputation for high energy and integrity in his role as European Commissioner for the internal market and services from 1994 to 1999—during which time he earned the nickname "Super Mario." Under Monti, DG-COMP once again began to exercise its powers. The fines by imposed by DG-COMP in 2001 totaled more than all the fines imposed by the EU in cartel enforcement up to that time, and between 2001 and 2003 DG-COMP issued an average of eight decisions per year, compared with an average of 1.5 decisions annually over the previous thirty years. Something of Monti's determination can be seen in his words in 2001:[45]

> I consider cartels to be a veritable cancer in an open market economy. Unlike other forms of anticompetitive behavior, they serve one purpose and one purpose alone: that of reducing or eliminating competition. They bring no benefit to the economy. . . . The managers and directors of companies engaging in such practices must be in no doubt that we shall leave them no respite, that they will be detected and that the penalties will be heavy.[46]

Yet Monti now faced far greater challenges in managing the Commission's competition policy than in his previous posts. His directorate-general, though the most powerful in the Commission, was systematically understaffed: year after year, national governments vetoed DG-COMP's requests for additional personnel.[47] This had been a chronic problem since the 1960s, and DG-COMP's growing caseload had made matters worse.[48] As we have seen, dissatisfaction with DG-COMP's performance had built up under van Miert, Monti's predecessor. To deal with the problem, Monti launched a sweeping modernization of the competition authority's structure and methods. Gone were the previous notifications, under which companies could seek advance clearance from the Commission for actions that might violate competition rules. Now the companies were on their own, and the Commission could go after them if they got something wrong. Monti also invoked the doctrines of subsidiarity and decentralization to engineer a wholesale devolution of jurisdiction to the national regulatory level.[49] At first, it appeared as though DG-COMP had surrendered power to member-states. But a closer look shows quite the opposite: Monti had actually harnessed national agencies, courts, and private litigants to extend the reach of EU competition policy. Kelemen called it "the most significant reform in the history of EU competition policy."[50] The result was to make DG-COMP more powerful than ever.

Neelie Kroes and the Third Package

As active as "Super Mario" was as head of DG-COMP during the first half of the 2000s (1999–October 2004), his successor, Neelie Kroes (2005–2009) went him one better. Under Kroes the total fines levied by DG-COMP (9.4 billion euros) were nearly triple the total levied under Monti.[51] Kroes took on some of the biggest corporate giants of the world—including Intel and Microsoft—and won. It was also under Kroes that the first energy sector inquiry was launched by DG-COMP, and although (as we shall see) Gazprom was not a target at first, it was Kroes's sector inquiry that prepared the ground for the investigation of Gazprom that followed under her successor, Joaquin Almunia, starting in 2011. In short, the shift from cooperation to conflict that began in the Russian-European gas relationship toward the middle of the 2000s has its origins to a substantial degree in the changes that took place in DG-COMP's approach to energy under Kroes.

Unlike Sutherland and Brittan, Kroes was not a lawyer by background; neither was she a noted academic economist like Monti. She had studied economics at Erasmus University in the Netherlands, but at the age of thirty she was elected to the Dutch parliament, and from that point on she was employed mainly as a working politician, serving for seven years as transportation minister. When she was first named competition commissioner by Commission president José Manuel Barroso, it was whispered in Brussels that he chose her because he wanted to increase the number of female commissioners, and all of her predecessors had been male. Kroes's confirmation hearings before the European Parliament did not go smoothly: her English was rusty and she knew little of the fine points of competition law. "She had a pretty rocky start," said one Brussels-based lawyer.[52]

But as a professional politician, Kroes had good survival instincts and a keen sense of politics, and these served her well as she learned her trade as commissioner. She was a tough and resourceful negotiator, and she ran a tight meeting, but with charm and skill. She developed a smooth working relationship with Lowe, her deputy. As Lowe commented, "At the beginning she was pretty unfamiliar with what competition policy was all about, but she rapidly caught up speed on the major cases involving companies such as Microsoft and Intel and on the follow up to the energy enquiry." Above all, she brought to the job deep promarket convictions, instilled partly by family business ties, that she expressed in tough and plain speech. "I feel myself come alive when I speak about our cartel work," she would say to audiences. "Nothing is more fundamentally wrong in our field than a cartel. I am not a shy person. . . . We are prosecuting more cartels and preventing more consumer harm than ever before."[53]

She brought equal passion to energy, as well as a sense that the field had been relatively neglected. As she told an audience in Vienna in the spring of 2006,

> Despite two waves of liberalization . . . a single competitive European energy market is still not a reality. . . . Shortly after coming into office, I was pleased that my Commission colleagues backed my suggestion for an in-depth assessment of the energy markets. Over the past nine months, we have used a new tool—the sector inquiry—to find out more about the barriers to free competition in energy.[54]

"The old incumbents," Kroes told her listeners—who happened on this particular evening to be the cream of the Austrian energy community—"appear to remain the top dogs." She was particularly scathing toward the Austrian gas and power industry, where "the largest player covers around half of the electricity and three-quarters of the gas supplies to final customers. . . . There is very little market entry." And she gave fair warning of what lay ahead: "The Commission will launch individual anti-trust investigations where this is appropriate, for example, where there is vertical foreclosure caused by long-term downstream contracts, or hoarding of capacity on pipelines, gas storage and interconnectors." "Vertical foreclosure," "capacity hoarding," and so on was new language for the established gas industry. With language and concepts such as these, drawn from the academic literature, DG-COMP was able to outflank the industry at the intellectual level.[55]

The sector inquiry lasted from mid-2005 to early 2007, focusing chiefly on electricity and gas. DG-COMP then proceeded, on the basis of the findings, to draw up a package of recommendations for new legislation—the famed Third Package. On the eve of its presentation, in September 2007, Kroes briefed an audience on the chief findings of the sector inquiry and the new package. Her chief conclusion was brief and to the point: "The results were deeply concerning."[56] The sector inquiry, the commissioner went on, had found five main problems:

> First, continuing high levels of concentration so incumbents maintain market power; second, vertical foreclosure, as the old monopolists continue to own the energy infrastructure; third, low levels of cross-border trade, due to insufficient interconnector capacity and to contractual congestion since spare physical capacity is not always released; fourth, lack of transparency about operations in the wholesale energy sector, which makes it difficult for new entrants to understand how the markets work in practice and the risks that they take on; and finally, lack of confidence that wholesale energy prices are the result of meaningful competition.[57]

However, these were, in a sense, generic problems, and previous commissioners would probably have come up with the same list. What made this speech different was that Kroes was now armed with facts, and there was a tone of moral outrage in her words as she described the abuses

DG-COMP had uncovered. For example, on the subject of discrimination in access:

> I can cite the case of a Transmission System Operator that granted its affiliated supply company substantial rebates that were not available to others. In another case a Transmission System Operator offered transport capacity to its affiliated company while refusing firm capacity on an almost identical route to other suppliers.

Certain of the abuses discovered in the sector inquiry came close to insider dealing:

> In some cases the top management of the supply branch have access to strategic business information of the transport company, either directly or as a result of their presence on the Board. Or on a more practical level, e-mails are copied to affiliated companies. In some cases it appears that central functions, such as legal advice, are still provided by the group holding company to all members of the group.[58]

Kroes concluded, "There are therefore limits to how far Chinese wall arrangements can actually achieve their function on the ground." The implication was clear: the only real cure would be to make the Chinese walls real by breaking up the companies involved.

Before we go on with the implications of such a breakup and what followed, it is important to note one singular feature of the sector inquiry: its main targets (and the only ones cited by name) were West European. Gazprom was not mentioned. In Kroes's 2007 speech, the only indirect mention of Russia was an endorsement of the Nabucco Third Corridor project to bring gas from Turkmenistan to Western Europe by way of Turkey. The chief villains of the inquiry, in contrast, were named explicitly—E.ON, Distrigas, GDF, and Eni, but not Gazprom. Indeed, the only mention of Gazprom in the sector inquiry report occurs in a footnote and is entirely neutral in tone: "New fields are also being developed in Russia, but these would appear likely to be marketed in the traditional way to the former incumbents or companies in which Gazprom has ownership or other links."[59]

The same silence on the subject of Gazprom and Russia during Kroes's tenure can be seen in the annual reports issued by DG-COMP. In the period between 2006 and 2009, neither Gazprom nor Russia was mentioned even once, despite the fact that 2006 and 2009 were the years of the two cutoffs of Russian gas to Europe. Only in 2011, the year of DG-COMP's first dawn raids in Eastern Europe, was the silence finally broken. One plausible inference is that as far as DG-COMP was concerned, the cutoffs were not competition issues. A more plausible one is that DG-COMP's sector inquiry had not covered Eastern Europe, where the toughest competition issues subsequently arose. But the 2009 cutoff changed the attitudes of the Europeans toward Gazprom. By 2011 DG-COMP was no longer in any mood to ignore the Russians, as we shall see in Chapter 13.

Persuasion Is the Best Weapon

At this point we need to circle back briefly to the 1990s to pick up an earlier story. DG-COMP looms so large in our account that we have left the other key directorate—DG-Energy—largely in the shade. It is true that DG-Energy lacks the legal, investigative, and enforcement powers of DG-COMP. It cannot batter down doors at dawn and haul away computers. But it has the power to persuade, and that is what it deployed in the mid-1990s. Even before the passage of the toothless First Gas Directive in 1998, DG-Energy had convened a pair of regular forums (the first in Florence and the second in Madrid) to discuss market reforms in the electricity and gas industries. The idea was to create a venue for all sides in the reform debate to be represented, explore the technical and political issues, and get the buy-in of the companies under the eye of government representatives. Both forums continue to meet twice a year, and their role has been invaluable in creating a common set of understandings throughout the power and gas industries.[60]

Thus as one visualizes the role of the Commission in the energy field from the 1990s to the present, it consists of two strands—that of competition law (embodied in DG-COMP) and that of regulatory persuasion and consensus building (embodied in DG-Energy). Both have been essential in building the Commission's role in the energy sphere. The complementarity between the two directorates-general is symbolized by the continuity represented by Lowe, who was director-general of DG-COMP from 2002 to 2010 and of DG-Energy from 2010 to 2013. Lowe commented about his own role in the collaboration between the two services:

> Before the end of the mandate of Mario Monti, I launched the idea of a competition enquiry into the energy sector. When she became Commissioner, Neelie Kroes enthusiastically took this up, with the active support of the Energy Commissioner, Andris Piebalgs. Quite apart from the antitrust investigations subsequently launched by DG-COMP into the practices of some energy companies, the authors of the proposals for the Third Energy Package were the services of DG-Energy, under Andris Piebalgs, and not DG-COMP, although we were of course among the main protagonists.[61]

Back in the USSR

Meanwhile, how did the neoliberal currents that were revolutionizing economic and political thinking in the Anglo-American world reach the Soviet Union, and what influence did they have on the rise of the reform movement in Russia? This is an important theme that we shall return to at intervals in the rest of this book, because it underlies several key questions to come: To what extent were the Russians aware early on of the revolutionary trends taking shape in the power and gas sectors of Europe? Did the Russians follow the rise of liberalization and competition doctrine in Brussels? Did they sense its implications for their business in Europe—and ultimately for the gas business in Russia itself? Or were they taken by surprise?

The answer, of course, depends on which Russians we mean. The future reformers of the Boris Yeltsin era could be found in ones and twos in the late 1970s in the academic institutes of Moscow and Leningrad, the media, and even the apparatus of the Communist Party. Yegor Gaidar—who as acting prime minister and then finance minister led the first radical reforms in 1991–1993 but at the same time (as we shall see) blocked the breakup of Gazprom—was an economic editor on the staff of *Kommunist,* the official monthly of the Central Committee of the Communist Party of the Soviet Union. He came from one of the most prominent families of the communist upper crust. "I started out as a convinced communist," Gaidar told me in an interview in 1996:

> In the late 1970s I began post-graduate studies, and realized that halfway measures or hybrid compromises—"market socialism"—simply wouldn't work. There were plenty of like-minded people to talk to, and it wasn't so difficult to find things to read. . . . If you read foreign languages and were willing to take the trouble, you could find a lot to read on market economies. The most important influence on me was the Hungarian economist Janos Kornai. His analysis of the economy of shortage, in the early 1980s, had a great impact on all of us. He was addressing *our* problems. Among Western writers, the most important influence on me was Friedrich von Hayek. He gave a very clear and consistent picture of the world, as impressive as Marx in his way.[62]

Anatolii Chubais—one of the first reformers in Saint Petersburg, the head of the Russian Privatization Committee in the 1990s, and the man often called the father of Russian privatization in the early 1990s—writes in a memoir, "Eurocommunism, market socialism, socialism with a human face—I believed whole-heartedly in it all at the end of the 1970s and early 1980s." But the formulas came and went, and nothing seemed to change: "The state of the Soviet economy became ever more hopeless, and along with the worsening of the economy our illusions melted away. . . . Gradually, step by step, we came to understand that the foundation of any healthy and successful economy is private property."[63]

Aleksey Kudrin, another early reformer in Saint Petersburg, went through a similar evolution, as did many other future reformers. Kudrin went on to be Vladimir Putin's finance minister for a decade and was the man most responsible for the modernization of the Russian fiscal and banking system in the 2000s, but he began life as a convinced Marxist. He recounts that his first doubts began in high school, but like many others, he was influenced primarily by reforms in Eastern Europe.

It was only as a graduate student in Moscow that he became convinced that a market economy was the only solution to Russia's economic problems.[64]

For Chubais, Kudrin, and other Russian reformers, a moment of revelation was the publication in 1987 of a book titled *Drugaia zhizn'* (Another life) by the economist Vitaliy Naishul'. Naishul' is widely acknowledged by the leading Russian reformers today as one of the spiritual founding fathers of the market economy in Russia. His book opens with a call to battle:

> Dear reader! Do you know that you get from the state only half of what you earn with your labor, live twice as meanly, eat two times worse, dress twice as badly, and live in worse conditions, than you might . . . in *another life?* And for this life, what is necessary is Economic Reform and a division of state property *from the people to the people!*[65]

Naishul' gathered around him a group of like-minded young people who came to be known (among themselves, at any rate) as the Snake Hill Gang, from the Moscow neighborhood where they met. In an interview he commented:

> I myself understood in 1979 that the Soviet Union would be dead soon. I worked in the State Planning Committee—Gosplan—and I observed huge economic discouragement, huge and growing. . . . The consensus was that we had failed. So I started to think about how to change from this system to a market economy. . . . I immediately began to think about privatization, about how it should be carried out. I thought about voucher privatization at the beginning of the 1980s.[66]

The members of the Snake Hill Gang were certainly aware of neoliberal ideas in the global economy. In fact, some of the members went to Chile to observe how market reforms were being carried out there under Augusto Pinochet.[67] But by this time it was late in the Mikhail Gorbachev period, when restrictions on foreign travel were eased.

As for the academic economists in Moscow, it is difficult to imagine two worlds more remote from one another than the University of Chicago and, say, the Institute of Mathematical Economics of the USSR Academy of Sciences, which had the reputation at the time of being the best institute for quantitative economics in the Soviet Union. The monetarist macroeconomics that was the central strand in the thinking of Milton Friedman could hardly have been more foreign to a system that had as its central ambition the elimination of money as anything more than a passive unit of account. It was not so long before, during the Joseph Stalin period, that Western-style economics had been brutally suppressed. As a noted specialist on the Soviet economy, Gregory Grossman, observed in his dissertation, "The economic literature disappeared with the economists themselves."[68] When modern economics was slowly reborn in the Soviet Union in the 1960s under Nikita Khrushchev, it was largely in the guise of abstract mathematical models, which were less controversial politically.

In sum, the exposure of different Russians to neoliberal ideas depended very much on who they were, what their main preoccupations were, and what period one is talking about. For some the turning point was in 1979, coinciding with the stagnation *(zastoi)* under Leonid Brezhnev, the end of détente, the invasion of Afghanistan, and a general depression that settled over the Russian intelligentsia in the late 1970s and early 1980s. But the key point about the reformers' life paths is that their early disillusionment with the Soviet system did not take concrete form as a program until late in the 1980s. By the beginning of the 1990s, the neoliberal tide had become a flood in Russia, but the main reason it poured in when it did and as powerfully as it did was the vacuum inside.

What then of gas? As the 1980s turned into the 1990s, what did the Russians make of events in Brussels and their potential implications for the Russian gas industry? Through what channels did they follow the trend toward the Single European Market in the 1980s and the Commission's efforts to apply neoliberal ideas to the gas industry? To what extent did they follow the gas regulation issues that brought such vehement opposition from the European gas industry, and what was their reaction to them? As the European gas market began to evolve away from the Groningen model and toward the development of spot trading and hubs, how early were the Russians aware of the trends and the politics surrounding them?

The answer depends very much on how close they were to the gas trade. A man such as Yurii Komarov (discussed in Chapter 10)—the first Russian cochairman of Wingas, Gazprom's joint venture with the German energy company Wintershall—was naturally at the center of the questions we have been discussing in this chapter and would have been informed to the last detail (although he may well have absorbed some of the skepticism of his German partners). The same would have been true of other foreign trade and financial specialists in the overseas offices of the foreign trade ministry and the USSR State Bank, such as the future head of Gazprombank, Andrei Akimov, and his partner Aleksandr Medvedev, until 2019 cochairman of Gazprom responsible for export policy, who were both investment bankers in Vienna during the late 1980s and 1990s.

But apart from such specialists, the broad answer is that until the beginning of the 1990s most Russians, even those at the upper levels of Gazprom, probably had very little knowledge of these questions and very little reason to be interested in them. But that changed practically overnight, starting in the early 1990s. There were several reasons. First, the collapse of the Soviet Union and the advent of market reforms inside Russia soon raised the question of the status of Gazprom itself. For the market reformers, Gazprom was the *sovetskii monstr,* a sinister remnant of the era of central planning, and as such needed to be broken up. In a remarkably short time, the reformers acquainted themselves with the basics of gas regulation, drawing directly from the Western literature and contacts with the Western gas industry. The World Bank was a particularly important channel of communication and learning.[69]

The second broad channel of communication was through Gazprom itself—or rather, GazpromExport. As we saw in Chapters 2 and 3, Russian gas professionals had been in contact with their Western counterparts since the earliest days of Soviet-European gas trade in the 1960s. These contacts intensified with the end of the Soviet system. Gazprom became a quasi-private corporation in 1989 and absorbed Soiuzgazeksport, the specialized gas-trading organization of the USSR Ministry of Foreign Trade. But more important, the breakup of the Soviet empire placed

Gazprom right in the middle of the European gas industry. Instead of dropping off gas exports at the West German border, as it had traditionally done for two and a half decades, Gazprom formed a trading alliance with a German company, Wintershall, and began participating directly in the midstream of the industry—that is, transportation and distribution. Through this connection, GazpromExport officials became acquainted with gas regulation issues in Germany and in other European countries where Gazprom created joint trading houses and became intimately involved in the European gas business as a direct player.

However, all this does not mean that the Russians were necessarily paying attention to the fundamental changes going on in the European Commission in Brussels. On the contrary, according to Stern, until the end of the Rem Viakhirev era Gazprom did not focus particularly on Brussels. For this there were, again, several reasons. First, the Russians have always been suspicious of the so-called European project and, more important, have tended to underestimate its importance, preferring to deal on a bilateral basis with individual European governments rather than the distant and (to Russian eyes) confusing bureaucracy of Brussels. For a brief period shortly after Putin was elected president, there was much talk of a strategic alliance between the European Union and Russia (the so-called Putin-Prodi process), but the Russians soon tired of the constant lectures they received from the European Union, and the process soon came to end as Putin returned to a more traditional bilateral approach, especially with the Germans.[70]

Until the mid-2000s there was little pressing reason for the Russians to pay more attention than they did. The Commission's efforts to reform the gas industry would come to nothing, their German partners assured them, and initially that appeared to be true. The Russians' own observations over the previous decade appeared to confirm this. One observation post for the Russians was the apparatus of the Energy Charter, a secretariat created to support the creation and adoption of a treaty to guarantee the security of Eurasian gas transmission. The Energy Charter never achieved its early aims. But it did serve as a vehicle for the education of a generation of Russian officials in gas regulation issues, especially those Russians who were assigned as staff members of the Energy Charter organization. When they returned to Russia, they became a resource for the dissemination of knowledge about gas regulation issues, in both Gazprom and its competitors.[71]

As the Russian-European relationship became more troubled in the mid-2000s, chiefly as a result of the Orange Revolution in Ukraine (see Chapter 11), the Russians began to focus more intently on what now appeared to Russian eyes to be the threatening progress of the Commission's gas and power directives, and the apparent crumbling of opposition to them from national companies and governments, particularly in Germany. This is the subject of Chapter 8.

CHAPTER 8

The Battle for Germany

We turn now to Germany. Germany's role as the center of the European gas industry reflects its central economic importance in Europe.[1] It is the largest gas consumer on the Continent, and it is the physical center through which much of Europe's imported gas must pass. Because of the increasing integration of Europe's gas transmission and trading systems, the commercial and regulatory relations that govern the gas flow through Germany affect every other player in the European energy sector—not to mention Europe's principal gas suppliers, Norway and Russia.

Germany has also been ground zero of the clash of cultures between the European Union's drive to create the Single European Market and the member-states' determination to maintain their sovereignty. The clash has been particularly dramatic in the German gas sector.[2] The result, after battles extending from the mid-1990s to the late 2000s, has been a wholesale transformation of the German gas market. The traditional basis of the industry—a close-knit family structure based on demarcated territories linked by long-term contracts—has practically disappeared. Gas and access to pipeline capacity are increasingly commodities traded through open spot markets and transparent trading platforms. As a consequence, the once-staid structure of the industry has been transformed as long-established companies have disappeared or been absorbed, and a new business culture, with new players, has taken hold.

Germany continues to surprise. Growing public concerns about climate change and other environmental issues have roiled German energy politics. More than any other country, Germany has chosen to confront environmental challenges head-on by phasing out nuclear power[3] (effective in 2022) and turning to renewables—chiefly wind and solar power—in a vast program the Germans call the *Energiewende* (energy transition).[4] But the combination of these technological changes and radical policy shifts has had unforeseen and highly disruptive side effects.[5] The German gas and power sector is in turmoil, as costs have skyrocketed and revenues have collapsed. Major utilities have been driven to the brink of bankruptcy. The very role of gas is being challenged.

Germany is normally known for its cautious style of policymaking, based on elaborate checks and balances among levels of government; painstaking consensus building among parties and interest groups; and slow, careful, and incremental change. On the face of it, the upheaval in the German energy sector seems an astonishing departure from form. In this chapter we argue that one important reason for the radical sweep of the *Energiewende* today is that the traditional structure of buffers and bulwarks that had kept the energy sector stable for decades and enabled it to resist change (one face of which was the celebrated German "Gas Club") was fatally weakened by the new business models and regulatory doctrines emanating from the United States, Great Britain, and Brussels—the ideological wind from the West that had begun to blow into Germany in the early 1990s. By 2010, the stout resistance to change by the gas industry had crumbled. As other pressures (notably, that of the environmental movement) mounted over the following decade, there was less and less to stand in their way.

What were the consequences for the Russians? To make a long story short, when the Russians entered the German gas market at the beginning of the 1990s, the rules they encountered there were already familiar and comfortable to them: they were essentially the Groningen model under which the gas bridge had been built, starting in the 1960s. By the beginning of the new century, through their alliance with the chemical giant BASF—via its oil and gas subsidiary, Wintershall—the Russians were established members of the German "Gas Club." They were winners, but mainly through their passive participation in the Wingas joint venture; the heavy lifting was done by their German partners, and the Russians were essentially along for the ride. They were the passive beneficiaries not of liberalization, but of the traditional order.

But within a few years, the rules had changed and the "Gas Club" had disappeared. As a result, the Russians found themselves subject to new rules and demands, which they had had no role in creating. By the end of the 2000s the scene was set for conflict, this time involving Russia and the European Union, with Germany as one of the main battlegrounds. This time, the Russians were passive losers. But to follow what happened next, we will need first to turn back to Russia and recount the events taking place in parallel in Russia and the key republics of the Former Soviet Union, above all Ukraine. (That will be the subject of Chapters 9–11.) The aim of this chapter is to weave together two themes: first, the long resistance of the German "Gas Club" and the reasons for its sudden collapse; and second, the implications of these events for the Russians and their response.

A Portrait of the German Gas Industry on the Eve of Liberalization

At the beginning of the 1990s, none of these dramatic changes could be foreseen. The structure of the German gas industry had evolved over the previous three decades into a complex but highly stable three-tiered structure, consisting of five gas-importing companies (which together were called the "interregional transmission companies") and six producing companies, ten regional transmission and trading companies, and around seven hundred regional and local distribution companies. Gas in the German system changed hands up to four times among these various levels before reaching the final consumer.[6]

Ruhrgas, the largest of the gas-importing companies, was by far the biggest player, with sales greater than those of the other four importing companies combined. The regional and local distribution companies were mostly small. Two-thirds of them were wholly owned by the municipalities they served, and municipalities had majority ownership of most of the rest (hence the name *Stadtwerke* [city utilities]). They did not compete with one another across their districts, and their trading was limited to buying gas from the tiers above them.

The companies were linked by an extensive and largely opaque network of ownership and sales contracts. Many of the regional transmission companies were at least partly owned by local or federal authorities, and the interregionals (particularly Ruhrgas) frequently owned some stakes in them as well. Despite the multitude of companies, they did not compete with one another but divided tasks among themselves through informal agreements. Peace was maintained through the use of vertical and horizontal demarcation contracts (a system that dated from the energy law of 1934, in the Nazi era). Contracts were long-term, frequently for up to twenty years. As German gas expert Heiko Lohmann describes the traditional system on the eve of liberalization: "Horizontally, gas companies agreed not to deliver gas into the network area of neighboring gas companies; vertically, gas suppliers along the delivery chain agreed not to acquire customers by directly bypassing distribution companies. This kind of demarcation was legally supported by the German competition law, which exempted the energy sector from the general competition rules."[7] Thus, for the most part the gas companies felt like part of one family and not potential competitors. Lohmann concludes, "None of them was really in a position to become a competitor."[8]

By the late 2000s, the picture had changed beyond recognition. Competition had arrived. New players had entered the midstream. Power utilities and other outsiders were taking over the gas companies. Trading on spot markets was beginning to appear. The rush of events was accelerating. But the battle for Germany had taken fifteen years, from the early 1990s to the mid-2000s, before the established structure began to show cracks.

By this time, however, Gazprom had arrived in Germany, and the Russians were firmly established in the German gas structure. How had this happened, and what were the consequences? We begin by turning to the early 1990s and the creation of the first Russian-German joint venture in gas, Wingas.

The Origins of the Wingas Partnership and the Arrival of Gazprom in Germany

The established gas community received a first jolt when a serious competitor entered the German scene at the beginning of the 1990s and launched gas-on-gas competition—a historic first in the German gas business—by building its own pipeline network and importing gas from Russia. It was at first only a plan on paper: it took five years before the new pipeline system was in place and the new company had signed supply contracts with local distributors. But it was a premonition of things to come. For the first time, a major consumer had rebelled against the de facto monopoly structure dominated by Ruhrgas.

In October 1990, Stepan Derezhov, head of international relations for Gazprom—the reader will remember him as the pioneer of Soviet gas exports to capitalist Europe back in the 1960s—signed an agreement with the CEO of Wintershall, Heinz Wüstefeld, to create a joint venture called Wingas to deliver and market gas in Germany. Wintershall had not previously been part of Germany's tight-knit gas family. The decision of its parent company, BASF, to go into the gas business was triggered by BASF's need to safeguard its fertilizer plants at its headquarters in Ludwigshafen, at that time the largest single-location gas consumer in Europe. The opening shot was actually fired in the Netherlands, where Gasunie had cut the price of its gas to the Dutch fertilizer industry. Suddenly BASF faced a crisis—the price it paid to Ruhrgas, its established transporter, was too high to enable BASF to compete with the Dutch. But Ruhrgas would not budge on its price. "Ruhrgas with its prohibitively high prices was stonewalling us," declared the head of gas purchasing for BASF. BASF decided to hunt for another source of supply, which meant challenging Ruhrgas.[9] BASF initially talked to the Norwegians and the Danes, who declined to take on the gas giant.[10] Then in 1990 BASF approached Gazprom, at that time still a Soviet enterprise.[11]

Wüstefeld was a visionary. He was determined to shake up the comfortable world of German gas by promoting open access to transportation to all comers, beginning with himself.[12] There was only one problem: BASF did not own a gas pipeline, and it had no gas customers. There was no history of gas-on-gas competition in the industry. But when the Iron Curtain went down, Gazprom—which as the Soviet gas ministry had long exported gas to the German Democratic Republic (GDR) and the West German border—began looking to expand into West Germany itself. Gazprom had reasons of its own to be dissatisfied with its longtime monopoly buyer, Ruhrgas.[13] BASF, aware of the Russians' dissatisfaction, approached Gazprom, and the two companies agreed to do something that had never been done before in Europe: to cooperate to build a new pipeline system from the ground up. Since the new company had as yet no customers, it would build its network on speculation—effectively, on a wish and a prayer—while touring the local *Stadtwerke* to sign them up as buyers.

Overlapping with BASF-Wintershall's effort to gain independence in West Germany was a battle for control of the East German gas system. As reunification suddenly loomed, the questions for the West German gas companies were: Who would control the East? Would the East Germans retain control, or were there opportunities for a takeover by the West?

Ruhrgas was first off the mark. In early 1990 Ruhrgas's CEO, Klaus Liesen, signed a series of agreements with VNG, the East German gas company, that led to Ruhrgas taking a 35 percent share in the company. Liesen made no secret of the fact that his ultimate objective was to take over VNG and to control the supply of Soviet gas to East Germany.[14]

The key to the arrangement was that VNG would continue to hold the contract with Gazprom as the sole supplier of gas to East Germany, but at this point Ruhrgas's plan ran into resistance from Gazprom. Gert Maichel, then head of the gas section of Wintershall, recalls:

> When in the summer of 1990 [Victor] Chernomyrdin and Derezhov found out that Ruhrgas wanted to take over VNG they felt very unhappy because they perceived the GDR as their home turf. That Ruhrgas would just carry on in Eastern Germany as in the West, with Gazprom delivering at the border and the rest of the gas value chain being under Ruhrgas' custody, did not please them.[15]

Moreover, Gazprom saw no reason to continue the privileged relationship of the past, based on low subsidized prices to a friendly socialist power. East Germany was now part of the West—why continue to offer it cheap gas? Here Wintershall saw an opportunity to steal a march on Ruhrgas. It approached Gazprom and proposed to take over the marketing of gas in East Germany, raising prices in the process while cutting out numerous local middlemen. This proved highly attractive to Gazprom, because it gave the Russians the opportunity to capture a portion of the value of their gas through direct marketing in East Germany. The Soviet political leadership gave its blessing when Wintershall arranged to prefinance $90 million worth of wheat deliveries from Germany to the Soviet Union, which Mikhail Gorbachev's government considered an act of friendship at a time of financial crisis. The Gazprom-Wintershall deal was signed within a matter of weeks.

In addition, however, Wüstefeld demanded a 25 percent share of VNG and then raised the stakes to 45 percent. At this point the German government stepped in to avoid the political embarrassment of a gas war between Ruhrgas and Wintershall. The Ministry of the Economy mandated that equal 5 percent shares be distributed to a range of interested players, including British Gas, Elf, Statoil, and Gazprom. Wintershall's share would be limited to 15 percent. Shortly afterward, Wüstefeld, who was widely considered to have overreached, resigned as CEO of Wintershall.[16]

It was left to his successor, Herbert Detharding, to turn the Wingas joint venture into reality. Detharding was a remarkable figure. Formerly the CEO of Mobil Oil Germany, he approached the new challenge with relish. Behind a mild-mannered exterior, Detharding was an entrepreneur who loved a good fight and a prize worth taking. As an external board member of Ruhrgas, he had seen the monopoly culture of Ruhrgas at first hand, and he had no love for the company. For Detharding, Ruhrgas's long-established business with Russia was the prize he was looking for. A senior Wintershall executive recalls: "Herb Detharding's time at Wintershall was a time of out-and-out, take-no-prisoners competition with Ruhrgas. We had a great time selling gas contracts in the 1990s. We were on the road all the time. Ruhrgas couldn't understand how we could cover so much ground. They thought we had a large sales organization, but there were only a handful of us."[17]

Ruhrgas fought back with all the weapons at its disposal. In early 1994, Wintershall thought it had won a twenty-year contract with a

Stadtwerk called Giessen, and Ruhrgas seemed to be out of the running. But at the last moment a telegram from Giessen arrived at Wintershall's company headquarters, informing Detharding and his team that Ruhrgas had offered a big discount and had won the contract. Detharding seethed: "To retain market access, Ruhrgas is underbidding us to hell and back." To which Ruhrgas's director for sales, Frederick Späth, retorted, "If Wintershall wants competition, then it had better not cry when it can't find any customers."[18] And so it went in the early years. It was an epic battle or, as Detharding put it, "Competition at the most primitive level."[19]

By the mid-1990s, the tide began to turn. A key victory for Wingas came in 1996, when it succeeded in taking over supplying gas to the largest industrial customer in Bavaria, the chemical company Wacker-Chemie. By now the entire German gas industry was becoming aware of the larger implications for competition in the German market, and in the following year or two several major regional distributors shifted part of their purchases to Wingas. BASF-Wintershall was finally winning its battle and breaching Ruhrgas's monopoly.

Detharding made it his mission to build not just a business partnership with Gazprom, but a solid friendship that ran through both companies.[20] Wingas had a joint Russian-German management, Russian and German teams worked side by side at the company's headquarters in Kassel, and on their time off the Russians and Germans played soccer together. Rainer Seele, the future head of Wingas and today the CEO of OMV, the Austrian oil and gas company, formed a close friendship with his Russian colleague Yuri Komarov, who was the protégé and emissary of Chernomyrdin—a tie that grew even more important when Chernomyrdin became prime minister in late 1992. But the key was Detharding. In Seele's words,

> When Detharding went to Russia he seemed to change personality. He enjoyed fairy tales and *anekdoty* [Russian jokes]; his whole personality seemed to soften and slow down. He had the Russian soul. He understood that the basis for business with the Russians is trust. He set his mind to building a close personal relationship with

> Chernomyrdin and [his successor Rem] Viakhirev and in this he was successful.[21]

The same close ties have been maintained in the next two generations of managers on both sides. These long-standing personal relationships are essential for understanding the continued growth of the Gazprom-Wintershall alliance, right down to the present participation of Wintershall and OMV in today's joint consortium to build the Nord Stream 2 pipeline from Saint Petersburg to north Germany, discussed in Chapter 12. The Wingas alliance also marked the beginning of the active presence of Gazprom as a stakeholder in Germany.

The entry of Wintershall into the German gas business, then, was a historic event, but it did not yet mark the dawning of the age of competition in Germany. Lohmann observes:

> The appearance of Wingas partly disturbed the comfortable scene [of the traditional German gas industry] but in fact the business model of Wingas was not so different from the rest of the industry. It was based on procurement contracts and its own investment in the pipeline system. It offered contracts to distribution companies which were similar to those of other importing gas companies. By the beginning of the liberalization era many market observers argued that Wingas had become part of the club.[22]

But for the Russians the Wingas partnership was crucial. Thanks to its stake in Wingas, Gazprom had a seat at the table and could observe at first hand how the German "Gas Club" functioned. More than anything else, Gazprom's long-standing partnership with Wintershall marked the Russian ideal of what a partnership with the West should look like. Equally important, as the Russians worked side by side with their German partners (Wingas as well as Ruhrgas, with whom they were soon reconciled), they absorbed from them the prevailing views of the traditional German gas industry. This was particularly the case where relations with the European Union were concerned. In all the battles of the 1990s and early 2000s, the Russians were largely spectators. But the lesson they absorbed was that the European Union could be successfully resisted. This may have led the Russians to underestimate the ultimate power of the European Union and the single-market movement in Eastern Europe. Once the European Union's membership expanded in 2004 to include the former satellites and members of the Soviet Union, the stage was set for the confrontations of the following decade.

But in Germany the battle over liberalization was barely beginning. We turn now to the story of how competition finally arrived and transformed the rest of the German gas industry.

The Clash of Cultures

As late as 2005, seven years after the adoption of the European Commission's First Gas Directive, it remained in effect a dead letter. In early 2004 the Commission wrote disapprovingly in its annual benchmark report that "progress in Germany and Austria is still very disappointing." The German Ministry of the Economy, on the whole favorable to reform, agreed with the negative assessment: "Competition in the gas sector has developed only in the market for big users and even in this sector only to a limited extent."[23]

The established large players, led by Ruhrgas, insisted that the German gas industry was unique, and therefore the one-size-fits-all model of regulation advocated by Brussels was inappropriate. How could there be a single state-run gas regulator model in Germany, the club's members asked, when there were over seven hundred companies, many of them—especially the largest—privately owned? As for competition, they pointed to the example of Wingas as evidence that gas-on-gas competition had arrived in Germany. Since the system was not broken, there was no need to regulate it. If there was to be third-party access, let it be on the industry's terms: negotiated rather than regulated.

So the incumbents dug in, but the old family feeling was gone. From 1999 on, bitter negotiations pitted the German industry associations against one another, the large gas transporters and distributors doing battle against groups representing the consumers and smaller distributors, and all without result. Finally, in 2003 the negotiations broke down altogether, and that is where matters stood until 2006, effectively at a stalemate.

Yet in the background other forces were at work that were causing the position of the incumbents to weaken, step by step. We turn first to the broad economic environment. These were the years of Germany's so-called lost decade, a fact that shaped the terms of the gas debate.

Germany's Lost Decade (1995–2005)

For the decade between 1995 and 2005, Germany appeared to have lost its way. Economic growth slowed to a stop and unemployment soared, while German capital migrated to cheaper labor markets outside the country—notably Eastern Europe and Russia. Germany during that decade was one of the worst-performing countries in Europe: in 2002 the *Economist* called it "a basket case." The most important reasons were the drain on the country's investment resources following reunification; growing competition faced by German exports from goods from emerging economies; and perhaps most important, the burden of German welfare and labor laws. As a result, the German economic model, which had been hailed as more successful than the short-termism of the US system, was suddenly out of favor. As the chairman of the influential Bundesverband der Deutschen Industrie (Federation of German Industries) the very symbol of German economic policy by association), put it, "Nobody wants our model anymore."[24]

The pattern of gas demand in the 1990s reflected the stagnation of the economy as a whole, as well as the influence of reunification. From 1990 to 1997 overall gas demand grew by over 50 percent, but much of that was due to a state-led effort to substitute natural gas for the low-quality brown coal called lignite that had been the chief fuel for the East German power and heating system in Soviet times. In the rest of the German economy, growth in gas demand had effectively reached a standstill (see Table 8.1).[25]

But after 2005 the picture changed radically. The German economy turned around, and over the following decade Germany prospered. Why?

Much of Germany's sudden revival has been credited to the reform policies of the Social Democratic Party (SPD)–Green coalition government headed by Chancellor Gerhard Schröder (1998–2005). One might have expected that a left-wing government led by socialists would pursue

Table 8.1 Natural gas demand in Germany by sector, 1990–2018 (standard billions of cubic meters, gross calorific value)

	1990	1995	2000	2005	2010	2015	2018
Residential	16.9	24.6	27.4	27.2	28.0	23.1	25.3
Industrial[a]	23.0	23.2	24.5	24.8	24.9	23.8	24.3
Electric power	9.6	9.6	9.7	13.6	16.2	10.7	15.7
Commercial	7.9	10.7	11.7	10.8	11.6	12.2	12.6
Hydrogen generation	0.0	0.0	0.5	0.6	0.9	0.9	1.1
Transportation	0.0	0.0	0.0	1.0	0.6	0.5	0.6
Agricultural	0.2	0.3	0.3	0.0	0.0	0.0	0.0
Other[b]	5.1	6.7	6.5	8.8	8.7	7.4	7.4
Total	62.7	75.0	80.7	86.8	90.8	78.6	87.0

Data sources: Rick Vidal, *2019 Update to the Rivalry Energy Data Set* (IHS Markit Global Scenarios Data, July 2019); historical data from the International Energy Agency and the US Energy Information Administration.

Notes: Numbers may not sum to totals due to rounding.

[a]"Industrial" includes feedstocks.

[b]"Other" includes energy sector uses, distribution losses, and statistical differences.

a demand-side strategy favoring labor and retired people. (Indeed, that was the policy favored by the SPD's left wing, whose leader, Oskar Lafontaine, was finance minister in 1998–2000.) But surprisingly, the Schröder government went the other way, opting for a supply-side strategy (called "Agenda 2010") aimed at rationalizing the welfare and pension system and liberalizing the labor market. It was, in short, a strongly pro-business policy, aimed at restoring the competitiveness of German industry by lessening charges and constraints—particularly those borne by German family-owned small and medium-sized businesses, the famous *Mittelstand,* that were so crucial to the prospects of German exports.[26]

This effort, which was controversial in much of German society and absorbed the time and political capital of the Schröder government, may help explain why avenues of reform in sectors other than those featured in Agenda 2010 received much less high-level attention. This was notably the case with energy, where the goals of liberalization and marketization were proclaimed in legislation—chiefly to bring German law into formal compliance with the single-market directives of the EU—but were backed by little action in implementation. As Lohmann observes in his detailed analysis of gas policy in the Schröder years:

> The government has not followed a coherent strategy of opening the market. From the start of the liberalization process launched by the [European] Commission, the German government employed a mixture of policy targets, which included environmental issues, employment in the energy industry, the creation of national energy champions, the security of supply, the protection of the coal industry and cheaper energy prices. There has never been a clear [state] commitment to energy marketization in Germany.[27]

This at least was the picture in 2005. At that point, the German gas industry had succeeded in defeating every major reform proposal advanced by the government to comply with EU legislation—as indeed it had since the passage of the Single Europe Act at the beginning of the 1990s. As late as 2005, the German gas industry was still defended by solid corporatist walls at the *Land* and federal levels of government, and above all at the municipal level.

Yet in retrospect, the Schröder years are crucial to understanding what followed. Underneath the seemingly solid ice of conservative resistance to change, cracks were already beginning to appear, even before Schröder came to power in 1998. It was during this period that key negotiations took place, laws were drafted, new institutions were created, and mind-sets began to change—all of which led in 2005–2006 to a flood of unexpectedly radical outcomes signaled by a whole new vocabulary—"third-party access," "entry-exit capacity contracts," "balancing regime," "hub trading," "transfer facility," and the like. Behind the technical terms was a whole new way of thinking about the gas industry.

In retrospect, perhaps the most important practical effect of the marketization breakthrough was to weaken permanently the traditional political power of the German utility sector. This in turn may have made the sector less able to resist the increasingly radical consequences of the *Energiewende*. These two trends acting together—marketization and the *Energiewende*—have devastated the middle of the value chain in the gas industry.

The situation has been described as a clash of cultures. This has been a central theme throughout Chapters 6 and 7, and it is worth reminding the reader of its relevance here. The neoliberal culture emanating from the European Commission sought to promote the single energy market in a competitive market environment. Its ideas and tools, as we have seen, were originally developed for the gas market of Great Britain. Whether the British model was the most appropriate for continental Europe is another question. But the point is that it became the central ideology of the Commission under Jacques Delors, and particularly of the Commission's competition directorate—which through its successive directives strove to impose a neoliberal regulatory framework on the EU member-states.

The clash of cultures was bound to be especially strong in the case of Germany. The German gas and power industry had grown up organically over more than a century. The resulting network of interlocking structures, linked by strong traditions of cross-ownership and shared information and voting power, had never been "designed"—and certainly not by state action. In this respect it is a misnomer to call the German gas industry ordoliberal, as it frequently is in textbooks. In its own way it was already responsive to the market, but the market to which it responded was the nontransparent structure of shared interests that had grown up organically over the years—within the gas sector, across competing fuels, and even extending to other network-based industries such as electricity.[28] The government's role in the gas sector was more reactive than proactive—at least until the government was prodded by the increasingly potent force of the European Commission and especially the Directorate General for Competition (DG-COMP), backed up by the European Court of Justice (ECJ).

The combination of this closed culture with the economic recession of the mid-1990s helps to explain why, when the Christian Democratic Union–Christian Social Union (CDU-CSU) coalition government under Helmut Kohl attempted to launch the first energy reforms, established interests reacted fiercely—especially the municipalities and the associated *Stadtwerke.* The mid-1990s were a time of crisis in municipal financing. When the coalition attempted to incorporate into national legislation the first EU energy directive of 1995, it raised a storm of protest from the municipalities. The new law infringed on their right to self-government, they said. It would create an uneven playing field. But above all, it would lessen their revenues and prevent them from passing on the profits from gas and power to other municipal services such as public transportation. The battle over the energy law lasted three years.[29]

There were two main issues in the implementation of the First Gas and Power Directive of 1998: network access and long-term contracts. Both were significant roadblocks from 1998 to 2003, but each began to give way between 2003 and 2008. The following is a brief summary of the changes that took place between 2003 and 2006, the key date of the breakthrough.

The Battle over Network Access

The main obstacle in the way of change was the issue of access to the gas network. At the time of the first Directive in 1998, the German government did not establish a regulatory authority and instead chose negotiated third-party access as its model. But industry associations could not agree on a system that really facilitated access for new market players and promoted competition. Lohmann comments, "Negotiations among the four major associations of network operators and network users were a nightmare."[30]

Associations representing the interests of producers and consumers are a characteristic feature of industrial politics in Germany, and one of the biggest points of conflict in any negotiation is which associations will be allowed to sit at the table and have a voice. In the battle over third-party access the associations representing the large incumbents (mainly the transmission companies) and the large industrial customers had the lead roles. Municipalities and other network users were also represented, but the association representing traders (who arguably had the greatest stake in seeing third-party access adopted) found itself excluded and ignored.

At issue was whether third-party access by negotiation could actually be made to work. While everyone agreed in principle that the goal was an agreement that would encourage "competition, fairness, transparency, and simplicity,"[31] the parties struggled to find workable formulas. The large transmission companies were opposed to any change. Particular points of contention included the establishment of a gas exchange, access to storage, and various competing formulas for transportation tariffs.[32]

The battle went on for five years, from 1999 to 2003, until talks finally broke down. Negotiation was clearly not working.[33]

In the meantime, however, Brussels had begun turning up the heat. In 2001 DG-COMP brought complaints against five transmission companies, three of them German. It took four years to reach settlements, but when the dust had settled a number of improvements had been made. A particularly important one was an entry-exit tariff system, which the industry and the German government had steadfastly resisted. Under an entry-exit gas tariff, a shipper reserves capacity in two separate steps: first, reservation of entry capacity, to transport gas from injection points to a virtual balancing point, and second, reservation of exit capacity from the balancing point to the exit points in the system. An entry-exit system is considered an essential prerequisite for a liquid, competitive gas market. Note that the role of DG-COMP was crucial, in view of the unwillingness of the German government to act.

Another noteworthy aspect of the German companies' settlements with DG-COMP was the increasingly prominent role of the internet and online services. In 2001 Thyssengas, the first of the five companies to settle, offered online balancing and an online map. By 2003, BEB, the next company to settle, had introduced free online balancing and an online bulletin board, as well as online publication of available capacity. The introduction of entry-exit tariffs by BEB in 2003 made possible the launch of the first over-the-counter trading system the following year, something that would have been impossible without advanced computing power and information technology capabilities. The advent of competition in the German gas industry was as much a technological phenomenon as a political one.

In 2003, the European Commission adopted a Second Gas and Power Directive that repealed the first one and made regulated access compulsory. The German Ministry for Economics, which up to this point had played a passive role, this time actively intervened.[34] German network operators fought hard against the new model and succeeded in delaying its implementation for three more years. Lohmann summed up the landscape as of 2006 in the following words:

> Customers and shippers were not able to achieve any significant improvement or meet any of their targets [for] network access. The

> system remained lacking in transparency and this created barriers to entry. . . . The number of transportation contracts remained limited and there was no over-the-counter trading of significant liquidity. Improvements were mainly enforced by cases settled between the companies and DG-Comp in Brussels.[35]

All these issues came together over a question that might seem obscurely technical but that turns out to be critical for the creation of a true market: the concept known as liquidity. To perform its function of signaling scarcity by linking supply and demand through the price mechanism, a market must have a sufficiently large number of active players buy and sell on the basis of real-time information, so that no single player or group of players can manipulate the market. This quality of high transparency and activity is known as liquidity. The classic example of a liquid commodity market is the crude oil market, in which a vast number of participants are linked together through instant communications.

Conflict over liquidity was at the core of the clash of cultures. The first liquid gas market in Europe, in the form of spot trading of natural gas, originated in Great Britain in the mid-1990s. As the number of players and the volume of spot trades on computerized platforms increased, so did the liquidity of the British market. The City of London was a major contributing force, as banks actively supported the new platforms. Greater liquidity created more confidence in the market, and this in turn fed greater liquidity. By about 2000, a confidence-giving level of liquidity had been achieved in Great Britain.[36]

However, as a result of the features we have discussed (demarcation agreements and interlocking companies resulting in limited market access), the German gas market at the beginning of the 2000s was precisely the opposite of liquid. Consequently, gas prices failed to perform their economic function. Without a liquid market and a transparent trading system, buyers and sellers could not gauge price or volume risk or hedge against it. The German gas industry continued to base its prices on indices of oil-product prices. To place their gas, German players still needed to have a direct bilateral relationship with a buyer. There was not yet a true market in gas.

The Struggle over Long-Term Contracts

The next battle was over contracts. Until the late 1990s the typical contract for gas imports was for 20–25 years.[37] The buyer bought both gas and pipeline capacity in the same contract to cover his entire demand, and the price included both—with the price calculated according to the traditional principle of *Anlegbarkeit,* that is, with reference to the closest competitive product, which was almost always heating oil.[38] The resulting structure of long-term contracts reinforced the overall pattern of close relations among companies at all levels. It also prevented third parties from breaking in.

Starting in 1998–1999, under pressure from the Commission and some domestic competition (driven mainly by Wingas), the picture began to change. The opposition from the traditional suppliers and distributors was not nearly as public as in the case of third-party access; rather, it took place at the level of individual companies in price negotiations with customers. Suppliers and distributors began to offer partial releases from long-term contracts (typically about 20 percent of the total), accompanied by rebates.

It was in 2003, however, that several circumstances came together to force more radical and rapid change. The movement was initially driven by the lower German courts, as legal experts came to perceive that long-term contracts were incompatible with European competition law. The first to feel the pressure was the power sector, where local courts began to throw out long-term contracts as early as 1999. Within a few years, long-term contracts in power had been swept away. Gas buyers were inspired to revolt, with a series of lawsuits beginning in 2001. The Ministry for Economics joined the fray, and by 2003 even Ruhrgas began offering partial releases and rebates—although at first it attempted to find a way out by offering multiple contracts to the same buyer, a tactic that came to be known as "stapling."

A major turning point came the following year, when the German Federal Cartel Office (Bundeskartellamt) conducted a sweeping investigation of 750 supply contracts with sixteen different companies, in which it found that over three-quarters of the contracts were anticompetitive. While initially it did not bring legal action but attempted to act by persuasion, by 2005 it had formulated clear guidelines for the courts. Stapling, in particular, was banned.

Not surprisingly, there was strong opposition from the incumbents, and especially Ruhrgas, who argued—much as Gazprom was to do a few years later—that long-term contracts were needed for security of supply. They also argued, on more theoretical grounds, that the Bundeskartellamt's rules amounted to an abridgment of freedom of contract. The Bundeskartellamt was unpersuaded, and in 2006 the Bundeskartellamt banned all of Ruhrgas's long-term contracts, effective in 2008. This was followed by a wave of challenges to long-term contracts all over Germany, and by the late 2000s the entire structure of long-term contracts had been decisively weakened—except for import contracts, where the battle remained to be fought.[39]

The Takeover of Ruhrgas by E.ON

While these developments were taking place, a second major jolt to the established system occurred in 2001–2003. Even more far-reaching than the creation of Wingas a decade earlier, this was the hostile takeover of Ruhrgas by Germany's second-largest power utility, E.ON. The consequences affected the entire German industry. "Ruhrgas was the head of the 'German gas family,'" writes Lohmann. "One of the results of the takeover was that the family began to break up."[40]

To understand why E.ON wanted to take over the German gas giant, one must go back to the context of the moment. E.ON was effectively making two bets. The first was prompted by the fact that the German government had just declared its intention to phase out nuclear power. E.ON, which owned five nuclear power plants in Germany,[41] felt the need to ensure its power supply in the coming postnuclear environment. Renewables were not yet part of the picture. Therefore, gas appeared to be the correct answer, especially since combined-cycle gas turbines (CCGT) had produced a revolution in the efficiency of gas-fired generation. (The British "dash to gas" was a particularly persuasive model.) But a secure supply of gas was needed. The second bet was that gas prices, still tied to the price of oil, would rise (this proved to be correct). No market in gas yet existed, and consequently it was not possible to hedge. Hence Ruhrgas was an attractive target, from both near- and long-term strategic perspectives, because of the long-term supply contracts it still appeared to control.

To that one might add that large corporate takeovers were fashionable at the time, and hence support from banks was readily available.

Before its takeover by E.ON, Ruhrgas had had a unique place in the German gas industry. It was in a sense the keeper of the overall economic interests of the German gas industry. But after the takeover, Ruhrgas was only the gas subsidiary of a market-listed company whose main business was power. Its loss of independence marked the beginning of a rapid fade—highly traumatic to those who had spent their lives at the company—that ultimately erased even the brand of the former gas giant.

What made the takeover so remarkable was that most observers beforehand would have said that it could not happen. Like other German companies, Ruhrgas was protected by a tight network of equity crossholdings with other firms inside and outside the gas industry. It took eighteen months before E.ON was finally victorious, and the battle was fought on two fronts.[42] First, in a series of purchases and exchanges, E.ON undid the complex shareholding structure of Ruhrgas and acquired a majority stake in the company. (E.ON's CEO at the time, Ulrich Hartmann, was particularly renowned as a corporate wheeler and dealer, and he lived up to his reputation in this case.[43]) Second, E.ON fought a bitter legal battle against the determined opposition of a coalition of private and public players, while appealing to the government for support. Though the government was divided, the Ministry for Economics ended up approving the merger subject to a number of conditions. Even so, the battle in the courts continued. But in a dramatic denouement, in January 2003 E.ON reached an out-of-court settlement with the last remaining opponents: "One hour before the ruling of the court was scheduled to be announced, the last complainant withdrew."[44]

In retrospect, two things are especially significant about the takeover. The first is that despite the complexity and opacity of shareholding relationships in German industry, it proved possible, with sufficient determination, to unravel them and achieve control. In the wake of the E.ON takeover, several further outsider acquisitions took place.[45] Second, neoliberal thinking was making its way into the German government, particularly the Ministry for Economics. The key moment in the takeover came when the ministry, through a special procedure, overruled the opposition of the powerful Bundeskartellamt and gave a qualified approval to the deal.[46]

The takeover of Ruhrgas by E.ON, concludes Lohmann, was "one of the most interesting episodes in German industrial history."[47] It opened the way for a cascade of takeovers, particular of gas companies by electric utilities.[48] The full implications took several more years to become visible. But for the Russians one consequence was immediately apparent: instead of Ruhrgas they were now dealing with the largest gas-and-power utility in Europe.

The Breakthrough: The Creation of an Independent Network Regulator

The breakthrough in the market liberalization of the German gas sector finally came in 2005, with the creation of an independent network regulator, the Bundesnetzagentur (BNA). Both the German business community and successive governments had long opposed the creation of an energy regulator. But by 2003 mounting pressure from the European Commission, together with a growing consensus that negotiated tariff agreements had failed to lower power and gas prices, had caused the German political leadership to change course and accept the principle of regulation. What finally opened the way was a compromise between the federal government and the German states *(Länder),* which had long opposed the creation of a regulator on the grounds that doing so would weaken their authority. In 2005 the two sides agreed to split the task: the states would oversee the local and regional level (especially the *Stadtwerke*), while the federal government would handle the interregional level.[49]

But the most remarkable feature of the German turnaround was the decision to make the BNA an independent agency. Independent agencies are unusual in German administration. Although the BNA is formally subordinated to the Ministry for Economics, its decisions, which are made by quasi-judicial ruling chambers, cannot be revoked by political authority. It has a strong legal mandate (the Energy Industry Act), and its independence is reinforced by the fact that its decision making takes place in public proceedings.[50]

Yet in the end the role of the BNA depended less on its formal status than it did on the people who staffed it and the ideas they brought with them. In this respect the BNA became a powerful agent for the transfer of neoliberal ideas into German practice, overcoming the resistance of the established utilities and their associations. Over the previous decade, the new doctrines of marketization and liberalization had gradually gained ground in German academic circles, competing with the older teachings of ordoliberalism. A younger generation of economists had been strongly influenced by colleagues from elsewhere in Europe, especially Great Britain. In the Ministry of Economic Affairs and Energy, midlevel officials had been exposed to the deliberations going on in various European forums such as the Madrid forum. They were also keenly aware that Germany, because it lacked a regulatory authority, was excluded from other key groups where the important regulatory issues were being discussed. In the Ministry for Economics, a split had opened between the political level, which remained opposed to regulation, and the professional level, which had become increasingly impatient with Germany's obstructive stance.

Thus, when the creation of the BNA opened an opportunity for people and ideas to migrate to positions of influence where they could make changes. The first head of BNA, Matthias Kurth, appeared to represent continuity. He was an experienced jurist and a long-time SPD politician and parliamentarian in Hesse, and since 2000 he had directed BNA's predecessor agency, which was responsible for the post office and telecommunications. Yet in reality he was one of the strongest voices for change. Post office and telecommunications had been two of the sectors where Germany had led the way in promoting liberalization and privatization.[51] Kurth had briefly been an executive in a private telecoms company, and he brought this varied experience to his agency's new responsibilities for gas and power. For several years he had waged a vigorous public campaign for the creation of an expanded network authority, calling for stronger regulation of the utilities.[52] After 2005, he delighted in positioning himself and his agency as the brave David, fighting for the public interest against the Goliaths of the established power and gas companies.[53]

Kurth was seconded by a team of experienced economists and ministry functionaries who shared his vision that change was long overdue. Martin Cronenberg, his deputy at BNA, had come from the Ministry for Economics, where he had been one of the leading drafters of the first energy law in 1998.[54] The head of the new Energy Department at BNA. Klaus-Peter Schultz, had held the same position at the Bundeskartellamt and before that had worked in the Ministry for Economics, where he had led a task force on network access and had also helped draft the first energy law.[55] Finally, Kurt Schmidt—the chairman of Beschlusskammer 7, the chamber that threw out the industry's defensive proposals for limited third-party access, a decision that "changed the German gas world"—had also served in the Ministry for Economics before moving to the predecessor agency of BNA, where he had been responsible for telecommunications.[56] In short, the four leading figures at BNA who played key roles in overcoming the resistance of the gas industry were people who had served in the trenches and who brought conviction and experience, as well as a decade of frustration, to the expanded agency. In their view, their time had come.

Thus, by 2009, a somewhat startled Lohmann wrote in a follow-up study, "Competition is actually taking place."[57] Burckhard Bergmann, the long-time head of Ruhrgas, in his retirement speech in February 2008—an event that marked the passing of an era—spoke of "a revolutionary change in competition in the German gas market."[58]

What followed next was indeed astonishing, Over the next ten years, the seemingly immutable world of the German "Gas Club" was turned upside down. The interlocking ownerships and agreements that had bound the three levels of the gas industry—the interregional importers, the regional transporters, and the municipal distributors—into what seemed the epitome of German Rhineland capitalism were gone.

The transformation had immediate repercussions on Europe's international trade, which were further catalyzed by a gas glut that suddenly appeared in Europe in the wake of the economic recession of 2008–2009. The shale revolution in the United States redirected to Europe liquefied natural gas (LNG), that would otherwise have gone to North America, and low EU demand caused excess natural gas to become available on the emerging spot markets. The traditional importing companies, such as E.ON Ruhrgas and VNG, found themselves trapped. Upstream, they were still locked into expensive long-term oil-indexed contracts with minimum take or pay clauses. Below them, however, there was a major realignment, as second-tier distributors were able to sell part of their long-term gas on the exchanges, while their customers were increasingly free to purchase much cheaper gas on a spot basis. The importers came under unsustainable financial pressure, and they clamored for changes in their contracts. There followed a wave of arbitration cases, which enabled importers to negotiate lower prices with Gazprom, as discussed in Chapter 10.

The impact on the old guard was traumatic. Over a lifetime in the gas business, the incumbents had been accustomed to a world in which the interregional transporters—especially Ruhrgas—were king, and customers did what they were told. But the seismic shift in power relationships could be felt in a conversation at the Madrid forum between Bergmann and one of his major customers, which was reliably reported by a witness:

> As Bergmann was explaining, convincingly and coherently as was his wont, the virtues of the way the German industry was established, his customer suddenly interrupted. "Mr. Bergmann, It doesn't matter what you say. What matters is the fact that you are talking, and I am your customer. You should be listening." Suddenly it was a world turned upside-down. Bergmann, who was not easily shaken, reportedly blushed.[59]

But the gas sector was not alone. By the middle of the decade the entire structure of the German energy sector—gas, power, nuclear, coal, and renewables—was in upheaval. Long-established companies had disappeared or been absorbed, and some of the biggest names in the industry teetered on the edge of bankruptcy. Energy policies lurched unpredictably, while the rules and structures by which energy was acquired, transported, and sold were changed beyond recognition. The German energy sector, previously so staid that nothing ever happened from one decade to the next, had suddenly turned into an experiment on a national scale.

By the end of the 2000s the battle for Germany had been won by a combination of pressure from the European Commission, action by the German courts, intervention by regulators, corporate takeovers, and the spread of new thinking in German ministries and even finally in the German utilities themselves. The victory of liberalization in the gas industry was absolute.

Indeed, as one looks back a decade or so later, a striking fact about the battle for the liberalization of the German gas market is how quickly it has become ancient history. The climax of the story and its denouement—the collapse of traditionalist resistance and the triumph of the new liberal order—seem curiously dated, even though they occurred only a short while ago. The reason lies in the very completeness of the victory. The war for open gas markets, one might say, abruptly ended for lack of combatants, when the liberalizers were left in sole command of the German battlefield.

The German gas industry is not more peaceful for all that. Rather, the lines of controversy have moved on, in three directions. First, the battle for liberalization continues, but the front has moved east, to Eastern Europe, the Baltics, and Ukraine. Second, in Germany itself, the contest is now all about the environment; and the battle cry is no longer competition but decarbonization. Lastly, security has returned to the fore, as the upheavals in Ukraine have revived long-standing geopolitical anxieties.[60]

Each of these has major implications for the Russians, Russian-German relations, and the continued viability of the Russian-European gas bridge. In the battle for liberalization, as we saw in this chapter and Chapter 7, the Russians were passive spectators in the battle for Germany, and they were not directly affected at first by the militancy of the European Commission. But the accession of the new East European members in 2004, the Commission's 2007 sector review of the state of the gas industry, the 2008 Third Package, and the 2009 interruption of gas supply to Ukraine and Europe changed all that. From 2010 on, Gazprom and Russia were in the line of fire, on both regulatory and security grounds. These developments, combined with a sharp decline in gas demand caused by the financial crash of 2008–2009, brought the first serious tensions in what had been five decades of serene relations in the Russian-German gas trade. As we shall see, the changes in Germany—indeed, in Europe generally—were received in Moscow with alarm, incomprehension, and hostility. Having taken no part in the internal German debates over gas policy and having paid little attention to them, the Russians were therefore taken by surprise by the changes that followed, soon after the arrival in 2005 of the grand coalition led by Chancellor Angela Merkel.

The irony is that it was the Schröder years (1998–2005) that had marked the high point of Russian-German entente, the moment when Russian-German relations, founded on a thriving two-way flow of trade and investment, seemed about to catalyze a new era of good feelings between Russia and Europe as a whole (see Chapter 11). Yet during those same years the coming radical changes in the gas industry, which were soon to shake the entire relationship, had been gestating quietly, largely out of view—not least that of the Russians. The awakening came at the end of the decade with two seismic events, the adoption of the European Union's third energy package in 2008 and the beginnings of DG-COMP's investigation of Gazprom in 2011 (see Chapters 12 and 13).

CHAPTER 9

Gazprom Survives and Gets Away

As he launched his gas campaign, Leonid Brezhnev hardly thought he was providing for the future of a capitalist Russia. Yet that is how it turned out. During the darkest days of the 1990s, Brezhnev's natural gas was crucial in keeping Russia's cities heated and lighted and its industry alive.[1] Gas in those years was the principal engine of Russia's virtual economy. Without Brezhnev's posthumous gift, Russia would have been even more disrupted by the shock of the Soviet collapse than it was. Gas was the bridge fuel that enabled the Russian economy to survive the painful transition.[2]

Russia entered the 1990s with the largest (and yet the youngest) gas industry in the world. It was a unique inheritance: 40 percent of global reserves, three West Siberian supergiant fields supplying over 80 percent of Russia's domestic and export needs, and a unique highway of twenty new pipelines with the world's largest diameters. Gas had played a key role in the last decade of the Soviet economy, but in the first post-Soviet decade its role grew absolutely critical, as other energy sources faltered. From 42 percent of Russia's total primary energy consumption in 1990, the share of natural gas grew to around 53 percent by 2000 (see Table 9.1). In 1990, domestic consumption accounted for 74 percent of Russian gas production, while in 2000, the share was 69 percent. In short, one of the biggest stories of the first post-Soviet decade is that Russia became

Table 9.1 Russian primary energy consumption by fuel and share of natural gas, 1990–2018 (millions of metric tons of oil equivalent)

	1990	1995	2000	2005	2010	2015	2018
Natural gas	384.7	334.1	330.3	354.0	383.1	362.8	390.6
Oil and petroleum products	267.0	147.5	122.3	121.8	127.8	119.7	144.6
Coal	185.9	120.6	111.5	114.8	107.9	114.5	119.2
Primary electricity	63.2	45.4	48.8	51.7	49.7	52.4	55.1
Other (peat, wood, etc.)	14.2	7.2	4.8	3.9	3.2	3.0	3.0
Total	915.0	654.8	617.7	646.3	671.7	652.4	712.6
Natural gas share of total	42.0%	51.0%	53.5%	54.8%	57.0%	55.6%	54.8%

Data source: IHS Markit.

Notes: Numbers may not sum to totals due to rounding. The table shows apparent consumption (production minus net exports).

increasingly a gas-fired economy,[3] and its giant gas company, Gazprom, became Russia's richest and most powerful corporation, thanks to its exports.

But we come here to the first of several mysteries surrounding the Russian gas story. In the cataclysmic breakup of the Soviet system and the free-for-all grab for the state's assets that followed, surely the gas industry was the greatest prize of all. The oil industry was torn apart and quickly privatized in the first half of the 1990s by eager nomenklatura entrepreneurs, overnight banking tycoons, grasping local politicians, and gangsters—but the gas industry was not.[4] It was never broken up. It was never even fully privatized. Why?

There are two reasons. The first and underlying one is that gas is inherently a more centralized business than oil. But that alone would not necessarily have kept the gas industry whole in the charged atmosphere of 1991–1992, when the Soviet Union was breaking up and its assets were up for grabs. The second reason is the role of a handful of key players during the brief critical months when the old system had melted down and a new one had not yet gelled. In fact, the answer to the question, "Why didn't Gazprom break up?" ultimately comes down to two people,

Yegor Gaidar, the acting prime minister who led Russia's radical turn to the market, and Victor Chernomyrdin, the former minister of gas who was the first head of Gazprom and who became Gaidar's successor as prime minister. Then, to explain the partial privatization of Gazprom, we introduce a third key personality, Anatolii Chubais, the father of Russian privatization in the 1990s.

Why Gas Is Different

Gas is different from oil. Unlike oil, gas is comparatively easy to produce, but it is difficult and expensive to transport, and worthless unless there is a distribution system in place to consume it. Well into the twentieth century, gas was treated as a nuisance by-product of oil, a troublesome substance that was either flared or reinjected into the ground, and gas-prone fields were avoided or abandoned. It was only once large pipelines were built to take gas to major markets that it acquired value and began to be developed for its own sake. Until technologies to turn gas into liquid form (liquefied natural gas, or LNG) began to emerge, the dependence of gas on pipelines was total. Even now, the economics and politics of gas are still largely those of pipeline transportation and distribution—although that is changing fast, as LNG takes a growing share of the global gas market.

A gas pipeline is an umbilical cord, linking the producer and the consumer in an intimate tie. Once a gas pipeline is in place, it can't be moved about in search of a better market, nor does it make sense to build a competing pipeline alongside it. Thus, a gas pipeline system, along with railroads and electric power grids, is the classic case of a natural monopoly.

Soviet planners never met a monopoly they didn't like, natural or otherwise, and in the Soviet Union the traditional issues of gas markets and regulation were never present. The massive set of trunk lines built in the 1970s and 1980s did not divide and subdivide, as in the West, into the myriad low-pressure pipes that feed individual homes and small businesses, but ran instead straight into large factories and power plants, which received annual quotas of gas deliveries to meet their assigned production plans. No one in the perpetually supply-constrained Soviet system ever worried about marketing gas, least of all the engineers who ran the gas industry. Their job was to produce and ship the gas; what became of it at the end of the pipe was not their concern. The only time the planners needed to think about marketing their product was when they exported it. Yet gas exports, like all foreign trade under the Soviet system, were handled by foreign trade professionals in the Ministry of Foreign Trade. They sold gas through long-term contracts with the major European gas companies, which did the actual marketing.

These basic facts about gas, and the quite different way in which the gas sector had grown up in the Soviet system, go a long way toward explaining why gas diverged from oil in the growing chaos of 1989–1991, as the Soviet system started to disintegrate. Russian oil producers and traders soon found ways of moving oil by the carload and tanker-load to the borders of Russia and onward into a well-developed global market with a multitude of ready buyers. As Chubais, Russia's arch-reformer, once said, "You can practically pour oil into a pail, carry it away from the field, and sell it."[5] But gas producers, hostage to the monopoly's control of the pipeline system, had no such outlets, and even if they had managed somehow to move their gas to the border, they would have found no one to sell it to outside the existing framework of long-term contracts, since no spot market for gas would exist in Europe for another decade. Thus, from the beginning the disintegrative force of greed, which quickly exploded the Soviet oil industry, was blocked in the case of gas.

But that didn't mean that market reform couldn't have occurred. A number of things could have happened. The trunk pipeline system might have been turned into a common carrier, open to all producers (as happened in the oil industry). The producers could have kept title to their gas as they put it into the pipeline system (as oil producers do). A single gas-trading agency, acting on behalf of all the producers, could have exported their gas under the existing long-term export contracts. In the domestic market, producers could have been allowed to find their own buyers at the other end of the pipe, at prices set by an open market.

All of these things would have been consistent with the radical market principles of the reformers who launched the so-called Russian shock therapy in late 1991 and early 1992. Indeed, they were discussed, as we shall see. But they did not happen. Why not? One reason, just suggested, is the structural properties of a gas system. The other was Chernomyrdin.

Victor Chernomyrdin, Father of Gazprom

Chernomyrdin was the key to the creation and survival of Gazprom. For two decades he had been the driving force in the Russian gas industry, as a senior Communist Party official and subsequently as gas minister. In 1989 he founded Gazprom as a state corporation and ran it until mid-1992 (Boris Yeltsin named him deputy prime minister then, and prime minister at the end of 1992). As prime minister until 1998, Chernomyrdin protected Gazprom from its numerous enemies. Only after 2000, when Vladimir Putin dispatched him to Kiev as Russian ambassador to Ukraine—in effect, an honorific exile where he remained for another decade—did Chernomyrdin's influence over gas and Gazprom finally end, as power passed into new hands.

Born in 1938 in a village in Orenburg, a rural province near the border of Kazakhstan, Chernomyrdin started out in the local oil refinery at Orsk and then worked on the staff of the Orsk Party city committee *(gorkom)* before being reassigned to the gas industry. In 1978, Chernomyrdin got the big break that launches a brilliant career. He was promoted to Moscow, to the apparatus of the Central Committee of the Communist Party in Moscow. This was the seat of power of the Soviet system, the policy staff that supported the top leadership of the country—the Politburo and its head, the general secretary. Suddenly, Chernomyrdin was on the fast track. His modest title, *instruktor* for heavy industry, gave no hint of the job's importance. In reality, he was the principal overseer for the gas sector behind the scenes, the eyes and ears of the ruling Politburo in this area, precisely in the years when Brezhnev was pushing natural gas as the highest domestic priority in the Soviet system. Central Committee instructors typically moved on to top jobs in the industry they oversaw. In 1982, Chernomyrdin was named deputy gas minister, moving up to minister three years later.

Thus, while Mikhail Gorbachev was developing his reform policies in the second half of the 1980s, Chernomyrdin was establishing his leadership over the fast-growing gas industry. These were the years in which gas was becoming the most important fuel in Soviet energy supply and the indispensable support of the Soviet economy. Like other senior Soviet managers, Chernomyrdin watched with growing apprehension as Gorbachev tinkered with the established system. The 1987 Law on Socialist Enterprise, in which Gorbachev sought to empower local managers and make their positions elected ones, struck people like Chernomyrdin as a death knell. "When they started to elect the enterprise managers," he recalled later, "it became 100 percent clear to me that we would collapse. With the first wave of elections of factory managers the best leaders were replaced and populists took over."[6] Chernomyrdin began taking steps to protect his unitary structure from the reformers: "We had to build a system such that, even if a fool came along, he would not be able to destroy it."[7]

"Chernomyrdin was no fool," recalls Gaidar. "He realized the old ministry system was collapsing. . . . As soon as coercion disappeared, it became impossible to manage through the old 'command methods.' Chernomyrdin saw that to preserve the gas industry you had to get people to work not by coercion but by self-interest."[8] Chernomyrdin began sending his deputies to Italy and Germany to look for models of Western management. The Italian energy giant ENI (today's Eni), with its blend of state ownership and private entrepreneurship, particularly appealed to him. He developed a plan to turn the gas ministry into a quasi-autonomous state corporation on the model of ENI, and after vigorous lobbying, he managed to persuade the USSR Council of Ministers to approve it.[9]

The gas ministry's change of status in 1989 to a state concern seemed at the time a mere switch of nameplates, but two years later it had its first far-reaching result. In the wake of the attempted coup of August 1991, all the members of the Council of Ministers were dismissed for having sympathized with the plotters. But Chernomyrdin was no longer a minister. As chairman of the Gazprom state corporation, he sailed on unscathed.

The Threat to Gazprom

In the fall of 1991 and early 1992 Russia teetered on the edge of chaos. After the failure of the attempted coup of August 1991, Communist Party General Secretary Gorbachev was sidelined, a powerless figure who with each passing day looked more and more like an anachronism. The Soviet government, under attack from all sides, disintegrated. Meanwhile, Yeltsin, Gorbachev's archrival, had turned the Russian republic government—until then hardly more than a fig leaf in the fictitious structure of Soviet federalism—into a rival political force, a kind of shadow government in waiting. This government now emerged as the dominant power in Moscow. By the end of 1991 the Soviet government had ceased to exist, and the Russian republic stood alone in Moscow as its triumphant successor.[10]

For the next few months the Russian government was in the hands of the market radicals, most of whom had never held leadership positions before. Few of them, indeed, had had any significant experience in government at all. They described themselves as "kamikazes" and had little expectation of staying in power long. But they were determined to use their brief turn—their moment in history—to destroy the basis of the Soviet system and prevent it from ever returning.

Typical of these market radicals—often dubbed "accidental people"—who rose like bubbles to the top, only to pop and disappear within months, was Valerii Chernogorodskii, a longtime minor official in the military-industrial complex who suddenly found himself propelled upward in the new Russian government as the head of a newly created Anti-monopoly Committee (AMC), with the mission of encouraging competition in the new private sector that was about to be created.[11] Chernogorodskii made it his personal mission to break up Gazprom. For people like him, Gazprom was the embodiment of Soviet evil. An all-powerful giant monopoly, overseen by the Communist General Secretary's Secretariat—for the radical reformers these were so many red flags. In March 1992 the AMC had its moment of glory, when it blocked the registration of Gazprom as a joint-stock corporation. The AMC objected on the ground that the assets of the gas industry were being handed over to a single monopolist.[12] The committee's main objections were that the process had been top-down, not bottom-up, and that employees of Gazprom had been excluded from owning shares in the new structure. Thus, the committee charged, it was the classic case of "*nomenklatura* privatization."

Another hotbed of radical reform, but a very different one, was the State Property Committee. A plan to break up Gazprom, which circulated in Moscow at the beginning of 1992 and was probably the work of this committee and its Western advisors, bears a strong resemblance to similar plans in Great Britain, where the longtime monolith of the gas industry, British Gas, had recently been broken up into a separate producer, transporter, and marketer and a competitive marketplace in gas had been created.[13] The Russian reformers planned to turn Gazprom into a holding company, make the individual producers into separate joint-stock corporate entities under a single state-owned holding company subordinated to the Ministry of Fuel and Power, turn the pipeline system into a common carrier, and encourage gas-on-gas competition by such means as transparent gas auctions.[14]

But while the market radicals held the front of the stage, in the background remnants of the Soviet bureaucracy pursued a very different agenda. In the wake of the failed August coup, Gazprom had been transferred to the jurisdiction of a newly created Russian Ministry of Fuel and Power, which at the time had been taken over by the reformers. For Chernomyrdin, this was a highly dangerous situation, which potentially put him at their mercy. As early as October 1991, just a few weeks after the failed coup, he approached the first deputy prime minister, Oleg Lobov, with a plan to turn Gazprom into a joint-stock corporation—but one designed by Chernomyrdin, which would preserve its top-down structure and monopoly control. In other words, apart from the term "joint-stock," nothing would change. Lobov obligingly signed the necessary decree.[15]

Chernomyrdin's moves in those chaotic last days of the Soviet Union were not only defensive but also offensive, revealing his wider strategic ambitions for Gazprom. In the fall of 1991, taking advantage of the wholesale transfer of Soviet institutions to the Russian Republic that took place in the wake of the failed coup, Chernomyrdin captured a long-coveted target, the gas-export arm of the Soviet Ministry of Foreign Trade (Soiuzgazeksport).

From December 1991 on, under this body (renamed Gazeksport) gas exports became an integral part of Gazprom.[16] This proved to be Chernomyrdin's master stroke, since gas exports would be the principal source of income for Gazprom throughout the 1990s and down to the present day. As a result of this history, GazpromExport (as it is known today) has a longer history than, and a distinct corporate culture from, Gazprom itself. As GazpromExport people like to joke, "We may be a daughter company, but we are the only daughter that is older than its mother."

Chernomyrdin's ambitions did not stop there. For over two decades the Russians had sold gas to Germany at the German border, watching in frustration as their trade partners, the giant European gas shippers, captured the lion's share of its value—or so they perceived. So long as exports had been the preserve of the Ministry of Foreign Trade, the *gazoviki* in the gas ministry had been helpless to do anything about it. But in 1991, as Chernomyrdin absorbed the gas exporters, he found an opportunity to do what no Russian exporter had been able to do in Soviet times—enter the German market directly and challenge the local middlemen for a share of the prize. The result was the Wingas joint venture, described in Chapter 8.

Chernomyrdin's role in creating Gazprom has been painted in dark colors by the Western media, which have portrayed what he did as the ultimate example of *nomenklatura* privatization. And so in one sense it was. But the motives of men such as Chernomyrdin were mixed. In the chaotic days of late 1991 and early 1992, his first thought was to preserve his life's work from destruction. Chernomyrdin had done as much as anyone to nurture the Soviet gas industry, and as an engineer and manager he not only took justifiable pride in it but also saw its functioning as essential to keeping the country running. In the immediate crisis, his main objective was to ensure that Gazprom remained intact. He strongly believed that the gas industry operated effectively as an integrated whole.

To counter the reformers' threat, Gazprom fought back, feeding the media with alarming pictures of the chaos that would result from breaking up the gas industry:

> The people supplying gas from Urengoy would have to monitor the weather conditions in, say, Kaluga or Tula, to determine if more or

> less gas should be supplied. Sixteen pipeline associations cater to 100,000 marketing organizations. Now each gas producer will have to negotiate contracts individually and deal with the whole mass of clients. Producers will be obliged to run vast commercial structures. Gas from different suppliers is pumped through the same pipe and no one can tell which supplier delivers on time and which fails. One fine day a dissatisfied supplier or a pumping station will go and turn off the flow.[17]

This is a revealing quote, in that it shows how far Gazprom was from understanding how a market system, based on hubs and backed up with modern computer and information technology, could work. But that is hardly surprising. The European gas industry—even Britain's—was at a similarly early stage.

Gazprom's leaders also tightened their internal control of the company. They faced nothing like the wholesale revolt of the so-called oil generals. When the Russian government approved the Lobov decree, Gazprom headquarters promptly absorbed all the gas production subsidiaries "at the request of their worker collectives." Unlike the hapless Russian Oil Company that the former head of the Soviet oil ministry was trying to construct at the same time, Gazprom from the first had total control over its enterprises, budgets, profits, investments, and indeed all significant cash flows. Against this there is no evidence that the gas generals were ready to revolt. There were some small wisps of smoke here and there, but nothing like the brush fire from the provinces that swept the oil industry.

The basic reason was simple: it was quite impossible for producers to market gas outside the Gazprom structure. There were some brief attempts: A joint-stock company to auction gas was created in December 1991 and held an auction in February 1992. Its initial shareholders included producers such as Urengoygazprom and six pipeline companies such as Tyumentransgaz (out of a total of nineteen)—as well as, revealingly, Gazprom itself. At the February auction contracts for 44 million cubic meters of gas (a negligibly small volume by Russian standards) were bought at prices 10–15 times the price fixed by the state.[18] (Such historical footnotes are of interest today, because after a decade Gazprom has returned to the idea of gas auctions.)

Gas producers, located in the far north of West Siberia, were totally dependent on the mother company not only for markets but also for all the necessities of life. For anyone who has experienced the frozen desolation of the Yamal-Nenets District in winter—not to mention the mosquito-infested and impassible swamp the place becomes in summer—it is plain that the *gazoviki* in the area cannot live or work without an elaborate lifeline to the Bol'shaia Zemlia (mainland). When I visited Tyumengazprom in 1992, the general director, Rem Suleimanov, proudly served his guests tomatoes and cucumbers from his greenhouses, but he acknowledged that except for those, all food was flown in from Gazprom's farms in southern Russia. According to former CEO Rem Viakhirev (Chernomyrdin's successor), Gazprom at that time owned over two hundred farms.[19] The handful of towns in Yamal-Nenets were literally company towns, to a more extreme degree than perhaps anywhere in the West—not only fed but also populated, employed, and financed exclusively by Gazprom. Indeed, the existence of permanent settlements at such northern latitudes was a legacy of the Soviet era and could be maintained only by Soviet-like expedients. The idea of independence, if it was ever entertained by the local gas generals, would have been dismissed as no more than a fantasy.[20]

In contrast, in Moscow in early 1992 there seemed to be no experiment too radical to try. Yet the reformers failed to break up Gazprom, even when they were at the height of their influence. The reason is supremely ironic: the key decision to keep Gazprom whole was taken by the leading kamikaze, the architect of shock therapy, and the man whose name will be forever associated with Russian market reform—Gaidar, who was then acting prime minister.

Gaidar was different from the other market radicals. Though he had never held office in Soviet times, he was familiar with the corridors of power. As economics editor of *Kommunist,* the official monthly journal of the Central Committee of the Communist Party, he had been an insider, and he was sophisticated about politics. He realized from the beginning that carrying out the market reforms required staying in power, if only for a few critical months—and staying in power required making compromises. From the beginning, for example, he had agreed as a tactical concession (a temporary one, he hoped) to leave a short list of strategic commodities out of the massive program to decontrol prices that was his main policy. At the head of the exempted sectors was energy.

But the way Gaidar chose to deal with Gazprom was quite different. In May and June 1992, Gaidar overruled the market radicals in his government and intervened decisively in Gazprom's defense. In six major documents, the Russian government reaffirmed the position of Gazprom as a state monopoly and gave its blessing to a Gazprom-designed privatization plan and a new charter that preserved Gazprom's centralized structure and its current leadership. It granted Gazprom the right to extensive access to foreign credits and to its own hard-currency revenues, together with major tax concessions. In short, Gaidar recognized the unique position of Gazprom and its strategic role as the supplier of gas to the wounded Russian economy, and he acted.

The timing of Gaidar's actions is significant. To be sure, he acted partly under duress. By the late spring of 1992 the tide of reform had weakened, the opponents of marketization had overcome their initial disarray and regrouped, and Gaidar was fast losing the support of President Yeltsin. By May he was forced to retreat from his radical reform program on several crucial issues—in particular the decontrol of energy prices, which he had hoped to achieve in a second phase. But in later years Gaidar never disavowed his decision to keep Gazprom whole, nor did he portray it as a retreat. Even after he left office, Gaidar remained adamant that breaking up Gazprom would have been a mistake. "The gas industry is a natural monopoly," he insisted in an interview in late 1995, three years after he had stepped down as prime minister. "It is not in our interest to break up Gazprom."[21] In the same interview, Gaidar went on to spell out his thinking: "But it is in our interest, of course, that the people's money—since gas is not just the property of the *gazoviki* but of the entire people—should be spent sensibly. . . . Look how much money has been transferred abroad for machinery that has never been delivered, and spent on contracts that have never been fulfilled!"[22]

Not demonopolization, then, but regulation was his goal. What is especially striking is Gaidar's description of gas as "the property . . . of the entire people"—a remarkably Soviet phrasing from the man considered the father of Russian market reform. By implication, oil is not, or at least not to the same critical extent. Oil is what keeps the budget balanced, but gas is what keeps the lights on and the homes heated. In the end Gaidar was always a pragmatist, not a neoliberal ideologue.[23]

Two Fateful Understandings

At the core of the deal between Gazprom and the Russian government lay two key understandings. The first was that Gazprom would keep Russia supplied with gas, come what might. The second was that Gazprom would be a Russian company. These two understandings have shaped the politics of gas ever since, for Putin as for his predecessors.

The supply agreement was crucial for both Gazprom and the Russian government. It amounted to this: Out of every five molecules of gas produced by Gazprom, four would be delivered to the Russian domestic market (and the gas-poor former Soviet republics, chiefly Ukraine and Belarus) at controlled low prices. The fifth molecule would be destined for export to Europe. In exchange for whatever losses Gazprom might incur in the domestic market, the Russian government would grant the necessary tax preferments on the export revenues to make Gazprom whole, without asking too many questions about where the money went.

It is difficult to say how explicit the supply deal was. It was certainly not made public, yet it was universally believed in Moscow to be true. Russian commentators speak of it as *neafishiruemyi* (unadvertised). A Russian article in 1996 summed up the terms this way:

> Under this agreement Gazprom supports the unrealistically low domestic gas prices, thus slowing the decline in [industrial] production and the inflation rate, while in addition supporting the interests of the government in the former USSR, by agreeing to practically free supplies to the gas-poor republics. On the other side, Gazprom does not include in the export price for gas the actual costs of maintaining a stable level of output and the functioning of the transportation system, as a result of which it obtains extraordinary profits on its exports. Until recently Gazprom was allowed to spend its entire

> hard-currency revenue on its own investment programs through a stabilization and development fund, instead of turning over 50 percent to the Central Bank.[24]

Subsequently Gazprom's obligation to supply the domestic market was codified in the 1999 Law on Gas, but the other side of the deal, for obvious reasons, has never been openly spelled out.

The supply agreement between Gazprom and the Russian government bound the two together at the hip as intimately as Siamese twins. The commitment to supply cheap gas at home ensured that the Russian economy would remain dependent on an underpriced resource. The indulgent and totally untransparent treatment of export revenues opened the way to illicit gains. Above all, the deal meant that Gazprom, however much it might subsequently be privatized in formal terms, would never become a truly private company.

Gaidar understood what the arrangement implied, and once again, his remarks after he left office suggest that he shared the same basic view that low gas prices were a blessing for Russia. "We have a natural advantage, a purely geographic property," declared Gaidar in 1995:

> Our closeness to the large fields, when transportation is the largest component of cost, insures a lower price both for gas and for gas-generated electricity in Russia, by a pure market mechanism, because that's the way our economy is structured, and not because of some kind of administrative decisions. It is not in our interest to strive at all costs to raise gas prices to West European levels. It would not be necessary, it would be unnatural, and more than that, it would be economically unjustified. But a gradual movement toward a market-clearing price—that's a normal process.[25]

Putin, two decades later, would put it no differently.

The second understanding, that Gazprom would remain a Russian company, also had far-reaching consequences. After the breakup of the Soviet Union, one of the most contentious issues was whether it was necessary to try to maintain the Soviet economic space—that is, the integrated structure of pipelines, power lines, supply relationships, and the like that had knit the command economy together. In the early 1990s this issue was particularly acute. Breaking up was hard to do, both economically and psychologically. Well into 1993, for example, the newly minted independent republics maintained the ruble as their common currency. However, the arrangement soon proved untenable, as their new central banks could agree on nothing except printing more rubles, fueling near-hyperinflation (which reached 2,500 percent in Russia in 1992 and remained in triple digits throughout the first half of the 1990s). Not surprisingly, therefore, the prevailing view in the Russian government soon switched to that of a group of so-called patriots, who argued for a rapid separation from the other republics.

Gas was the ultimate symbol of Soviet integration. While West Siberian production dominated, the Central Asian Republics were also major producers, supplying gas to the southern provinces of Russia and the Russian Caucasus, as well as the newly independent Transcaucasus republics. Above all, Russia was acutely dependent on Ukraine, not for gas but for transit and procurement. Some 40 percent of Gazprom's machinery and pipe and 70 percent of its R&D and design services came from Ukraine. What would become of Gazprom assets located on Ukrainian soil and, most crucial of all, the pipeline network and underground storage reservoirs connected to the export system? The threat was real and immediate. As soon as the Soviet Union broke up, the newly independent republics started taking control of assets on their territories and raising barriers to the flow of goods across their borders. In early 1992 the Turkmen gas industry walked away from Gazprom, raising prices and squeezing deliveries to show it meant business. Ukraine began siphoning gas from the export pipelines for its own failing economy, sparking rancorous disputes that would go on for more than a decade (see Chapter 11).

Chernomyrdin's first instinct was to try to preserve Gazprom as a multinational company. Gazprom's first plan, drafted in January 1992, envisioned a multinational company, in which the republics of Ukraine and Belarus would own a small minority of the shares. Decisions would be taken by unanimity. This plan was rejected by the Russian government, on the ground that it would give the two non-Russian republics veto power.[26] The next effort, in the summer of 1992, dropped the multinational concept in favor of a purely Russian company, and Gazprom's former assets in the non-Russian republics were quietly relinquished. On this point Chernomyrdin and Gaidar had little difficulty reaching agreement; the market reformers too, in the main, were "Russia firsters."

But by this time Gaidar's time on the stage was running out. At the end of June 1992 he was forced to jettison his energy minister, Vladimir Lopukhin, when the Siberian oil generals revolted against him, and to accept Yeltsin's appointment of Chernomyrdin as deputy prime minister responsible for energy and natural resources.[27] With Chernomyrdin in the government, the radicals' efforts to break up Gazprom were definitively quashed. In July Chernogorodskii was dismissed as head of the AMC, and while the committee continued through the fall to protest against the privatization of Gazprom, its voice was hardly heard.[28] In November 1992 a presidential decree codified the founding deals and Chernomyrdin's victory. It declared Gazprom a Russian Shareholding Company and authorized a stabilization fund. It also opened the way for Gazprom's formal privatization.[29]

The last act for Gaidar came in December 1992, when he was forced out as acting prime minister and replaced, fittingly enough, by Chernomyrdin. Chernomyrdin's accession was something less than triumphal: he had been put forward by President Yeltsin only after Yeltsin's first two candidates had been defeated by an increasingly hostile legislature, and Chernomyrdin's appointment was never more than the compromise reluctantly accepted by both sides. But for the next five and a half years, the creator of Gazprom would serve as prime minister.

The Privatization of Gazprom

As Chernomyrdin took office, one more task remained. Gazprom had been protected from breakup. Now the founders of Gazprom would strive to become its owners.

One of the great ironies of the end of communism in Eastern Europe and the Soviet Union is that upon the breakup of the Soviet system Russia—which had spread the Soviet-style command economy throughout its empire and maintained the most extreme form of it at home—adopted an extreme form of fast-track privatization. Moreover, the Russian privatization campaign did not begin until after Gaidar, the symbol of market reform, had been replaced by Chernomyrdin, one of the most emblematic members of the Soviet *nomenklatura.*

The key to these paradoxes lies in the design of the Russian privatization campaign and its primary architect, Chubais, who at that time was the young chairman of the State Property Committee. Chubais was one of the original band of Saint Petersburg economists who had landed in Moscow in the late 1980s, and he shared their radical free-market convictions. Chubais proved to be a fearsome bureaucratic infighter and a brilliant political tactician. He quickly rediscovered the oldest secret of politics—that the best way to win is to turn your opponents into allies. But he certainly had his enemies. Gazprom's Viakhirev in particular was unsparing in his condemnation of Chubais: "There was no man more harmful for the Russian state than Chubais, nor will there be any time soon. . . . They all wanted to tear Gazprom apart. They only knew how to subtract and divide, but not add or multiply."[30]

But in 1993–1995 all that still lay ahead. The State Property Committee—under the dynamic leadership of Chubais and an idealistic squad of young Russians and backed by Western consultants—carried on where Gaidar had left off, driving forward the most massive giveaway of state assets in history. Between 1992 and 1996, over 114,000 state companies passed into private hands, and by the end of 1996, 90 percent of Russian industry was private.[31]

Chubais and his team designed the privatization campaign following three pragmatic principles: speed, simplicity, and transparency. To avoid lengthy and costly delays, companies would be privatized first and restructured later. To prevent state assets from falling into the hands of foreigners or criminal elements, they would be given away, not sold. And to make the whole process legitimate in the eyes of the public, the Chubais team drew on the example of voucher privatization in Czechoslovakia and issued free privatization checks to every man, woman, and child in Russia that could be traded for shares at public auctions. But to get the privatization campaign through the increasingly resistant legislature, at that time still called the Supreme Soviet, Chubais agreed to a formula for privatization that reserved most of the shares for company employees. Whether he realized it or not, this practically guaranteed that the Soviet-era *nomenklatura* insiders would stay in control. By the time the first phase of privatization was over, only five of the top fifty companies privatized in this period ended up in the hands of outsiders.

Insider control is exactly what happened with Gazprom. The internal distribution of Gazprom shares began in 1993 and voucher auctions in the spring of 1994. By the end of the first phase of the privatization process in 1995, the shares had been allocated in this way:

- *Sale to Gazprom Management:* 10 percent of Gazprom shares were acquired by the company.
- *Closed subscription:* a 15 percent stake in Gazprom was distributed to 282,000 current and past employees of Gazprom, including retirees, partly for cash but mainly in exchange for vouchers.
- *Voucher auctions:* in auctions held in sixty regions of Russia, 32.9 percent of Gazprom shares were auctioned in exchange for vouchers (747,000 people became shareholders in this way).
- *The government* initially remained in control of over 40 percent of Gazprom's shares; subsequently this number fell to 38.7 percent.

At the end of this first phase, over a million Russian citizens had become shareholders of Gazprom. At first glance, it seemed to be precisely the sort of mass ownership that Chubais and the market reformers had hoped for.[32] All told, some 40 percent of the company's shares had been acquired for vouchers (in contrast to an average of 20 percent for all voucher auctions nationwide). Moreover, the auction of Gazprom shares was carried out relatively transparently and honestly, compared to the scandals that discredited the process for many other companies, especially in the oil industry. The general idea was that shares would be made available for voucher auctions in rough proportion to the number of Gazprom employees in any given region, and by and large this criterion was observed.[33] Thus the Gazprom auctions stand out as a relative model of the way privatization was meant to happen but so seldom did.

However, mass ownership did not mean shareholder control. Gazprom imposed tight restrictions on the resale of its stock, to prevent any possibility that outsiders could gain a toehold in the company by buying up stock on the secondary market. Those who obtained Gazprom shares at the voucher auctions did not receive an actual stock certificate but only an affidavit that shares were on deposit in Gazprombank in the shareholder's name, which the shareholder was forbidden to sell without Gazprom's consent. The affidavit did not carry voting rights—in other words, Gazprom management voted the shares.

This arrangement not only kept Gazprom's managers in power, but it subsequently enabled them to concentrate Gazprom shares into their own hands. Over the next few years, many Gazprom shareholders sold their certificates, and the number of individual shareholders declined from over one million to half that number. A decade later, individuals owned only 15 percent of Gazprom shares, while 34 percent were held by various "juridical entities" (to use the Russian legal term), some of them presumably affiliated with management.[34]

This approach to privatization had several further consequences, which affected Gazprom along with all the others. First, as a result of the decision to give away rather than sell, neither the government nor Gazprom received any capital infusion from the initial transfer. For most other companies in the same situation, this was a serious problem. However, thanks to Gazprom's protected "fifth molecule," it enjoyed such hefty revenues from exports that it hardly noticed. Only a decade later, when the government went back on the fifth-molecule arrangement and raised taxes on Gazprom and the company needed capital to invest in the next generation of gas, did it begin to feel the pinch.

Second, and ultimately much more serious for Gazprom, the decision to privatize before restructuring meant that Gazprom began life as a private corporation loaded down, like a ship encrusted with barnacles, with the noncore assets that had accumulated over the Soviet decades—an astonishing 72 million square feet of housing, 245 kindergartens and daycare centers for 34,000 children, 17 hospitals that collectively had 2,852 beds, 20 clinics with enough staff to accommodate 4,700 visits per shift, 48 resorts and retirement homes, several dozen construction companies, and the over 200 farms. At the time it was privatized, Gazprom employed over 400,000 people.[35]

The last consequence was the remaining 38.7 percent stake in Gazprom still held by the state. Again, this was a typical outcome: at the end of 1995, when voucher privatization ended, the state ended up with an average one-third stake in the fifty largest companies. This set the stage for the shares-for-loans scandal two years later, in which the state's stake in the oil industry was sold to the emerging oligarchs for a fraction of its worth.[36] But Chernomyrdin protected Gazprom from the same fate. In March and April 1996, he engineered two presidential decrees that barred the state from selling off any of its stake in Gazprom and instead entrusted the management of that stake to Gazprom.[37] The victory of the insiders was complete.

Before one concludes that the privatization of Gazprom was simply a variant on the inside jobs of the 1990s, one should also take account of the specific features of Gazprom as a gas company. The entire gas industry of Europe at that time was still dominated, as we have seen, by similar monopolies. The system of gas exports that had been constructed since the 1960s, linking Europe and Russia in a highway of steel, was governed by large long-term contracts between the Russian gas export monopoly, Soiuzgazeksport, and the major European gas companies, notably ENI, Ruhrgas, and Gaz de France. That meant there was no room for small players. There was not yet a spot market through which smaller lots of gas could be exported to consumers. It was not until a decade later that the first spot gas was sold from Russia to consumers in Western Europe.

The contrasting fates of the two key Soviet-era trading bodies—Soiuznefteeksport for oil and Soiuzgazeksport for gas—are revealing of the fundamental differences between the two commodities and the two external markets. Soiuznefteekeport disintegrated into dozens of small trading bodies, competing against one another and the dozens of other oil-trading bodies that quickly sprang up all around them. There was nothing to keep established oil traders together and everything to drive them into business for themselves. Nor was there anything to prevent anyone who could get their hands on a quota and a tank car or a river tanker from jumping into the business. In 1991–1992 there was chaos in the Russian oil export trade. In contrast, Soiuzgazeksport had no possible existence outside the familiar world of the long-term take or pay contracts, and in 1992 the logical outcome happened: it was absorbed by Gazprom and became simply the gas monopoly's export arm, another trump in the hands of Gazprom's Soviet-era management. Thus, unlike oil, in which the state monopoly disappeared (despite Putin's efforts to salvage at least some of it), in the gas sector the Russian state today

Table 9.2 Gazprom's equity structure in 2018

Entity	Shareholding (%)
Federal Agency for State Property Management[a]	38.37
Rosneftegaz[a]	10.97
Rosgazifikatsiia[a]	0.89
ADR holders	25.20
Other registered holders	24.57

Data sources: Andrew Neff, *Gazprom: Upstream Strategy Assessment* (IHS Markit Upstream Companies & Transactions Profile, November 2018); Gazprom.

Note: ADR = American Depositary Receipt.

[a]Companies controlled by the Russian government.

retains a de facto monopoly over gas exports (except for LNG), thanks to the state's retention of control of Gazprom overall (see Table 9.2).

But even that is not the end of the mystery. Granted that there was no possibility of spot gas exports, three other things could possibly have happened in the gas sector that would have made it possible for the gas producers to turn into independents. First, they could have exported gas to Ukraine and the western part of the Former Soviet Union and learned how to make money in that market. This is effectively what an independent start-up favored by Gazprom's management, Itera, learned to do in the second half of the 1990s. (But that is precisely the point: the support of Gazprom's management was a key part of Itera's initial success.[38]) Second, gas producers could have sold gas on a spot basis to Russian commodities producers, who could use gas to make fertilizers, metals, and petrochemicals for export and could afford to pay in dollars. (But that kind of domestic spot market, as of this writing, is still struggling to be born.) Third, gas producers might have found new long-term buyers in the European market, as Gazprom did in Germany. But at the time, the established European buyers were not willing to do business on any other basis than the long-term contract.

It is only in retrospect that we can see that those opportunities even existed in theory. At the time, all three were bridges too far. They represented too radical a break with the established way of doing business. One must remember that the gas generals never had the same degree of autonomy and privilege as the oil generals in Soviet times. They had always been more tightly subordinated to Moscow and their ministry. At any rate, for a variety of reasons, none of these avenues to liberalization was available in the 1990s.

The next stage in the insider privatization of Gazprom began when Chernomyrdin was named prime minister and promptly appointed his longtime protégé, Rem Viakhirev, to run Gazprom in his place. The Viakhirev era lasted until Putin came to power and marked the ultimate stage in the de facto insider privatization of Gazprom. It also marked the beginnings of a vast expansion of Russian gas exports to Europe, particularly Germany.

The Odd Couple: Chernomyrdin and Viakhirev

They made an unlikely looking couple. With his stocky frame and square jaw, Chernomyrdin looked like the classic Soviet bureaucrat. Viakhirev, his protégé and heir at Gazprom, could not have been more different. Short and squat, with wavy hair and a round, pockmarked face, Viakhirev looked like a Mediterranean shopkeeper, and only his eyes, constantly moving behind steel-framed glasses, gave a hint of the man's wary cunning.[39]

They both had roots in the Russian village (Viakhirev's parents were schoolteachers), and they shared a common start as oil engineers from the Volga province of Samara (then called Kuybyshev). They both switched to the nascent gas industry in Orenburg, where they first met. Viakhirev rose through the ranks to become chief engineer of Orenburggazprom. His career nearly came to an early end when the first well at Orenburg exploded, and he was almost expelled from the Party and arrested. However, he was cleared when the gas minister, Sabit Orudzhev, visited the site and absolved him of responsibility for the accident.[40] In later years, Chernomyrdin and Viakhirev, together with a small handful of protégés who rose with them and formed the core of the Gazprom leadership, were known, as already mentioned, as the "Orenburg mafia."

As Chernomyrdin rose, Viakhirev came right behind. From 1983 on, his career matched Chernomyrdin's in lockstep. Whatever post Chernomyrdin vacated, Viakhirev immediately filled, from deputy gas minister to first deputy and then deputy chairman of Gazprom. In 1992, when

Chernomyrdin became energy minister and deputy prime minister, Viakhirev smoothly took command of Gazprom. It was only when Chernomyrdin became prime minister, in December 1992, that the pattern was finally broken. According to a common rumor at the time, Chernomyrdin invited Viakhirev to take over his former post in the government, but Viakhirev elected to stay put at Gazprom. For the first time, Viakhirev had his own seat of power. Viakhirev remained Chernomyrdin's man, but over time, a subtle change began to take place as both moved into new roles. That is one key to the story of Gazprom in the 1990s.

Initially, the patron-protégé relationship between Chernomyrdin and Viakhirev not only symbolized the tie between Gazprom and the Russian government, it *was* that tie. Chernomyrdin defended Gazprom against its enemies, and in exchange Viakhirev provided Chernomyrdin with ready cash for emergencies, natural disasters, and election campaigns, a priceless backdoor treasury for a government that was perpetually short of funds.

No one would have called Chernomyrdin charismatic, although in later life his habit of blunt speech and his trademark malapropisms won him a certain popular affection. Some of his sayings have passed into the Russian language, such as his exclamation, "We wanted the best, but everything turned out like always." His attempt to develop political charisma led to disaster when he tried to lead a political party into the parliamentary campaign of 1995. His organization was positively clueless. A political poster created for that campaign showed the prime minister sitting with his hands joined, fingers touching in the shape of a roof. He meant the gesture to be reassuring, even paternal—the sign of a man who would protect Russian homes. But as every Russian knows, the word "roof" *(krysha)* stands for the special kind of protection provided by organized crime, and Moscow wags seized upon the slogan of Chernomyrdin's party with delight, turning it from "our house is Russia" into "our home is Gazprom." Chernomyrdin and his party went down to defeat.

Chernomyrdin was not perceived by Russians as personally corrupt. That in itself is remarkable, in a decade in which state officials were openly on the take. As President Putin once mused in a press conference, "At that time . . . very many of the so-called elites were busy pilfering whatever could be pilfered and squirreling it away somewhere" or setting up their children in some "cushy spot."[41] Indeed, Chernomyrdin's children did well in the 1990s, as did Viakhirev's. But at the time this was not considered particularly scandalous. The Western concept that a high official should put his assets in trust or divest himself of them in areas of potential conflict of interest was, to say the least, a novel idea in post-Soviet Russia. By that yardstick, most of the business and political elites would have considered it surprising—indeed, somewhat unfair—if the man who had played such a key role in the creation of Gazprom had not been a stakeholder in his own company.

Chernomyrdin was more than an oligarch in government. The term "oligarch," as it came to be used in Russia in the 1990s, meant a nongovernment figure whose private fortune was created by taking over state assets, multiplied many times over by favorable backdoor treatment from high government officials, and whose contribution consisted of extracting rents rather than investing in his property. But Chernomyrdin was different. He had created Gazprom virtually single-handedly, in battle against the failing government of the time, to hold together an industry to which he had devoted his whole life. Later on, as prime minister, he continued to defend the terms of Gaidar's founding compact with Gazprom—the export monopoly, the sole control over transportation, and above all the delegation of the government's Gazprom shares to the company. Thus, in these ways, Chernomyrdin was not an oligarch in the classic sense of the term—whereas his successor, Rem Viakhirev, conformed much more to the type.

Managing the Initial Gas Bubble and the Cash Crisis

Over the past quarter-century the Russian gas balance has shifted twice, from surplus to shortage and then to surplus again, and this is an important clue to Gazprom's behavior in the 1990s and afterward. In the immediate post-Soviet years there was a Russian gas bubble, caused mainly by the combination of giant legacy fields, imports from Central Asia, and a decrease in domestic consumption both in Russia and the so-called near abroad countries of the Former Soviet Union. Then, from the mid-1990s to the mid-2000s, the Russian gas surplus disappeared, as Gazprom's legacy fields declined and gas from Central Asia vanished, while domestic consumption bottomed out and began to recover. This led to a vigorous response in the second half of the 2000s, as both Gazprom and independent gas companies began investing heavily in the next generation of gas in the Yamal region and the Yamal Peninsula (see Chapter 10). As a result, the Russian gas balance today is once again in surplus. By some Russian estimates, Gazprom alone had nearly 150 billion cubic meters a year of developed capacity available in 2017, which has been reduced to about 100 billion cubic meters per year. In short, there is again today a Russian gas bubble, although Gazprom is no longer its sole master.

But in the first half of the 1990s all that lay far in the future. As it exited from the Soviet regime, Gazprom had more gas than it knew what to do with.[42] Thanks to the massive investments of the previous decade, its three main West Siberian supergiants were still at their peak, and its massive pipeline system was still relatively new. Yet domestic demand was falling as the economy went into a severe depression. By the time it hit bottom, domestic demand had dropped by just over 100 billion cubic meters per year, or 21 percent, from around 477 billion cubic meters in 1991 to a low of about 377 billion cubic meters in 1997 (as measured according to the apparent demand definition; production less net exports). That was the origin of the first gas bubble, which lasted from the early 1990s to nearly the end of the decade (see Table 9.3).

This surplus of gas generated massive amounts of rent. Despite chronic losses in the domestic market, Gazprom's "fifth molecule"—the export molecule—was enough to make it the most profitable business in Russia. Gazprom's managers were besieged with favor-seekers of all sorts. As Viakhirev recalled a decade later:

> The contracts with the Europeans, especially the joint venture with the Germans, saved Gazprom. In Russia nobody was paying anything to anyone, or in the former Soviet Union either. But thanks to Europe, Gazprom had money. And long lines stretched out for that money: we began to pay taxes; we paid wages; and there was even something left over for capital investment. And the state was constantly rip-

Table 9.3 Russian Federation gas balance, 1990–2018 (billions of cubic meters)

	1990	1995	2000	2005	2010	2015	2018
Production	640.6	595.5	583.9	641.0	650.7	635.5	725.2
Exports	205.6	190.6	217.9	222.7	220.7	211.7	264.8
Pipeline exports	205.6	190.6	217.9	222.7	207.3	197.2	237.9
Non-FSU	109.3	117.4	129.0	154.3	138.6	159.6	202.3
FSU	96.3	73.2	88.8	68.4	68.7	37.7	35.6
LNG exports (all non-FSU)	0.0	0.0	0.0	0.0	13.4	14.5	26.9
Imports	36.4	2.9	37.3	25.2	36.2	19.2	16.3
Net exports	169.2	187.7	180.6	197.6	184.5	192.6	248.6
Consumption	471.4	407.8	403.4	443.4	467.4	443.4	477.5

Data source: Matthew J. Sagers, *Eurasian Gas Export Outlook, May 2019* (IHS Markit Market Briefing, May 2019).

Notes: Numbers may not sum to totals due to rounding. The table uses the Russian measure of 8,850 kilocalories per cubic meter (in gross calorific value). Production excludes flared volumes. The former Soviet Union (FSU) includes the three Baltic Republics. Consumption is apparent domestic consumption (production minus net exports). LNG = liquefied natural gas.

> ping us off. Everybody was asking us for money—customs officials and generals, everybody. All my friends came around too, including the movie director Nikita Mikhalkov. . . . I had a clever little arrangement in my office—I kept a little round table by the door. Anybody new who came to ask me for something—I used to sit them down at that little table. We'd negotiate and off they would go, good-bye.[43]

But the domestic market was a bottomless sea of red ink. The first problem was that domestic gas prices initially remained tightly controlled in the midst of near-hyperinflation. While general price levels rose over twenty-fivefold during 1992, wholesale gas prices rose only fourfold. By the end of 1992, the posted domestic price of gas fell to only 2.3 percent of the export price. But domestic prices hardly meant anything anyway, because consumers failed to pay them. Even though prices subsequently recovered a bit by mid-decade, Gazprom continued to bleed money in the domestic market.[44] Its domestic revenues did not even cover operating costs. By November 1995, Gazprom was owed

$13.5 billion by domestic consumers. In that year Chernomyrdin attempted to provide some relief by signing new rules authorizing Gazprom to cut off delinquent consumers, but electoral politics intervened. In the run-up to the critical parliamentary election of 1995 and the presidential election of 1996, neither Chernomyrdin nor Viakhirev were prepared to alienate influential industries and gas-consuming voters by cutting them off for nonpayment. But Gazprom was not allowed to shut off supplies to delinquent consumers.[45]

Rather than accumulate IOUs that would probably prove worthless—since they amounted to dubious promises to pay rubles that in any case were melting away—Gazprom began accepting payment in any commodity that could ultimately be exported or otherwise turned into cash or property.[46] Within a few years Gazprom traders had become virtuosos at constructing elaborate daisy chains of barter, designed to extract at least minimal value from the domestic market.[47] As a result, Gazprom accumulated a large portfolio of nongas assets in a wide range of industries (especially power) and other properties, including real estate—a good deal of it outside Russia. In addition, Gazprom systematically used the accumulated debts of local gas distributors to acquire equity in those as well. By the end of the decade, Gazprom had evolved from an upstream company with trunk lines to a company with near-total control over gas transportation and distribution, and over many gas-consuming industries as well.[48]

The regional governors were an influential part of the chain of barter and debt and quasi-monies. Most of the local distributors were tied into the regional power structures. To stem its losses at the local level, Gazprom was constantly forced to negotiate with the governors to get paid for its gas. As Gazprom took over local assets, it needed the local governors' blessing to blunt local opposition to the takeovers. Regional authorities connived with nonpaying consumers. "All the local governments do their utmost to make sure that local debtors don't pay us," complained one high-ranking Gazprom official in early 1995.[49] Even when consumers paid, distribution companies—under pressure from local authorities—sometimes diverted the payments back into the local budget or local projects.[50] When Gazprom, under growing pressure itself, attempted to get tough, governors pushed back, hitting Gazprom offices with inspections, tax audits, and other forms of harassment. There were even instances in which regional governors threatened to send special troops to dispatching points to prevent Gazprom from turning off the gas to nonpaying consumers. From the mid-1990s on, therefore, Gazprom's senior managers began systematically traveling through the provinces, signing mutual cooperation agreements with regional governments. In all of these, the basic idea was the same: Gazprom agreed to invest in local pipeline projects (particularly the regional gasification programs) and buy from local equipment suppliers, in exchange for the local authorities' support in collecting gas debts. Life as a "gas cow" did not come cheap.

Gazprom maximized its export revenues by shifting its own exports from the non-Russian Former Soviet Union to the more lucrative European market. In the process, Gazprom closed the European export market to Central Asian gas producers and encouraged them to export to the Former Soviet Union market instead, thus displacing Russian gas. In the first half of the 1990s, these were junk markets from Gazprom's perspective, because it was even more difficult to get paid there than in the Russian domestic market. This was a rational enough strategy, given the situation in which Gazprom found itself at the beginning of the 1990s. By seizing the higher-value export stream for itself (that is, exports to Europe) and forcing the Central Asians into the lower-value stream (exports to the non-Russian Former Soviet Union), Gazprom in effect forced the Central Asians to absorb the consequences of Russia's excess gas.

During this period Gazprom cut back its own investment in new gas and allowed its own production to decline. From a peak of 642.9 billion cubic meters in 1991, Russian gas production declined to 571.1 billion cubic meters in 1997; but within that envelope, Gazprom's own production declined even more sharply. It was less a strategy than a forced retreat. Gazprom's so-called Big Three fields had peaked and were beginning to decline. The worst years were 1999–2001, during which Gazprom's production declined from 546 to 512 billion cubic meters. No new gas fields were developed for a decade, until the start-up of Zapolyarnoe—the last of the conventional giants in the major gas-producing region of West Siberia known as the "Nadym-Pur-Taz" triangle in November 2001.

This lack of investment proved fateful. Toward the end of the 1990s, economic growth resumed in Russia, domestic demand began to recover, and the gas bubble abruptly vanished. Russia, the country with the largest gas reserves in the world, suddenly found itself acutely short of gas.[51]

The State Goes After the Fifth Molecule

Despite its problems, Gazprom was fortunate, compared to other Russian businesses—as long as its revenues from exports remained relatively immune to taxation. But in the summer and fall of 1994 came the first signs that Gazprom's privileged status was weakening. Gazprom-bashing had always been a popular sport among liberals in the media and the government,[52] as we have seen, but Gazprom's support in the prime minister's office had seemed unshakable.[53] Yet in December 1994, to the astonishment of observers and apparently of Gazprom itself, the government abruptly quadrupled the export duty (tax) on gas *(eksportnaia poshlina).* There were reports that the prime minister's office had initially intended to raise the export tax tenfold, but desperate last-minute lobbying by Viakhirev won a concession.[54]

Why did Chernomyrdin weaken his support for Gazprom? The short answer is that Chernomyrdin had evolved. Over the years as prime minister he had come a long way from his initial worldview as a Soviet technocrat, who referred to the market economy as a "bazaar," to a more prime ministerial—and indeed market-oriented—perspective. He never understood the finer points of market economics, but he did come to realize that inflation was harmful and that balancing the budget was the first essential step to restoring both the government and the economy to health. In the end, his conversion proved incomplete, and his administration too weak, to halt the government's addiction to deficit spending. But for Gazprom the implications of the prime minister's evolution were ominous: it had been the one agency that the taxmen could not touch in 1992–1994, but by 1995 it had become a prime target.

Throughout the summer of 1995 the Russian government narrowed the loopholes that had made Gazprom the wealthiest company in Russia in the first half of the 1990s. The tax breaks granted to Gazprom under the so-called stabilization fund were eliminated as of January 1996, at the insistence of the International Monetary Fund. A 25 percent excise tax on production was instituted and then quickly raised to 35 percent.

The export tax was extended and increased.[55] Along with the new taxes came fresh attacks. Critics charged that Gazprom had grossly misused the funds it derived from the stabilization fund. One investigation showed that Gazprom had systematically paid for equipment purchased abroad but not taken delivery on it, implying that the sellers were enjoying interest-free loans—and presumably paying someone for the favor.[56]

To pressure the government to rescind its increased take, Gazprom cut back programs it had been funding on the government's behalf, such as defense conversion, rural gasification, and the so-called Yamburg Accord, under which local infrastructural investments were paid for by exports of gas on behalf of regional governments.[57] But this attempt to play hardball did not work. How, for example, could Gazprom not contribute to Yeltsin's reelection campaign in 1996? Gazprom's protection from the Kremlin might have weakened, but it still cost money. Even before the financial crash of 1998, the combination of higher taxes and nonpayments exceeded Gazprom's export revenues. Then the crash pushed Gazprom even deeper into the red.[58]

In the immediate emergency, there was only one option—to turn to Central Asia.

Viakhirev's Kowtow in Ashgabat

"If there is a gas bubble in Russia," the crusty chief of Gazeksport, Stepan Derezhov, once said in my presence in the early 1990s, "we will make the Central Asians swallow it."[59] But that was when Russian demand was declining and Gazprom had plenty of gas to spare. By the end of 1999, the situation had changed dramatically, and Viakhirev was in trouble. After a decade of noninvestment in new gas, Gazprom's production was declining. Viakhirev had lost a battle to curtail the gas quota for electrical power, and suddenly he could not cover the anticipated call for gas for 2000. In desperation, Viakhirev flew to Ashgabat, the capital of gas-rich Turkmenistan, on an emergency mission to negotiate a short-term supply contract from its eccentric and unpredictable dictator, Saparmurat Niyazov, the self-styled Father of the Turkmens.

Niyazov relished the irony of the moment. For the past six years Gazprom had blocked Turkmen gas exports, forcing the Turkmens to swallow the Russian surplus or export it to the low-value market of Ukraine and the Transcaucasus. Now the head of Gazprom had come to his capital, hat in hand. Niyazov played the visit for all it was worth. He put Viakhirev live on Turkmen television, and the extraordinary dialogue was watched by the entire Turkmen viewing public. The following excerpt, probably unique in the annals of gas negotiations, is quoted directly from the transcript released by the Turkmen government:[60] "I will tell you frankly: Turkmenistan was offended by you," Niyazov began. "You said things in press conferences like 'The Turkmens will eat sand.' We kept silent." "I am getting older and will soon leave for the second world," Viakhirev replied, clearly rattled. "I have a big sin left behind me, which is to provide a market for Turkmen gas. . . . Please forgive me if I said something wrong." But Niyazov was relentless: "I will tell you what you said: 'They have no choice; they will crawl to us. . . .' You know these words were excerpted from your interviews. . . ." "No, no—I don't want to say such words," floundered Viakhirev desperately. "I don't speak with journalists anymore." With that Niyazov softened, and the televised conversation turned to prices. Viakhirev, presumably much relieved, returned to Moscow with a one-year gas contract in his pocket.

Viakhirev's visit marked the beginning of a sea change in Russia's gas relations with Central Asia. Within a few months of Viakhirev's visit Gazprom was back, this time to negotiate a longer-term contract. Niyazov initially played hard to get. He was being courted at the same time by the Ukrainians and Western companies, which encouraged him to believe that Turkmen gas exports could bypass Russia, going westward through a Trans-Caspian pipeline to Azerbaijan and on to Turkey, or southward via a pipeline through Afghanistan to Pakistan. But by 2000 Niyazov was losing hope in the prospects of such new export routes, and the election returns in Russia in 2000 and the rise of Putin refocused his attention northward.[61] Within days of his inauguration as president, Putin made a brief ceremonial visit to Ashgabat, and though his mission was ultimately the same—to shop for gas—the context was already different. Having been forced to swallow the Russian gas bubble in the 1990s, now the Central Asians would be invited to regurgitate it. But the Central Asian breather was only a stopgap. It took nearly another decade before Gazprom's investments in the next generation of Russian gas from the

Yamal began to come in, and during that time Gazprom was chronically short of gas.

Conclusion: Gazprom in the 1990s

The battle for survival and control of Gazprom absorbed nearly all the energies of Gazprom's management for a decade. Their attention was necessarily selective—survival came first, and everything else was secondary. As the 1990s ended, the company faced a long list of challenges that had been neglected or postponed that now needed to be dealt with. First among these was the financial crisis caused by the government's rescinding of Gazprom's unique fiscal privileges. The company had spent little on upstream investment or pipeline maintenance in the first half of the decade; in the second half it spent practically none. But in addition, as we have seen, Gazprom had become the "gas cow" for the rest of the economy. These problems came to a head in 1998–1999, when the global recession brought lower gas prices in Europe and more unpaid gas bills in the domestic market, topped off by the state's default and crash.

The 1990s were equally important in shaping Russia's future gas relations with Europe. Gazprom did nothing to challenge the existing gas regime in Europe—with which in any case it was quite comfortable. Even the Wingas joint venture with BASF, though it challenged Ruhrgas's monopoly position as Gazprom's partner, soon settled back into the established order of things. Gazprom's early experiments with trading houses, designed to enable Russian gas to reach consumers directly, faded away when it met with resistance from incumbents.[62] In a book written in 2014, with the benefit of a decade and a half of hindsight, Jonathan Stern summed up the main consequence of Russia's experience in the 1990s for its gas relations with Europe: "The Russian Federation was preoccupied with establishing a post-Soviet order in relation to gas, and the EU as well as the gas companies were struggling to achieve first steps toward liberalization of the gas industry."[63] In other words, both sides were focused inward. The conflicts that arose later in the 2000s were aggravated by the fact that when both sides began to look more closely at each other, they were surprised by what they discovered.

By the late 1990s, the Viakhirev era was coming to an end. As long as Chernomyrdin was prime minister, he continued to provide Viakhirev with essential (if declining) protection from political and fiscal pressures. But in 1999, Chernomyrdin stepped down and Viakhirev committed a blunder that cost him his job: in the election of 1999 he supported the opposition—Evgenii Primakov and Yurii Luzhkov—against the Kremlin's Unity Party and its candidate, Vladimir Putin. When his contract came to an end in May 2001, he was summoned to the Kremlin and informed that it would not be renewed. From that moment on, the de facto head of Gazprom was Putin. But that is the subject of Chapter 10.

CHAPTER 10

Gazprom under Pressure

In the first two decades of the new century, Gazprom, and the Russian gas industry as a whole, came under new pressures. There were basically three. The first was the threat to Gazprom's gas supply, as the Soviet-era generation of fields began their inevitable decline. The second was the challenge, but also the opportunity, stemming from the process of market liberalization in Western Europe. The third was the rise of new domestic competitors, weakening Gazprom's traditional monopoly position.

Over the following twenty years, Gazprom met the first two challenges successfully. By 2020, by developing the huge gas resources of the Yamal Peninsula, Gazprom had ensured a comfortable surplus for the next generation of supply. In Western Europe, its prime export market, Gazprom had adapted to the radical changes in market structure and business models that have transformed the European gas industry. But the third challenge still looms, and it is growing. Domestic competitors have captured over half of the domestic gas market and have seized the lead in LNG development and exports, breaching Gazprom's traditional monopoly of gas exports. Gazprom's historic agreement with the state—the "Gaidar deal" discussed in Chapter 9, under which Gazprom supplied the loss-making domestic market in exchange for monopoly control of transportation and exports—is gradually weakening, and might soon disappear.

Gazprom is often portrayed, both in Russia and outside, as a monolithic and unchanging dinosaur, a direct and immutable descendant of the USSR

Ministry of the Gas Industry. If the period from 2000 to 2020 changed anything, this narrative runs, it was that Gazprom reverted to Soviet form, as Gazprom's quasi-autonomy in the era of Boris Yeltsin was undone under Vladimir Putin, and Gazprom was turned back into an instrument of Kremlin policy. Moreover, Gazprom is depicted as a highly centralized, unitary actor. This view has shaped the prevailing view of Gazprom and of Russian gas policy in the West throughout the past two decades.[1]

This view, although it has some basis, is nevertheless a caricature and does not capture what has changed. Large organizations such as Gazprom are complex and made up of many parts, each with its own history and culture. Gazprom is more like a feudal barony, in which the various departments are led by strong personalities, each of whom has political ties to the elites in power, and the control of the chief executive is imperfect and often contested. To make a rough approximation, there are at least three Gazproms: the first consists of the engineering departments, in charge of production and transportation; the second is export and external relations; and the third is the central management, in charge of overall policy, security, and relations with the Kremlin. In addition, there is the part of Gazprom that handles domestic distribution and sales, which we do not discuss here. The three Gazproms are often at odds over issues of policy and personal ambition, and the responses of Gazprom as a whole to the pressures upon it are shaped by the different behavior and complex interaction of these three parts.

Yet overall Gazprom remains firmly attached to its traditional vocation: to produce natural gas and deliver it by pipeline to European Russia and its preferred export market, Western Europe. As a result, Gazprom tends to react slowly to pressures for change, following what two perceptive observers describe as "a strategy of minor acceptable adaptation steps"—with the important exception of Gazprom's activity as a trader, where the company's adaptation, according to the same two observers, has been marked by "huge flexibility and creativity."[2] Consequently one should not overlook the very real process of evolution that has taken place over the years. Time passes; new leaders take charge; new technologies arise; and slowly, one step at a time, Gazprom evolves.

GazpromExport, which handles Gazprom's export and trading activity, is the prime example. It is striking that in the half-century since

Soyuzgazexport, then an arm of the USSR Ministry of Foreign Trade, began exporting gas to Europe on behalf of the USSR Ministry of Gas (subsequently Gazprom), there have been essentially only four heads of Russian's international gas exporting operations. Stepan Derezhov, a pipeline engineer by background, practically founded Russia's export business in the 1960s and presided over Gazprom's international operations into the era of Rem Viakhirev, Gazprom's chairman until 2001. Yurii Komarov, followed briefly by Viakhirev's son Yurii, headed Gazexport in the transitional 1990s. Alexander Medvedev, who succeeded the younger Viakhirev in 2002, was an investment banker by background who spent thirteen years in finance in Vienna before he was promoted to be head of GazpromExport.[3] Though he was a new entrant to the gas business at the time of his return to Moscow, he quickly became expert in all its commercial aspects, especially the arcana of pipeline contracts. At the same time, facing internal resistance to the development of liquefied natural gas (LNG), Medvedev led the creation of Gazprom Marketing and Trading (GM&T), a vehicle through which Gazprom could begin to learn how to trade LNG, even though it did not yet produce it, and could also learn the skills of trading pipeline gas in the liberalizing markets in Western Europe, as we shall see later in this chapter. Elena Burmistrova, who has been head of GazpromExport only since 2014, symbolizes the rise of a fourth generation. She did not come from the pipeline business but, as a protégée of Medvedev and his Vienna partner Andrey Akimov, directed Gazprom's growing trade in LNG and liquids. As a member of the latest generation, she brought a whole new set of skills and experiences to a position of potential power inside Gazprom.[4]

Because of this internal diversity, Gazprom's adaptation to the new world of gas happened more slowly and in varying ways than strict commercial logic might suggest. Thus, the objective of this chapter is to attempt to sort out the mixture of change and nonchange, of stagnation and dynamism, and of commercial and political motives as the different parts of Gazprom relate to one another, Gazprom's management, the markets, and the political center.

The critical period on which this chapter focuses is the twenty years from 2000 through 2020, beginning with the advent of Putin as president and the purge of Rem Viakhirev as chairman of Gazprom. These two decades were perhaps the most momentous time in the history of Gazprom, surpassing even the key developments of the 1990s. The previous management team was replaced. Assets that had fallen into the hands of friends and relations of Victor Chernomyrdin and Viakhirev were recovered. The Kremlin came close to merging Gazprom with Rosneft, with the aim of creating a single state-owned oil and gas giant. Central Asian imports waxed and then waned: Central Asia was no longer a major supplier by the end of the period, as Gazprom grappled with the increasingly urgent problem of offsetting the decline in its West Siberian Soviet-era giant fields.[5] Independent companies emerged in the Russian gas sector, posing new competitive challenges as Gazprom's monopoly of production ended, its control over the domestic market weakened, and its monopoly of exports was partially breached.

For much of the period Gazprom was internally divided over strategy, as powerful players inside the company did battle over which road to follow—whether to head north to the Yamal Peninsula and develop the next generation of conventional pipeline gas or, at the time, go west via LNG to the United States and other global markets. Gazprom's leaders argued over export strategy, as the company took its first hesitant steps into the fast-growing spot market of Europe. Chronic conflicts with Ukraine led to two interruptions of Russian gas deliveries to Europe as Gazprom attempted to raise its prices. The conflicts gave new impetus to Gazprom's policy, begun under Viakhirev, of bypassing Ukraine through alternative pipelines to Europe. Lastly, Russia's leaders faced the new challenges posed by the rise of global LNG and then of shale gas in the United States, as they struggled to develop a strategy for LNG. It is fair to say that nearly everything that is happening in the Russian gas industry today has been driven by the decisions made—or not made—during the period from 2000 to 2015.

The chief catalyst of change was Putin. Almost as soon as he was elected president, it became evident that he too had a list of priorities, and it was very different from Chernomyrdin's and Viakhirev's. The first task was to change the management. The second was to regain control of the gas industry. Yet to say that Putin sought to regain control is very far from saying that he achieved it, and still farther from saying precisely what he would do with it. Putin is unique among world leaders in the active interest he has taken in the gas industry and in the depth of his knowledge. Yet his own power is subject to constraints, and his motives are mixed. He too has evolved over time.

One of the main points of this book has been that the Russian gas industry, despite its geopolitical significance, is a business, and a highly technical and a highly complex one. A state-owned gas company may be an instrument of government policy and even of geopolitical ambitions, but it is also interested in profit and market share as well as its commercial reputation, the implementation of its engineering skills, and the management of such a large and complex system. Putin is clearly the chief decider in Russian gas policy. But in the everyday conduct of business Gazprom, like any large organization, has the capacity to delay, resist, and reshape the Kremlin's commands, if they run counter to Gazprom's commercial objectives, business models, and core competences. Russian gas policy has been at its most effective when the two levels mesh smoothly. The prime example of this over the past twenty years has been the strategy of bypassing Ukraine by building a new system of pipelines to Europe. In contrast, Putin has had much more difficulty persuading Gazprom to carry out his Far Eastern program for LNG, and he has increasingly turned to other champions, chiefly a fast-rising new company called Novatek. One of the chief challenges of this chapter is to try to understand this interaction of motives and behaviors.

This chapter consists of three parts. The first describes Gazprom's long hesitation before embarking on the conquest of the Yamal Peninsula and the next generation of Russian supply. The second recounts Gazprom's response to the opportunities arising from the liberalization of the European gas market. The third tells of the reluctant response of Gazprom to the advent of the shale gas revolution in the United States and its implications for its LNG strategy. Gazprom's response in the first case could be summed up as traditional, in the second as innovative, and in the third—initially at least—as denial. Each has had far-reaching consequences for the present day.

Finally, in the background, after an initially positive period the first decade of the new century was marked by a serious cooling in East-West relations. This too became a major factor in Gazprom's behavior, but we shall deal with it only peripherally in this chapter, reserving a fuller discussion of it for Chapters 11 and 12.

First, we begin by describing the change in Gazprom's leadership.

The Kremlin Turns against Viakhirev

From the beginning of Putin's presidency, members of his administration had been thinking about what to do with Viakhirev. He showed no respect to government ministers, even the prime minister. Moreover, he was too obviously out of step with the Kremlin's emerging priorities, especially Putin's. For example, in June 2000, shortly after Putin's inauguration, Viakhirev came up with a plan to amend Gazprom's charter to allow him to sell off various Gazprom subsidiaries—presumably to interests related to his family and friends. His plan was blocked by the Gazprom board of directors as a transparent ploy to benefit the existing management.[6] But the more serious strategic problem with the Viakhirev plan was that it aimed at decentralization, whereas Putin's intention was to recentralize.

This was not immediately obvious. When Putin first came to power he seemed inclined toward liberal reform, and he had formed a liberal think tank headed by a reform-minded economist, German Gref, to study various proposals for the liberalization of the economy.[7] Breaking up Gazprom was at the top of their list. But to their surprise, Putin had already decided that the solution to the gas industry's problems was not to break it up but to reconcentrate it under the direct control of the Kremlin—and that meant, first of all, getting rid of Viakhirev.[8] Putin's decision to keep Gazprom whole was one of the first indications that he would turn out to be more statist than liberal.

Viakhirev's contract as CEO was up the following year, and the search began for a replacement. So long as Chernomyrdin was prime minister, Viakhirev had been protected. But with his patron gone, Viakhirev could feel his position weakening. He was not part of the inner circle around Putin or a member of "the Family" that had brought Putin to power,[9] and he knew he was vulnerable. Viakhirev tried to flatter the new president personally, but Putin treated him with mocking contempt.[10] In June 2000 Chernomyrdin retired as Gazprom board chairman and was replaced by Dmitry Medvedev, a lawyer from Saint Petersburg who was chiefly known at the time for having been the manager of Putin's successful presidential campaign. The following spring, Chernomyrdin was packed off as ambassador to Ukraine, a form of honorable exile. Dmitry Medvedev,[11] the new chairman of the board, was in no doubt about what Putin wanted:

> Do you remember how much Gazprom was worth and how it was being managed? There was no oversight by the government; the market for its shares was in complete disorder. That's when we decided that the state had to regain control. A company like Gazprom, when you consider its roles and functions, needs to be managed by just one owner: the state.[12]

As far as Putin was concerned, the last straw came in the summer of 2000, shortly after his election as president, when Viakhirev attempted to coerce the government into raising domestic gas prices. Otherwise, Viakhirev claimed, Gazprom would be unable to supply enough gas to the electricity sector to get through the winter: "This winter we'll survive . . . but in the winter of 2001–2002 we'll have serious problems. And a year after that we won't have anything to heat with. But that won't be any of my concern—by then I'll be retired."[13] His tone was defiant, altogether a demonstration of what the president's men objected to about Viakhirev. Viakhirev had to go. When Putin decided to replace him, he reached into the ranks of his former Saint Petersburg associates—as he so often did—and came up with a relative unknown, a man named Aleksey Miller.

When he was named by Putin to head Gazprom, Aleksey Miller knew nothing about the gas industry. That may have been one of his attractive qualities in Putin's eyes: he knew no one at Gazprom, he was unconstrained by prior ties or loyalties, and he was untainted by previous deals. He was the perfect choice to head Gazprom for a man who intended to be the actual boss. Even before he became president, Putin showed a fascination for the details of the gas industry, and over the past two decades he has intervened repeatedly to take the tiller into his own hands. Visitors are always impressed with the Russian president's mastery of his brief on gas matters.

Miller had the gift, priceless given his circumstances, of being underestimated by all who knew him. Initially a member of Anatolii Chubais's reform team in Saint Petersburg in the late 1980s, he was left behind when Chubais moved to Moscow to run the privatization program. Miller gravitated to the Committee on External Relations of the Saint Petersburg mayoralty, whose head, Putin, was rapidly becoming the de facto mayor of Saint Petersburg. But Miller was a minor member of the team, and when Putin too left for Moscow to join the presidential administration, Miller was again left behind. At the committee, he was the proverbial man in the back room, a shy, uncertain figure whom nobody noticed. He came to work early and stayed late. He worked hard, "but he lived like a shadow, and he worked like a shadow, he slunk along the walls," said one former colleague from the mayoralty. "He was the ideal deputy," concluded another.[14]

Yet it was his qualities of hard work, discretion, and reliable loyalty that led Putin, as he left Saint Petersburg for Moscow, to appoint Miller to his first job with real responsibility, head of development for the Port of Saint Petersburg. In 1999 he was named director-general of the Baltic Pipeline System, an oil pipeline that Putin was backing strongly. It was shortly after that, in 2000, that Putin brought him to Moscow as deputy minister of energy. It was widely believed at the time that Putin had already settled on Miller to replace him.[15]

Over the past two decades, Miller has grown into his role of CEO. In contrast to Viakhirev, who came up as an engineer and almost seemed to know every centimeter of Gazprom's pipeline system, Miller leaves the engineering details to his deputies, but on all other matters he is known as a hands-on manager, with a policy wonk's obsession with detail. Western gas professionals, after initially underestimating Miller, have learned to respect his knowledge of even abstruse commercial points and his capacity for grasping issues. Above all, they have come to appreciate his qualities as a survivor. The media have reported Miller's impending dismissal more times than one can count, and ill health has been a repeated theme—yet in 2016 Putin renewed Miller's contract as CEO for a new five-year term.[16] Thus, as of this writing it would seem that Miller's career at the head of Gazprom is not yet over, although rumors continue to fly.

The Post-Viakhirev Purge and the Great Recentralization

Between 2001 and 2005 the size of Gazprom's management committee was cut in half, as Viakhirev's veterans were pushed out to make way for Miller's incoming team. While the technical specialists in production and transmission tended to remain in place, or at most were replaced by other Gazprom specialists from local production subsidiaries, the more strategic positions—in the sense of those in charge of money, property, and security—were staffed with people Miller had recruited from his previous jobs in Saint Petersburg. Elena Vasil'eva had been his chief accountant at the Port of Saint Petersburg; she took over the same position at Gazprom. Mikhail Sereda had been financial director of the Baltic Pipeline System; he became chief of staff for the management committee. A thirty-two-year-old rising financial expert from the same background, Andrei Kruglov, became chief financial officer. Kirill Seleznev had been head of external relations at the Saint Petersburg Seaport and head of tax policy at the Baltic Pipeline System; he became Miller's chief of staff and then head of gas distribution and sales at Gazprom. There were only two true outsiders, presumably nominated by the Kremlin: Alexander Medvedev, as noted, returned from Vienna to take over GazpromExport, while his partner in Vienna, Andrei Akimov, became head of Gazprombank. For much of the ensuing two decades, this team remained remarkably stable. It is only now, in 2019, that a new generation of managers, who do not share Miller's Saint Petersburg background, are rising to top positions in the company.[17] What the impact on Gazprom's behavior may be is as yet unknown.

In parallel with the purge of Viakhirev holdovers, Miller embarked on a campaign to recover Gazprom assets that had been taken over by friends and relatives of the previous management. In May 2001 Putin presided at the ceremonial start-up of production at Zapolyarnoe field, the last of the big four conventional supergiants of West Siberia and the first major field to begin production since the end of the Soviet Union. Turning to the newly appointed Miller, and with Viakhirev standing by in attendance, he demanded, "Where's the money? How is it that we're selling gas so cheaply? Where's the difference?"[18] Putin was convinced that various middlemen were capturing half or more of the value of Russian gas.

Miller's orders were clear: find the middlemen and recover the value that had been leaking from the company, both inside Russia and outside.

Putin's hand was clearly visible in the purge that followed. Within days of Miller's appointment special agents from the Federal Security Service (FSB) appeared in the key offices of Gazprom and took over the books. Key members of Viakhirev's team were summarily dismissed.[19] Friends and relatives of Viakhirev and Chernomyrdin, of which there were a number, were likewise quickly fired, beginning with Viakhirev's son Yurii, who had briefly headed GazpromExport. Others quickly followed—Viakhirev's brother Viktor (who headed Gazprom's drilling directorate, Burgaz), his daughter Tatiana (a major shareholder in Gazprom's construction contractor, Stroytransgaz), and two of Chernomyrdin's sons (who likewise held major stakes in Stroytransgaz)—all of whom had been in positions overseeing large cash flows.[20]

As for Viakhirev and Chernomyrdin, between 2001 and 2002 they quickly fell off *Forbes*'s annual list of billionaires.[21] The *Forbes* list makes an interesting barometer. Before 1996, it reported no Russian billionaires at all. The first Russians appeared in 1997, following Yeltsin's reelection. There were five that year, including Viakhirev in fourth place. Eight Russians made the list in 2001. Mikhail Khodorkovskii, the head of the oil company Yukos, was reported to be Russia's richest man, with $2.4 billion; Viakhirev was third, with $1.5 billion; and Chernomyrdin was eighth, with $1.1 billion. But Vikahirev and Chernomyrdin did not appear on the list in 2002. Ironically, it appears that their shares and options had been next to worthless only a few years before, and now that they had made the two men rich, they were stripped away.

A long list of former subsidiaries that had broken away from Gazprom's empire was quickly recovered by Miller, clearly again with Putin's assistance behind the scenes. In short order Miller regained control of Sibur, Purgaz, and Zapsibprom, important gas and petrochemical assets that had wandered off into the hands of well-placed friends and relatives of the previous management. In May 2002 it was the turn of Stroitransgaz, Gazprom's main pipeline construction contractor, which had acquired 4.83 percent of Gazprom shares in 1995 for a paltry $2 million. A few years later the market value of the block was estimated at $1 billion. With each reacquisition, the closer the state came to recovering majority control of the company. This opened the way for the state to sell dollar-denominated shares on the open market, increasing the attractiveness of Gazprom shares for foreign investors. In parallel, Gazprom sold off a number of small banks that had become part of its portfolio, frequently in exchange for small blocks of Gazprom shares. Each little bit helped Gazprom regain majority ownership on behalf of the state.[22]

Some of those who had taken over Gazprom assets proved harder to convince than others. Yurii Viakhirev, Rem Viakhirev's son, surrendered without a fight. However, Vyacheslav Sheremet, a childhood friend of Chernomyrdin and deputy chairman under Rem Viakhirev, spent three nights in jail, and Yakov Goldovskii spent several months behind bars before he surrendered his holdings in Sibur. Arngolt Bekker, the former head of Stroitransgaz, was reportedly persuaded by Alisher Usmanov, acting on behalf of Putin, to return his shares to Gazprom and reconsolidate the state's equity control over Gazprom.[23]

All of this activity took place largely out of sight of Europeans, except for the few energy companies that did business in Russian gas—principally Shell (active at Sakhalin), Wintershall (Gazprom's joint venture partner in Wingas), and BP (active in East Siberia)—and even they could only follow from afar. A handful of specialized Western media reported on the goings-on inside Gazprom, but few in Brussels, in particular, or in the capitals of the European member-states, were more than distantly aware of the fact that Gazprom had been effectively renationalized under the control of the president.[24] The *Financial Times*, for example, warned in an editorial: "Mr. Putin must avoid the temptation to reimpose full control from the Kremlin. Gazprom cannot prosper as a state-run company. And if Gazprom does not prosper, nor will Russia."[25] But ironically, with those very words the *Financial Times* had summed up the essence of Putin's program and his rationale for it. By the time of the May 2001 board meeting, the reimposition of "full control from the Kremlin" was already well under way.

Meanwhile, the liberals in Putin's administration—beginning with Gref, his chief reformer—imagined that with Gazprom once more under the government's control, they could proceed to modernize it along Western lines, much as the market reformers had been recommending since the beginning of the 1990s. The company would be broken up; the upstream would be privatized, and different producers would compete against one another; and the pipeline system would be turned into a regulated public utility.[26] But Putin had different plans for Gazprom. He kept Gazprom's export monopoly intact while encouraging the company to resume its expansion, this time as Russia's national energy champion.[27] But none of this was visible in the early years of the decade, and the liberals, still believing that they had the president's support, continued laying their plans.[28] (I remember attending a Christmas Eve meeting in 2004 in the office of one of the top reformers at which he vowed, "Next year we will break up Gazprom.") That hasn't happened yet.[29]

Expanding Gazprom's Empire: The Near Takeover of Rosneft

From the moment he became president, Putin had encouraged Rosneft and Gazprom to work together. That was the trend worldwide. In contrast to oil, gas had long remained a regional industry, but that was rapidly changing, as technological advances—chiefly the growth of LNG—made gas increasingly global. All the major international companies did both oil and gas. In Russia, the separation between oil and gas seemed an anachronism. From the first, Putin showed signs of wanting a single champion.

But Gazprom and Rosneft were unlikely partners. In the 1990s, under Viakhirev and his associates, Gazprom had hewed to its traditional Soviet-era profile: dry gas from West Siberia, shipped through pipelines to western Russia and Europe. Gazprom had no connection to the eastern half of the country and, despite occasional press releases to the contrary, showed no real interest in creating one. As for the valuable gas liquids that are a by-product of the extraction of natural gas, known as condensate, Gazprom had traditionally treated them as a nuisance by-product, for which it had little use. Gazprom had inherited no significant oil production and showed no interest in acquiring oil licenses or investing in oil exploration or development. In the 1990s, indeed, Gazprom invested in everything except hydrocarbons, hardly investing even in its own gas resources, let alone oil.

Up to this point, Rosneft and Gazprom had had practically no contact with each other. But in late 2004 it became apparent that Putin had a completely different plan in mind—to merge Gazprom with Rosneft. The story begins on a day in mid-September 2004, when Miller and Prime Minister Mikhail Fradkov paid a call on Putin at his presidential dacha at Novo-Ogarevo. Fradkov and Miller had brought Putin a draft proposal on a subject so innocuously technical that it hardly seemed worth a presidential visit. The issue was Gazprom's "ring fence" problem for its shares. Gazprom and the government were stumped over how to undo the unforeseen consequences of a law passed five years before that limited foreign ownership of Gazprom shares to no more than 20 percent.[30] This led to a situation in which there were, in effect, two classes of Gazprom shares, the dollar-denominated shares (ADRs) that the government had sold on the London and New York stock exchanges and were traded outside Russia, and those that circulated in the domestic market only. At various times since 1997, the domestic shares had been worth as little as 10 percent of the foreign shares. As a consequence, the market worth of Gazprom was far lower than it would have been had Gazprom shares traded freely. But if the ring fence could be removed, Gazprom would suddenly become (in terms of market capitalization) one of the largest companies in the world—not to mention the fact that a number of influential Russians would see their net worth grow substantially.[31] In addition, Gazprom would have access to more capital to develop the next generation of gas supply.

But Putin did not want to remove the ring fence until he was certain that the state had secured majority control of Gazprom. By the end of the 1990s, Gazprom and the government controlled only a little over 38 percent of the company's shares. Aleksey Miller, acting under Putin's orders, had acquired an additional block amounting to 17 percent, giving Gazprom majority control.[32] But the cautious president, wary of repeating the past, wanted full legal ownership in the hands of the state itself. The difficulty was how to pay for it. A commission had been working on the ring fence problem since 2001 but had bogged down amid competing proposals.[33]

The plan that Miller and Fradkov brought to the president was simplicity itself. The Russian state owned 100 percent of Rosneft. All it had to do was swap its Rosneft shares for a block of Gazprom shares of equal value. Gazprom would then own Rosneft, while the state would gain direct ownership of the combined Gazprom-Rosneft entity. Miller calculated that this would require a little more than 10 percent of Gazprom stock, since the market value of Gazprom at that time was far higher than that of Rosneft. The move would finally make Gazprom transparent, attractive to investors, and adapted to market conditions, Fradkov told the president.

To Putin the idea was attractive for several reasons. First, it eliminated the ring fence. But even more important, it created at one stroke the national energy champion that Putin had come to believe was essential. From his point of view, it made no sense to have two national champions. Rosneft was still perceived at that time as weak and ineffective. By itself, it was a company with little acreage and a limited future. Equally, Gazprom by itself was only the "West Siberian Pipeline Gas Company," with no presence in the east and little capability to deal with liquids. But united, Gazprom and Rosneft could provide the kind of unified total hydrocarbons approach that appealed to Putin.

Yet to Putin's evident surprise, what had seemed at first to be a purely financial maneuver to consolidate state control over a unified energy champion turned into a bitter political struggle between the two principal Kremlin clans of the day, both predominantly from Saint Petersburg. Because of the support of Igor Sechin, Putin's longtime chief of staff and confidant, Rosneft had come to be viewed in Moscow as the silovik oil company (from the name *silovik*, meaning members and former members of the security forces), a link that was symbolically confirmed in July 2004 when Putin named Sechin chairman of Rosneft's board of directors. Gazprom, conversely, was viewed as the company of the liberal Saint Petersburg business elite, with one of its members, Dmitry Medvedev, as the board chairman. Considered from this angle, the Miller-Fradkov proposal to merge Rosneft into Gazprom was anything but the narrow technical idea it seemed on the surface. To the Moscow media, always inclined to see politics behind every move, it amounted to taking the prime energy asset of the *siloviki* and turning it over to a rival clan.

The whole plan ultimately foundered on the opposition of Rosneft and its CEO, Sechin.[34] But three aspects of this episode are of lasting interest in understanding the evolution of Gazprom as a corporation and the Kremlin's approach to it, as well as Putin's attitudes. The first, mentioned above, was Putin's apparent enthusiasm at the time for a single oil and gas champion. Second, when confronted with determined opposition from Sechin, Putin was ultimately willing to drop the plan. Third, from this point on Putin increasingly came to accept the presence of multiple players in the gas sector. The result, unintended at the time, has been the rise of domestic competition in the Russian gas industry and a weakening of Gazprom's traditional export monopoly. We return to this topic below.

Ensuring the Next Generation of Gas Supply

Once Miller, acting at Putin's behest, had purged the Chernomyrdin-Viakhirev management and built a new team of his own, he turned to other issues. The most pressing by far was the future supply of gas. As we look ahead, it will help to recall that in the early 1990s Gazprom had a large surplus, but by the end of the 1990s the surplus had turned into a deficit. Each of these phases brought different responses from Gazprom's management. In what follows, we look at the decisions made between 2000 and 2006, when Gazprom faced an impending shortage, leading up to the decision to go to the Yamal Peninsula.

For much of the post-Soviet transition, Russia had been coasting on the inheritance of the Soviet gas industry, thanks to the massive investments made during the final twenty years of the Soviet era. The core of the industry now lay in the northern part of the West Siberian mainland. This major gas-producing region is known as the Nadym-Pur-Taz region, from the names of the three rivers that border the territory. The three supergiant fields located there—Medvezhye, Urengoy, and Yamburg—had been the mainstay of the Russian gas industry for two decades. But the first two had been declining for years, and in 2000 the newest of the three, Yamburg, finally peaked, and production began to fall there as well.

Gazprom relied heavily on these Big Three: in 2001 they still supplied over 71 percent of the company's output. The company had plentiful undeveloped reserves, but it faced rising costs and it lacked capital to develop new fields and bring them on line. Gazprom's strategists were pessimistic: from 410 billion cubic meters in 1999, they forecast that the Big Three would be down to 90 billion cubic meters by 2020. But instead of attempting to slow the decline with additional spending (such as additional drilling and repressurizing existing wells) Gazprom concentrated what resources it had on developing the last of the Soviet-era giants, Zapolyarnoye, while delaying a decision on its next move.

The ultimate prize lay well to the north of the Big Three, on the Yamal Peninsula. It contained a huge treasure trove of gas, plus an abundance of condensate, located along the coast of the Arctic Sea, jutting several hundred miles northward into the Kara Sea. In Soviet times, it had been explored but remained undeveloped.[35] All told, twenty-five large fields had been discovered on the Yamal Peninsula, with recoverable reserves exceeding 10 trillion cubic meters—enough to supply a large part of Russia's and Europe's gas demand for a generation.

But working on the Yamal Peninsula was a big step up in difficulty and cost. The first problem was the challenges of the climate and the landscape. As one travels north on the railroad that now links the mainland to the peninsula, one crosses an invisible climatic and geological divide. Just under the lakes and swamps that form during the brief Arctic summer, the ground underneath remains permanently frozen (hence the name "permafrost"), which vastly increases the difficulty and cost of laying pipelines. Gravel and stone for roads and drill pads become scarce. The ice-free season for shipping in supplies shrinks to a few weeks. Work frequently stops in winter, when winds reach gale force and the temperature drops to –50° F and lower.[36] Once one moves up the coast, to the northwest, the going gets even tougher.[37] There, on the exposed western coast, where the biggest supergiants are located, wind-driven tides flood the flat landscape with freezing seawater. Intermittent permafrost becomes continuous, and instead of solid ground there is a frozen mixture of one part sand to four parts of ice, shot through with salt water. At greater depths one encounters "cryopegs"—lenses of liquid salt water that slide under pressure, further weakening the load-bearing capacity of the soil. As a Soviet-era official once famously remarked, "Urengoy is 'little flowers'" *("Urengoy—eto eshche tsvetochki")* in comparison with the Yamal Peninsula.

For a generation, the Yamal Peninsula had been the Holy Grail of the Russian gas industry, forever beckoning and forever receding. In 1988, Chernomyrdin, then still the Soviet gas minister, announced that Bovanenkovo, the biggest field on the peninsula and the nearest to the mainland, would begin commercial production in 1991. But as the Soviet central planning machine weakened and then collapsed, funding and supplies for Bovanenkovo collapsed with it. Gazprom tried to keep the project alive, but by 1994 it was clear that it had neither the funds nor the demand to proceed, and the Viakhirev management postponed the project. Between 1994 and 2002, the Yamal Peninsula was largely abandoned, while studies and desultory debate continued inside Gazprom over the best way to tackle it. It was only in 2002, at President Putin's direct request, that Gazprom returned to the subject with an economic feasibility study for the development of Yamal.

Thus, the essential obstacles at Yamal were logistical and economic rather than technological, and they particularly came down to transportation. There were two parts to the problem—getting equipment and manpower to the fields, and then getting the gas and liquids back out to market. The first problem required a railroad, which did not yet exist. Before large-scale construction could begin at Bovanenkovo, a 520-kilometer railroad connection from the existing railhead at Obskaya (near Labytnangi, on the western bank of the Ob River) had to be completed. Construction of the railroad had begun in 1986, in the last five-year plan of the Soviet era, but stopped when the Soviet Union broke up. As late as 2006, after twenty years and a reported $1 billion in investment, only 268 kilometers of track had been laid.

The railroad, like many Viakhirev-era assets, had fallen into the hands of local notables, who had grown rich on the state subsidies for construction. The contractor on the project, Yamaltransstroy, had been privatized in the early 1990s. A father and son, Vladimir and Igor Nak, owned 41 percent of the company, and there was a scattering of outside investors, mostly connected with the local construction industry. The state owned 10 percent, which it had twice unsuccessfully attempted to sell at auction. The Naks were powerful, well placed in the local leadership of the pro-Putin United Russia party. In May 2005 Putin approved a plan to accelerate construction of the railroad, but the Naks replied that additional funding would be needed before the line could be completed. Even if all the necessary money was available immediately, they said, construction would require five more years. In short, progress on the Yamal Peninsula required clearing the Naks out of the way. Persuasion from the Kremlin was duly applied, and work finally resumed in late 2006 at a stepped-up pace.[38]

Even more problematic was the choice of a route for evacuating the oil and gas produced. Over the years, Gazprom considered and rejected a number of options, ranging from developing two other fields up the coast, Kruzenshtern or Kharasavey, as LNG projects, to building pipelines southeast down the length of the peninsula and under the Ob River to connect to the existing system at Yamburg. The most ambitious pipeline route would go southwest across the Baydaratskaya Gulf and around the northern end of the Ural Mountains to connect to the main network of trunk lines at Ukhta. For years Gazprom's planners and engineers had debated the choices.[39] Ultimately, they chose the third option. Even though it was the most costly, it was by far the shortest route to the European market, and this decision has played a major role in making Bovanenkovo today the prime source of low-cost gas for Europe.

These, then, were some of the obstacles facing Gazprom as it considered its supply strategy in the first half of the 2000s. Initially the Yamal Peninsula simply appeared beyond reach, both financially and logistically, and despite Putin's calls in 2002 for Gazprom to move north, the final decision to undertake the conquest of the peninsula was not taken until 2006.[40] In the meantime, however, with its core assets declining, how was Gazprom to meet demand?

A Fateful Turn: Gazprom and the Independents

Until the mid-2000s, Gazprom's monopoly position in the Russian upstream and domestic market had been unchallenged. As the gas analyst James Henderson describes it:

> Gazprom dominated the Russian gas sector to such an extent that the concept of "independent" third-party suppliers was a misnomer. Any producer of gas in Russia, or supplier of imports into Russia, relied on Gazprom to allocate room in the Annual Gas Balance, to provide space in the pipeline, and usually to buy the gas itself. Russia's state-owned gas company enjoyed an effective monopoly as distributor of gas to the domestic and export markets.[41]

However, beginning in the late 1990s three things gradually changed the situation. First, as we have seen, Gazprom's legacy assets began to decline. Second, gas demand began to increase, as the economy started to recover from the post-Soviet decline, and the growth continued after the crisis of 1998–1999. Third, as already observed, Gazprom lacked the capital to undertake a major investment in the Yamal Peninsula, and it repeatedly postponed the beginning of the northern offensive. For these reasons, as the 2000s proceeded Gazprom's supply balance was coming under growing pressure.

Under its new management, Gazprom was internally divided over how to proceed, and it turned first to interim solutions. The first was to replace the declining output from the Big Three with new gas from Zapolyarnoye, plus a handful of smaller satellites, located in the same Nadym-Pur-Taz triangle of West Siberia. The second was to rely on increased imports from Central Asia, chiefly Turkmenistan. But it was the third that proved fateful, a decision that has had momentous consequences for the future of the company: to meet domestic demand, Gazprom turned to "independent producers."[42]

The independents at the beginning of the 2000s were a diverse collection of mostly oil companies that produced associated gas, as well as a handful of small private gas companies. As Henderson observes, to call them "independent" was a misnomer. They produced small volumes that Gazprom accepted on sufferance in its pipeline system and sold to domestic consumers. As demand increased and real domestic prices began to rise, the independents managed to realize small margins, but they were essentially at Gazprom's beck and call. By inviting them to supply more gas to the domestic market, Gazprom was repeating a familiar maneuver, just as it had done in directing Central Asian gas to the Ukrainian market in the 1990s: Gazprom aimed to preserve its export flow to Europe while consigning the Russian independents to the less attractive domestic market.

Gazprom thought it was on safe ground: after all, it controlled the pipeline system, and no independent could deliver gas to a customer unless Gazprom granted access. Although third-party access was recognized in Russian gas law, there was no way to enforce it. Gazprom had only to argue that there was no room in its system and it could not provide delivery, and who could contradict it, since information on the capacity of the system was a corporate secret?[43] Thus, initially the policy of relying on the independents seemed a secure bet, and throughout the first half of the 2000s their output was allowed to grow steadily. The day when Viakhirev snapped at independent producers—including the oil companies—and say "they produce more stink than gas"—was now a distant memory.

In retrospect the policy of relying on the independents proved to be a strategic mistake for Gazprom. In the early years all seemed to go well. From only 45 billion cubic meters in 1999, the independents' output grew slowly. But by 2005, the threat could already be glimpsed: in that year, the combined gas output of the non-Gazprom producers reached 94 billion cubic meters, or about 15 percent of the Russian total of 641 billion cubic meters, and it was now growing fast. Moreover, the independents were actually making money. This undermined Gazprom's claims that the domestic market was a money loser and thereby weakened the foundation of its gentleman's agreement from the previous decade with the Russian state that it should continue to be given free rein in the export market in exchange for continuing to deliver gas at a loss to the domestic market.[44]

By the end of 2005 the independents and Gazprom had come to a crunch point. The three largest independents—LUKoil, Novatek, and TNK-BP—announced that they planned to produce, as a group, over 100 billion cubic meters of gas in 2006 (in comparison with 552 billion cubic meters for Gazprom). But Gazprom announced that it would accept an increase of no more than 1.5 billion cubic meters of gas into its system.[45] It banked on forcing the independents to sell their gas to Gazprom "at the top of the pipe"—that is, at the entrance to the Gazprom pipeline system and at a low price. Alternatively, Gazprom declared that it was prepared to grant access, provided the independents shared in the investment costs of building additional pipeline capacity. What it was not prepared to do was lower transportation tariffs for the independents to broaden their geographic reach in the domestic market. In other words, the latter would carry the growing burden of servicing the domestic market, while Gazprom would continue to enjoy its export monopoly.

Yet in the end, Gazprom was forced to make concessions—partly as a result of political pressure from the Kremlin, and partly because of the simple fact that it did not have enough available gas to meet all of the

Table 10.1 Gas production in the Russian Federation, 2010–2018 (billions of cubic meters)

	2010	2011	2012	2013	2014	2015	2016	2017	2018
Total for Russia	650.7	670.5	654.5	668.2	640.2	635.5	640.2	691.1	725.2
Gazprom	508.7	510.1	478.8	476.2	432.0	405.9	405.6	470.8	497.6
Others	142.0	160.5	175.7	192.0	208.2	229.6	234.6	220.3	227.6
Oil companies	58.2	63.7	66.7	76.2	81.6	83.4	87.1	93.7	92.9
Joint ventures or PSAs	37.7	28.7	36.9	37.8	37.8	27.3	36.4	19.0	26.2
Independents	46.1	68.1	72.1	78.0	88.9	118.9	111.1	107.5	108.4
Novatek	37.8	53.5	51.3	53.0	53.6	51.9	49.9	45.5	49.9
Other independents	8.4	14.6	20.9	24.9	33.3	67.0	61.2	62.0	58.5

Data source: IHS Markit.

Notes: Numbers may not sum to totals due to rounding. The table uses the Russian measure of 8,850 kilocalories per cubic meter (in gross calorific value). Production excludes flared volumes. PSA = production-sharing contract.

demand. Yet Gazprom still did not see the problem as life threatening. Its long-range planners anticipated that by 2030 the independents' share of total production would be no more than about one-quarter.[46]

From that point on, however, the barn door was open. The independents' surge continued into the next decade (see Table 10.1), and by the second half of the 2010s, their share of the domestic market exceeded 50 percent. Their secret lay in the fact that they had learned how to make money in the quality end of the domestic market—the part that was located closest to production and paid the highest prices (that is, when sold to large industrial enterprises)—leaving Gazprom with the loss-making dregs and essentially making it the swing producer serving the most distant and lowest-paying end of the market.[47] In addition, the independents developed an increasingly profitable product: condensate. The leader in this trend was a remarkable start-up called Novatek.

The Rise of Novatek

The rise of Novatek as a leading independent, and more recently as the leading player in Russia's LNG industry, is closely tied to the name of its founder, Leonid Mikhelson. Mikhelson started his career as a builder of gas pipelines. When Mikhail Gorbachev's reforms began in the late 1980s, Mikhel'son was the head of a pipeline construction trust in Kuybyshev (now Samara): "My father was the head of the Kuybyshevtruboprovodstroy, the largest construction unit in the Gazprom system. He took me everywhere with him, wherever the company built pipelines. I used to go to gas and oil fields before I went to school."[48] He quickly realized the opportunities presented by the Chubais privatization reforms. In 1991 his company, Kuybyshevtruboprovodstroy, was one of the first Soviet enterprises to privatize, under the name Nova—subsequently changed to Novatek.

Since there was no money to be made from pipe building in those years, in the mid-1990s Mikhel'son went into the gas business instead, focusing on Tyumen Province. He was familiar with West Siberia (his first job after engineering school had been to work on the Urengoy-Chelyabinsk pipeline, where he quickly rose to chief engineer), and through connections in Tyumen he began acquiring licenses to undeveloped gas fields. At that time licenses to even relatively large prospects could be had virtually for the asking, particularly from local geological groups. Because there was so much gas in the ground, and because there were as yet no pipes connecting it to the main pipeline system, the gas was virtually worthless. The fact that Novatek concentrated on buying undeveloped opportunities rather than Soviet-era producing properties was Mikhel'son's trademark, and it helped him avoid problems with Gazprom later on, when Gazprom began recovering assets previously controlled by it.

The business concept developed by Novatek and the other independents in the early 2000s rested on the fact that the independents were allowed to sell their gas at unregulated prices. This created the possibility of selling gas at an attractive profit to factories and municipalities along the main pipeline routes—particularly in the situation that emerged in the early 2000s, when Gazprom was short on gas and was limiting deliveries to industrial consumers.[49] In addition, Novatek kept tight managerial control over its contractors and its own operations and consequently was a low-cost producer.[50] Mikhel'son also proved adept at cultivating relationships, particularly with local political leaders such as the deputy governor of the Yamal-Nenets region Yosif Levinson, who joined the board of Novatek.[51] Mikhel'son was also successful at cultivating smooth relations with Gazprom, settling any potential disputes with well-timed compromises and joint projects.[52] It helped that Novatek's main producing assets were located close to Gazprom's declining fields, which enabled Novatek to take advantage of available space in the pipeline system with a minimum of friction with Gazprom.

In the early 2000s, Novatek also began carving out a profitable niche as a producer and exporter of condensate. This was an increasingly attractive business opportunity. As one moves deeper in West Siberia's sedimentary layers, natural gas grows wetter—that is, the concentration of condensate increases. Gazprom, accustomed to working in "dry gas" formations, had traditionally treated condensate almost as a waste product. The amounts it produced were gathered up in a pipeline and delivered to a gas-processing facility located in Surgut. As the independents' potential resources of condensate increased, this arrangement was increasingly inadequate for them: the Surgut facility was too small, and it was under Gazprom's control.

Thus, the challenge for Novatek's management was to find its own outlet to stabilize and export condensate to Europe, where refineries saw it as a valuable substitute for declining North Sea production. In 2005, Novatek achieved a breakthrough by building its own processing plant for condensate, followed by an enlarged railroad connection to a marine terminal on the Barents Sea.[53] It was an early preview of the capacity for innovative risk taking that the company demonstrated a decade later in the development of LNG. It was also a welcome buffer for the company. When gas demand declined during the recession of 2007–2009, income from condensate (even at lower international prices) enabled the company to maintain its cash flow.

Novatek's success in developing and selling condensate and its plans for LNG attracted the interest of Total, the French oil and gas company. At this time Total had limited exposure in Russia, apart from a small joint venture in the Russian northwest (Nenets Okrug). Its major foray into Russia, an extended negotiation with Gazprom to develop the offshore Shtokman field, had ended in frustration when the project was dropped (see below), and Total was looking for a partner. By this time Novatek was the largest independent gas producer in Russia: in 2010, its production of 38 billion cubic meters represented 10 percent of the domestic supply.

The two companies were familiar with each other, since Total had come close to acquiring a stake in Novatek in 2004. In 2011, with Putin (then premier) looking on, Christophe de Margerie, the CEO of Total, and Mikhel'son, Novatek's founder, signed an agreement under which Total acquired a 12 percent interest in Novatek.[54]

As part of the agreement Total also bought a 20 percent stake in Novatek's subsidiary, Yamal LNG, to develop the South Tambey field on the Yamal Peninsula, which held 1.25 trillion cubic meters of gas reserves and was strategically located on the northeast coast of the peninsula. The license to the field had had a succession of owners since the early 1990s before being acquired by Novatek in 2009. Its potential as an LNG project had been apparent from the beginning, but earlier development plans by previous owners had targeted the North American market pre-shale, when it was thought that the United States was going to be short of gas. The innovation of Mikhel'son and his partners was the vision that LNG from South Tambey could be exported both east and west—west to Europe during the winter, and east to Asia during the summer via the Northern Sea Route, with the assistance of nuclear-powered icebreakers. This was a daring concept, since the eastward route through the Arctic had never been used for LNG. Nevertheless, from 2014 on, thanks to the partnership between Novatek and Total and the strong support of the Russian government—which provided tax breaks and financed the development of a port at Sabetta on the Yamal Peninsula—progress was rapid. In 2017, the first LNG tanker, named after the late CEO of Total, Christophe de Margerie, who had recently died in an airplane accident in Moscow, was dispatched to market in Asia.[55]

Gazprom's Response to Gas Liberalization in Europe

We turn now to the response of Gazprom to the changes taking place in the European market. Gazprom's managers under Viakhirev had reacted with alarm to the first EU Gas and Power Directive in 1998 and the beginnings of liberalization in Europe. In this first phase, the emphasis in European liberalization was very much on cutting gas and power costs to the final consumer. To anxious Russian ears this sounded very much like a plan to gang up on producers to force down supply prices.

Viakhirev's first reaction was to bluster. Fuming about "European foolishness," he suggested that Gazprom would be more than willing to see its market share decline rather than sell its gas cheaply "at the bazaar" *(razbazarivat')* to support European liberalization and low gas prices. A more businesslike—though equally alarmed—reaction came from Yurii Komarov, GazpromExport's leading expert on the European gas trade, who worried how the next generation of gas development could take place in the absence of the familiar structure of long-term take or pay supply contracts to provide the foundation for the scale of investment required. This concern was widely shared at the time by most West European gas companies, so Gazprom in this regard was not out of step with mainstream thinking in the European industry.

Gazprom's initial alarm subsided when the Miller management took over, as it became apparent that incumbent players in Europe, especially in Germany, were successfully resisting change in the continental European gas market. European market reforms, the Russians concluded, was something they could safely ignore. In this respect, they proved to be mistaken. Even as Gazprom lowered its guard, the European Commission was preparing its Second Gas Directive, and was preparing to undertake the Energy Sector Inquiry, which ultimately led to the Third Directive and a new and more serious round of conflict between the Russians and the Commission.[56]

However, in parallel, by 2002 GazpromExport's strategists had also begun to see an opportunity in the evolving European spot market, especially in Great Britain. Gazprom began experimenting with spot sales to Great Britain in alliance with Wintershall. By 2004–2005, Gazprom had developed a trading structure in Great Britain and was poised to take advantage of the opening presented by the decline of North Sea gas production. In 2004, Yurii Komarov stated that Gazprom was considering renting gas storage in Great Britain and planned to sell as much as 8–10 billion cubic meters on the spot market in Great Britain by 2010.[57]

Gazprom Adapts: Two Faces of Gazprom

As we observe the behavior of Gazprom in the period from 2000 to 2008, we see two faces. There is the conservative Gazprom, which continued to be dominated by the engineering culture of the Soviet period that focused on dry gas and pipelines. Broadly, this is the face of Gazprom that is still dominant on domestic matters and for the most part sets overall investment policy. But this is far from the whole story. The parts of Gazprom that specialize in exports and foreign relations have a different history and culture. This is particularly the case with GazpromExport, which as we have seen was captured by Chernomyrdin. Even today, GazpromExport sees itself as a distinct organization from the rest of Gazprom, with skills and knowledge that the rest of the company does not have. GazpromExport prides itself on its sophistication about global gas trends and markets. It is, on the whole, the progressive end of Gazprom, the first part to become aware of challenges and opportunities in the global market and the first to adapt to them.

So which face prevails on any given question? In the next sections of this chapter we look at two early cases. The first is the creation of GM&T (Gazprom Marketing and Trading). The second is the battle between two internal factions over the development of Gazprom's policy toward LNG. In the first case, the progressives had a clear field and made the most of it. In the second, the conservatives won after a sharp conflict.

Gazprom Goes for the Market:

There are specialized departments in Gazprom that have an explicit mission to be innovative. The most striking example is GM&T, a wholly owned subsidiary of GazpromExport. GM&T was created in 1999, at a time when the British spot market was growing fast. It seemed to offer a golden opportunity for Gazprom to fulfill its decade-long ambition of becoming a direct player in distribution and sales. Its main source of gas was via Gazprom's 10 percent capacity in the Interconnector pipeline between Belgium at Zeebrugge and Great Britain at Bacton.

The challenge was to learn how to trade gas in the emerging liberalized gas market in Great Britain. At first, with a total staff of two, GM&T was hardly more than a beachhead and lacked trading skills. Andrey Mihalev, who went on to become managing director of GM&T, had a degree in petroleum engineering from the prestigious Gubkin Institute and held a postgraduate degree from the All-Union Academy of Foreign Trade. In other words, he was among Gazprom's best and brightest, but he had never traded gas on a spot market. In 2001, Mihalev was joined by GM&T's first foreigner: Daniel Gornig was originally from Germany, where he had earned a graduate degree in business. GM&T's London base quickly grew to have a staff of over three hundred people, all but a handful of them foreigners. They were veteran traders of leading European oil and gas companies, and some of them were quite senior veterans, such as Keith Martin from Shell and Frédéric Barnaud from Gaz de France.[58]

As the British and continental spot markets grew, GM&T grew along with them. Over the next five years, it moved quickly to obtain the necessary licenses and memberships to trade gas on Europe's nascent spot market. In 2001, GM&T acquired a shipper's license to use the British National Transmission System, the network of pipelines supplying natural gas from coastal gas terminals to forty power stations, industry, and twenty million British households. In 2002 it marketed its first gas on the UK National Balancing Point, Great Britain's gas hub. In 2003 GM&T gained access to the Zeebrugge hub in Belgium, which enabled it to move gas between Bacton and Zeebrugge via the Interconnector. In 2005 GM&T entered the Dutch spot market on the Title Transfer Facility hub. In 2006 came the first direct sales to end users in France and GM&T's first trades on a German hub.

Meanwhile, GM&T's business in Great Britain was expanding rapidly. In March 2006 Alexander Medvedev, Komarov's successor as head of Gazeksport (as it was then still called), proudly reported to the Gazprom board of directors that GM&T had sold 4 billion cubic meters on the British spot market the previous year and had recently acquired six hundred retail consumers in Great Britain, among them the William Hill betting shops, Debenhams department stores, and the Sunderland football club.[59]

Medvedev saw GM&T as an instrument to carry out his mission of learning "new gas"—first the spot trade and then, as Gazprom became more proactive, LNG. But London was only the headquarters. Soon GM&T opened offices in Houston and Singapore, and with nine hundred employees it was trading LNG around the world.

As GM&T's missions and operations expanded, Medvedev appointed a protégé, Vitaly Vasiliev, as worldwide CEO in 2004. Vasiliev at the time was a fast-rising star. He had majored in international economics at the Moscow State Institute for International Relations, Moscow's training ground for international diplomacy, and in 2003, the year before he joined GM&T, he earned a master's degree in management at Stanford University. Over the following thirteen years, under Vasiliev's leadership, GM&T built a cadre of experienced executives who felt at home in the UK and European trading environments.[60]

Medvedev and Gazprom's management now faced a strategic choice. Since the 1990s Gazprom had opened trading houses all over continental Europe. But the trading-house business model was proving unprofitable, and it was persistently dogged by rumors of corruption. Medvedev argued forcefully for a switch to outright acquisitions of foreign subsidiaries as the fastest and safest way to build business in Europe. His first target was Centrica in Great Britain—the sales arm of the now defunct British Gas, with seventeen million gas and electricity customers.

But at this point, Medvedev's strategy ran up against the suspicions of the British. Prime Minister Gordon Brown warned that his government would look hard at any acquisition by a Russian company.[61] Parliament was equally skeptical. On a sweltering day in July 2006, Vasiliev, accompanied by his head trader, Keith Martin, attempted to persuade the Trade and Industry Committee of the House of Commons that Gazprom was not a geopolitical arm of the Russian state and was as commercially driven as any other company with a high degree of state control, such as Electricité de France or Statoil. Gas from distant West Siberia was perfectly safe and economic, they insisted,[62] and Gazprom's ambitions were limited to an ultimate share of only 10 percent of the British market.

All of this was to no avail. It did not help that the first cutoff of gas to Ukraine (discussed in Chapter 11) had occurred the previous January, and by this time British-Russian relations, despite the heat of that summer, had turned distinctly cold. But there was worse ahead. In November 2006 a former Russian FSB agent, who had sought political asylum in Great Britain, was poisoned under circumstances that pointed to the involvement of the Russian secret services. By the beginning of 2007, Russian-British relations were in a deep freeze.

In retrospect, the abortive attempt to acquire Centrica marks the high point of Gazprom's ambitions to become a fully integrated player in Europe's evolving gas market by taking equity positions in major European companies. The experiment has not been repeated since.[63] Yet the historic contribution of GM&T was that over the years it has been an important vehicle for Gazprom to learn how to operate in a liberalized market and sell new products, eventually including LNG.

The Battle between LNG and Pipeline Gas

LNG in the early 2000s was a concept that was at once familiar and distant to Gazprom's managers.[64] Russia (and the Soviet Union before it) had watched the LNG industry grow up beyond its borders and at intervals had studied the idea of developing part of its gas resources as LNG. The Soviets had made plans for building LNG facilities in the Far East as early as the mid-1960s. Soviet scientists had even developed cryogenic technologies (cryogenics was considered a military industry and was treated as part of the military-industrial complex). However, their application in the gas industry had remained largely experimental. Yet the Soviets were attracted by the idea of shipping LNG from Russia to the United States. The North Star project—a plan to liquefy West Siberian gas and ship it to North America—was proposed in the 1970s to do this, and much basic study was carried out then. However, the emergence of a temporary gas bubble in the United States at the time, coupled with the Soviet invasion of Afghanistan and the subsequent cooling of US-Soviet relations, halted the project in the 1980s.[65]

Some Gazprom planners had long thought of using LNG to export gas from the Yamal Peninsula. In the mid-1990s, V. E. Brianskikh, head of long-range projects for Gazprom, had commissioned a study of the possibility of building a floating gasification facility to develop one of the Yamal gas fields (Kharasavey) as an LNG project, but the project was ignored by Viakhirev and his colleagues, who had more urgent matters on their minds.

Even after the arrival of Miller and the new management team at Gazprom, the concept of LNG was still a low priority. In the spring of 2003

Miller mentioned general plans for entering the LNG business, including in a speech to the European Parliament in March and a presentation at the Tokyo World Gas Conference in May, where Gazprom showed up in force with a delegation of over a hundred people.[66] By mid-2003 Miller had been persuaded that LNG needed to be part of Gazprom's supply strategy and ordered the creation of a special LNG committee to study the issue of global LNG markets and Gazprom's potential role in them. By the end of 2003 the committee had come up with two key recommendations. The United States was identified as the key target market for Russian LNG. At that time—ironically, in light of later developments—it appeared that the United States would soon run short of gas, and it seemed to be the future market of choice for LNG. As for the source, the committee pointed to a large deepwater field in the Barents Sea, 550 kilometers out to sea from Murmansk, called Shtokman.

Since Gazprom had no experience in such projects, a joint venture with foreign companies seemed indispensable. Talks had already gone on for several years over alternative strategies for Shtokman, including an undersea pipeline to the Russian mainland near Murmansk and on to Europe, but there seemed to be no way to make the field economic, and the talks had been suspended. However, at the end of 2002 and in early 2003, discussions with foreign companies on Shtokman resumed. This time the new theme for Shtokman was LNG and exports to the United States. By this time two things had changed. The United States was now short of gas—or so it seemed at that moment—and LNG technology had matured, making it cheaper by half than twenty years before. The American market looked like a sure thing. Indeed at that same moment the Americans were preparing to build large numbers of regasification terminals, in anticipation of the LNG imports that would be needed soon. The idea was further buttressed by discussions in the United States of a "U.S.-Russian Energy Dialogue."

Inside Gazprom's management, one senior manager emerged as a champion of LNG and the Shtokman project. Alexander Ryazanov, the vice-chairman in charge of liquids and other noncore products, was enthusiastic about the prospects for the new technology and correspondingly skeptical about the economic viability of pipelines from the Yamal Peninsula. In the first half of 2004, at his urging, Gazprom publicly endorsed the need to form a partnership with Western companies to develop LNG. Thus, in May 2004, Ryazanov first announced Gazprom's intention to complete negotiations with foreign partners on Shtokman and to form a consortium by the end of the year. Meanwhile, according to Ryazanov, Yamal had been shelved in favor of Shtokman.

But Ryazanov spoke too soon. His enthusiasm for LNG (and indeed for liquids generally) ran counter to the views of the more traditionally minded senior technical gas specialists, of whom the leader was the deputy chairman in charge of conventional production, Alexander Ananenkov. In addition, Ryazanov was a relative newcomer to Gazprom, whereas Ananenkov had risen from the core of the Soviet-era gas and pipeline business in West Siberia. There were bad feelings on both sides. Ananenkov and the technical wing of Gazprom's management had no intention of giving up on the Yamal Peninsula and betting the future of the company on an untried technology at a remote and complex offshore project. But in Ryazanov's view, members of the old guard were simply too slow in perceiving the threat from the rapidly declining costs of LNG. They thought of LNG as something long-range when in reality, Ryazanov warned, it could become a threat to Gazprom's traditional pipeline business.[67]

By the summer of 2004, Gazprom remained divided over LNG. This led to a compromise division of duties within the company. It would take years for Gazprom to develop the necessary technical expertise to produce LNG, but in the meantime it could learn how to sell it. One way to get into the LNG market quickly was through swaps of Russian pipeline gas to Europe in exchange for LNG cargoes that could be redirected to the United States. This part of the program was entrusted to GazExport and its new head, Medvedev, with the endorsement of Ryazanov and Miller. The idea was to purchase several cargoes of LNG from Western companies and deliver them to North America by 2005. In the fall of 2004 Gazprom signed a series of memoranda of understanding with foreign companies including Total and two Norwegian companies, Statoil and Norsk Hydro.

Another way of gaining experience with a new and unfamiliar technology was to partner with an established player. Shell, at that time the most experienced LNG company in the world, had created a consortium to develop oil and LNG on Sakhalin Island—on the basis of a production-sharing contract with the Russian government—in which Shell was the majority shareholder and the operator. In the early 2000s, Putin convened an international conference on Sakhalin, at which he gave what seemed to be a ringing endorsement to foreign investment based on production-sharing contracts, especially in the Far East. But at the same time, he reiterated the principle of equal access for Russian and foreign companies and the importance of using Russian personnel and equipment—a principle that had been part of the original PSA's in the 1990s but had been neglected in practice. Responding to Putin's message, Gazprom signaled that it wished to join the Shell-led consortium. Over the next several years, the two companies conducted negotiations, with no result. But by 2006, it became clear that the Kremlin was prepared to bring pressure against Shell to surrender its majority stake. Taking advantage of a series of cost overruns at Sakhalin, the government announced that it intended to sue Shell for over $30 billion in environmental damages and that it was considering bringing criminal charges against Shell executives. For Shell and its Japanese partners, it was clear that the game was up. On December 21, 2006, they signed an agreement to turn majority control over to Gazprom for $7.45 billion in cash.[68] The environmental suit evaporated.[69]

In acquiring a majority stake in the Sakhalin project, Gazprom was acquiring a ready-made platform through which to learn the LNG business. Shell had spent the previous several years signing supply contracts for Sakhalin LNG all over the Asian rim of the Pacific Ocean. Its mastery of the technology of producing and shipping LNG made it an ideal model for the inexperienced Gazprom, as well as a basis for further expansion. But for the next decade Gazprom showed little sign of following through. To this day, Sakhalin is Gazprom's only source of LNG.

The same impression of low priority came from Gazprom's internal structure. Until early 2005 there was no specialized LNG unit inside Gazprom, suggesting that the company was still not quite serious about LNG. This changed in April 2005, when Gazprom announced the creation of an LNG department. Yet even then there was an air of ambiguity about the decision: it was announced that the new department would be headed by Yurii Komarov, who had just resigned from the Gazprom management team to make way for Medvedev when the latter was promoted to head of the company's foreign relations in addition to his duties as head of Gazeksport. Komarov was one of Gazprom's most knowledgeable and energetic managers and had recently shown keen interest in LNG, but it was hard not to think that his appointment was a form of honorary retirement and that the LNG committee would lack clout.

Indeed, LNG remained controversial inside Gazprom. The two camps—the liquid and the dry gas factions—were now divided not only over LNG but also over the role of oil in Gazprom and the role of liquids generally.[70] In the dispute over the takeover of Rosneft, Ryazanov had been firmly on the side of acquiring the oil company, and when the plan was defeated, he was equally in favor of a consolation solution: to acquire another oil company, Sibneft, instead. The old guard wanted to split up Sibneft (now renamed Gazprom Neft) and distribute it among the various departments of Gazprom. But Ryazanov defeated Ananenkov and the dry-gas faction on that plan as well. Indeed, he was so closely identified with the acquisition of Sibneft that Miller named him to head it. But at this exact moment, Ryazanov became embroiled in an unrelated scandal involving assets in southern Russia. Suddenly his political position within the company was weakened. The old guard seized the opportunity and prevailed upon Miller to dismiss him. By mid-November 2006, Ryazanov's contract had been revoked, and he left the company.

This unexpectedly sudden outcome left Ananenkov in sole command of the field, and the company decided to take the plunge into Yamal instead of giving high priority to LNG or Shtokman. From 2006 on, the Shtokman project lost ground inside Gazprom, even though negotiations with Western companies continued. Indeed, by 2007 the two foreign partners had teamed up with Gazprom in a joint operating company. But Gazprom was increasingly ambivalent. Over the next three years the partners quarreled over key issues such as the location of the planned liquefaction plant and the technology to be used at the field. The plan had been to reach a final investment decision by the spring of 2010, but when the time came the partners were no closer to an agreement. By the summer of 2012 both sides gave up, and Shtokman was postponed indefinitely.[71]

The battle between the LNG camp and the Yamal-firsters was in effect a battle for supremacy between two cultures and two organizations within a single company. In this case the progressive side lost and the conservative side won, but what remained was a compromise that gave control over selling third-party LNG to GazpromExport's subsidiary, GM&T. Note in passing that the decision to give priority to Yamal preceded the first word of the shale gas revolution in the United States by about four years. The news of US shale gas did not begin to penetrate Gazprom until about 2010. Thus, it was not US shale gas that defeated the LNG faction; it was ultimately the view that the future of the company rested with West Siberian pipeline gas, a conviction that represented continuity from the Soviet times. The "shale gale" in North America is our next topic.

"Slanets-Uraganets": The "Shale Gale" Reaches Russia

In 2007–2008 the US gas industry awoke to the fact that something was happening to domestic gas production, which up to that point had been declining. It was the beginning of the extraordinary "shale gale," the revolution in gas technology that in only a few years brought the United States back to the position of the number-one gas producer in the world.[72]

But the news of the turnaround in the US gas industry did not reach Russia immediately. When it did, it took time to penetrate successive tiers of Russian policymakers, depending on their distance from the gas sector. Gas export professionals were aware of the news coming from the United States as early as 2008, but it was not until October 2009, coinciding with the World Gas Congress in Buenos Aires, that the Russian media began to take notice of the shale gas phenomenon and its implications for Russia's LNG strategy, and that Russian gas policymakers began commenting on it publicly.

Their initial reaction was dismissive. In Buenos Aires Miller restated Russia's long-range target for LNG production and sales of 80 million–90 million tons per year by 2020, and during this period, Gazprom officials downplayed the significance of shale gas. Medvedev, the head of GazpromExport,[73] stated, "Producing shale gas requires constant drilling; if you don't drill, production falls immediately by 80 percent. I am convinced that LNG will remain absolutely competitive in the [global] market."[74]

The news of the shale gas revolution in the United States finally reached the senior levels of the Russian government in mid-February 2010 from a source outside Gazprom, at a high-level conference on energy chaired by President Dmitry Medvedev. Significantly, the speaker designated to make the presentation on shale gas was Viktor Vekselberg, at the time head of TNK-BP's gas business. As Miller looked on, Vekselberg described the advent of shale gas in the United States as nothing less than a "revolution," with far-reaching implications for Russia's entire gas export strategy—not only for Russia's traditional gas market in Europe, but also in the East.[75]

As the scale of the shale revolution was confirmed, Gazprom's interest in the Shtokman project finally evaporated, although, as noted above, its interest had dwindled even before the news of US shale had arrived in Moscow. In February 2010, Yurii Komarov, then the chairman of the Shtokman development group, acknowledged publicly that the shale gas revolution had forced "new decisions."

Thus, in the space of a few months the shale gas phenomenon and its possible implications for Europe grew from a distant rumor that was initially dismissed in Moscow to a subject of public discussion and an action item for the top leadership. By April 2010, the Russian media had taken up the shale gale. (Moscow wags, looking for an equivalent to the English expression "shale gale," came up with *"slanets-uraganets"* (the little shale hurricane).[76]

By June 2010, top-level Russian policymakers had had time to digest the shale gas phenomenon and weigh its likely implications for Russian gas exports. At the Saint Petersburg Economic Forum that month, Deputy Prime Minister Sechin delivered a major address on gas before a high-level audience that included most of the CEOs of Russia's and the world's leading energy companies. In a thoughtful and thoroughly documented speech, Sechin argued that shale gas was real, but it was too soon to say whether it amounted to a real revolution. If it was one, it was likely to be confined mainly to North America. Talk of the shale revolution spreading to Europe and causing a long-term oversupply of gas was greatly exaggerated.[77]

The significance of Sechin's speech was that it was the first time a senior Russian heavyweight admitted publicly that Gazprom's strategy of exporting LNG to the United States was dead. Yet even then senior Gazprom decision makers clung to the view that the shale gale would soon blow over. Acknowledging that "we must temper over-optimism with a dose of realism," Alexander Medvedev wrote in a position paper in March 2010 that "once North America works through its modest deliverability surplus created by the price-driven shale boom, supply / demand unbalances will be largely self-correcting, with LNG competing on price with conventional and unconventional domestic gas production."

Other Gazprom spokespersons likewise expressed skepticism about the long-term significance of the shale gas phenomenon in the United States. Shale gas was expensive, they insisted, and economic only for supplying local demand. The sharp fall of US gas prices that followed, the Russian argument continued, had left some of the shale gas pioneers in financial difficulty. In addition, the Russians believed that environmental opposition would slow the spread of shale gas production. In short, they concluded, by 2020 the shale gas phenomenon would have played itself out. It was really only in the fall of 2012, when Putin convened a meeting of the presidential energy commission,[78] where gas export policy was among the top agenda items, that the shale gale story was fully integrated into Russian strategy.[79] Such skeptical views, it is only fair to add, were expressed in the United States as well. However, in the ensuing years the US shale gale has only blown harder, with no end currently in sight.[80]

Gazprom Adapts to "Black Swans," 2008–2016

In February 2019 at an investor conference in Singapore, Oleg Aksiutin, a board member of Gazprom and head of the company's prospective development department, was asked, "What keeps Gazprom managers awake at night?" He replied, "In particular the black swans, and trying to understand [how] we can whiten those swans and expect them to appear."[81] His use of the term "black swans" was revealing: it is a metaphor widely used by scenario writers to describe an event that comes as a major and unpleasant surprise.[82]

In truth, Gazprom has been dealing with "black swans" throughout the last decade. New and unexpected challenges began with the critical years of 2008 and 2009. European gas demand declined as Europe weathered the Great Recession caused by the collapse of the real-estate bubble in the United States and the subsequent financial shakeout. As a consequence, the European gas market was oversupplied, and spot prices began to drop. Compounding the surplus was the arrival in Europe of homeless LNG, diverted from Asia. In 2009 Russian gas exports declined for the first time since their beginning in the 1960s. Russia's market share in Europe, which had been 30 percent in 2007, dropped to 23 percent in 2009. Between 2009 and 2012, Russian gas exports dropped by 12 percent, and they did not regain their precrisis levels until 2016. Russian gas prices were still linked to oil, which made them unattractive to European gas buyers. Looming over these market factors was the political hangover from the interruption of gas supplies to Ukraine in 2006 and 2009, discussed in Chapter 11.

Meanwhile, the rapidly growing role of spot and hybrid markets in Europe was creating a new price benchmark and increased price transparency.[83] Starting in 2009, European gas buyers, increasingly aware of what other buyers were paying and under growing pressure from a shrinking market, began taking advantage of reopener clauses in their contracts with Gazprom to demand lower prices. In public, Gazprom vehemently defended its traditional model—Medvedev once went so far as to denounce the EU Commission–led regulatory developments that underlay Europe's emerging spot-based hubs as "communism"—but behind the scenes Gazprom reluctantly faced up to reality and entered into negotiations to preserve its market share. According to the gas experts Tatiana Mitrova and Tim Boersma, "Analysis of Gazprom's official reports demonstrates a much more flexible negotiating position than has commonly been thought to be the case. In 2013 Gazprom started to implement a new price discount model with so-called retroactive payments, which actually provided customers with partial compensation of the gap between spot and oil-linked prices."[84]

Over the following seven years, nearly all of Gazprom's export contracts with its top 40 buyers were revised, with Gazprom conceding an average discount of 25 percent, introducing a spot component into the price formulas, and relaxing take-or-pay requirements. Gazprom's step-by-step concessions have played a major role in changing the basis for gas prices throughout Europe, as the share of spot- and hybrid-indexed gas supplies went from 20 percent in 2008 to nearly 70 percent in 2015. This led to a sharp drop in Gazprom's export price levels and a decline in its revenues and contributions to the Russian budget.[85] One consequence was that Gazprom cut its capital investments and throttled back the pace of development at Bovanenkovo for the next several years.[86]

This was the situation in 2012, when Putin returned as president. His return, whether coincidentally or not, brought a sharp change in tone in gas policy, as Gazprom's deteriorating financial results led to growing criticism from domestic industrial consumers and the government. The Federal Antimonopoly Service began scrutinizing reports of violations of third-party access by Gazprom, as political pressures mounted on the company. Meanwhile, Gazprom's share of gas production continued to decline, from 85 percent in 2005 to 68 percent in 2017, and the share of domestic sales by independents exceeded 50 percent. A major symbolic turning point for Gazprom came in December 2013, when a law on LNG export liberalization broke Gazprom's monopoly of exports—a development so far confined to LNG but possibly leading to wider repercussions.[87]

Continued Slow Progress on LNG

It took another several years for the Russians to develop an alternative policy for LNG, and when they did, the initiative came from outside Gazprom, and the focus was East Asia, not the United States.[88] Russia's "pivot to the East" has been driven both by political factors—the falling out with the West—and economic factors—the size of the market, particularly in China. A full account of the pivot lies outside the scope of this book, but suffice it to say that when it happened, it was the result of a strong and sustained pressure from Putin, together with the growing ambitions of Russia's two leading independents, Rosneft and above all Novatek.[89] As soon as he returned to the Kremlin as president in 2012, Putin called for a sharp increase in gas investment in East Siberia. He had two main reasons. The first was his growing concern over the lagging economy of the eastern third of the country. But the second was his growing conviction that LNG was the new future of gas. At times, in fact, he appeared to put LNG ahead of pipeline gas. Thus, in October 2012, discussing Russia's Eastern Gas Program with Gazprom CEO Aleksey Miller, he said, "We will create a new gas export channel oriented towards the Asia Pacific . . . and develop LNG options first and foremost." It was, to say the least, a striking change in language, compared to Russia's traditional focus on pipelines. Since that time, Putin's prioritizing of LNG has only increased.[90] In December 2017, he presided over the launching of the first shipment of Yamal LNG. Significantly, it was produced not by Gazprom but by Novatek. At the same time, Putin issued a "List of Orders for the Development of LNG Projects," featuring a checklist of measures designed to stimulate the growth of Russian LNG, with the aim of ensuring the entry of the Russian Federation in the medium term into the ranks of world leaders in the production and export of LNG."[91]

Gazprom, in contrast, remained a reluctant recruit, both on LNG and on the pivot to the East—except for its Gazeksport wing, led by GM&T. It was only after several years of urging by the Kremlin that Gazprom's senior management finally committed itself to a major project in East Siberia, and when it did, it was a pipeline to China (the so-called Power of Siberia pipeline), not LNG. When Putin pushed Gazprom to build an LNG liquefaction plant in Vladivostok, Gazprom held back, and the project was ultimately shelved. To this day, Gazprom has trailed behind its rivals. Its sole production of LNG, as noted above, is from the Sakhalin-2 project, which it took over from Shell in 2006 (see Map 10.1).

There are several reasons for Gazprom's reluctance. One of them is the financial demands of its multiple-pipelines strategy to Europe, aimed at bypassing Ukraine (discussed in Chapter 11). A second is that Gazprom's core expertise remains its traditional business of dry gas by pipeline to Europe. It has taken several years to build a team with experience in East Asia. But perhaps the most important reason is Miller, who has been a consistent voice for traditional approaches and technologies. There has yet to be a fundamental shift in Gazprom's fundamental views on LNG and East Asia.

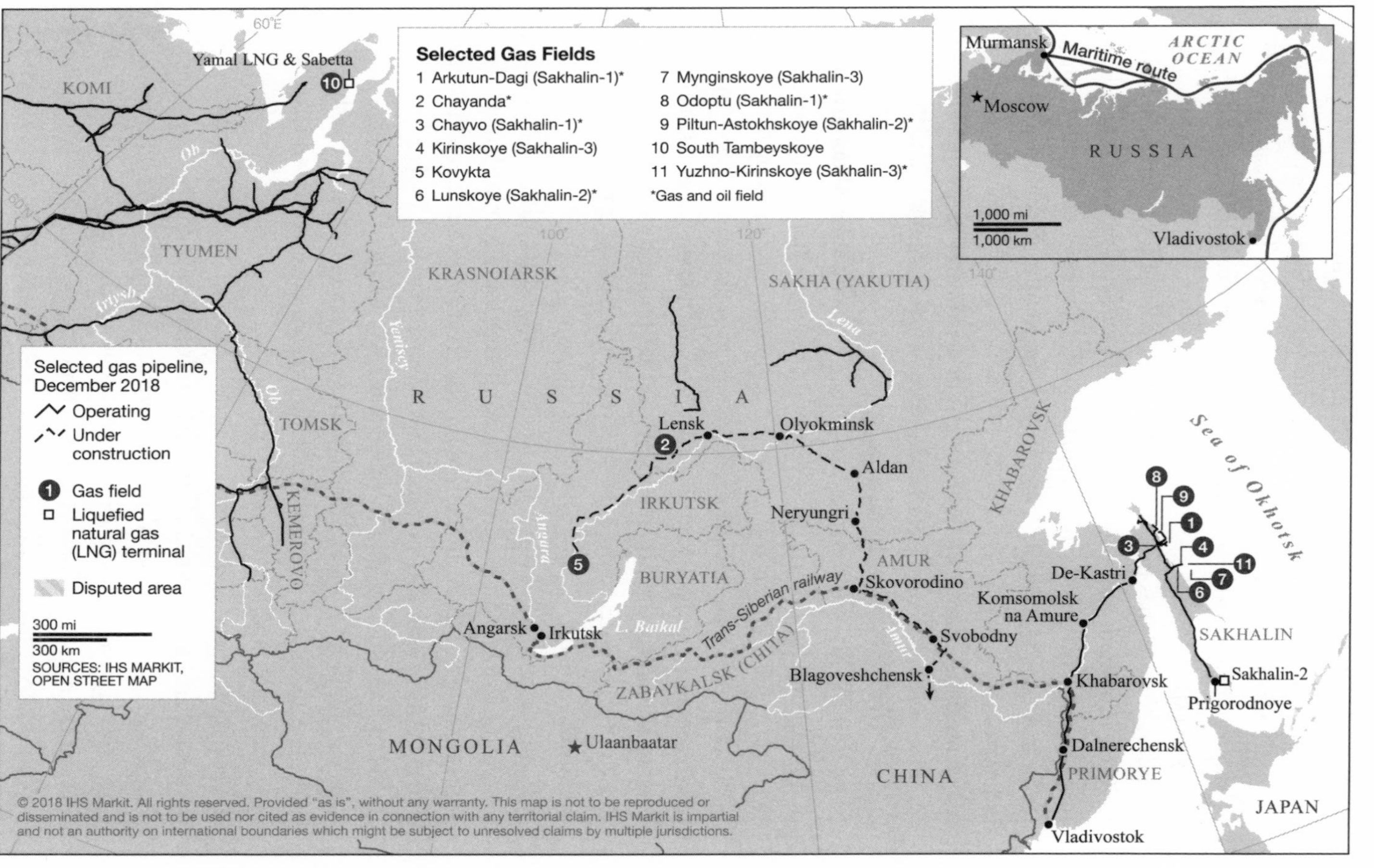

Map 10.1 Selected eastern Russian gas fields and infrastructure.

Gazprom's long resistance to LNG and its continued focus on pipeline gas as its core business are already having serious consequences for its position inside Russia. Gazprom, as noted above, has already lost its monopoly of LNG exports. Moreover, as the priority of LNG in Russian policy continues to rise, Gazprom is increasingly out of step with the Kremlin. Novatek has become, in effect, Russia's national champion for LNG development and export, with Putin's evident backing. Putin has come a long way from the man who promoted a merger of Gazprom and Rosneft to create a single state-owned monopoly of the oil and gas industry.

Gazprom has shown increasing signs of alarm. In the spring of 2019, Gazprom vice-chairman Vitaliy Markelov wrote an angry public letter, charging that Novatek's LNG exports were displacing Gazprom's pipeline gas in the European market, and since Novatek enjoyed tax benefits that Gazprom did not, this was depriving the state budget of revenue.[92] Novatek's CEO, Mikhel'son, answered diplomatically, stressing that Novatek supported the principle that the European market was reserved for pipeline exports, and consequently that Novatek had no intention of challenging Gazprom's dominance there. Yet he pointedly noted that Novatek's LNG exports enjoyed the support of the Russian government and of Putin personally.[93]

Looming on the horizon is a further potential threat. In March 2019, Vladimir Litvinenko, the rector of the Saint Petersburg Mining University, addressed a letter to Putin, calling for the creation of an "LNG cluster" on the Yamal Peninsula as part of an "Arctic National Project" to put Russia in a leading position in the global LNG market.[94] Litvinenko occupies a unique position as part of Putin's past career in Saint Petersburg: in the late 1990s he led a study group on energy policy at the Saint Petersburg mayoralty and he supervised the doctoral (in Russian, "*kandidat*") theses of both Putin and Rosneft chairman Igor Sechin. He continues to have access to the president as an advisor. It remains to be seen whether his proposal will be adopted by the government, but Miller has pushed back strongly, emphasizing that Gazprom needs the Yamal Peninsula to support the next generation of pipeline exports gas.[95] A major battle may be coming over LNG policy and the respective priorities of LNG and pipeline gas, as well as over the control of the gas riches of the north.

Despite these clouds on the horizon, Gazprom by 2020 could claim considerable success in adapting to the challenges facing it. It was in a much improved competitive position in Europe, which remained its prime market. It had built a new generation of gas supply in the Yamal Peninsula and a new set of pipelines to bring it to Europe. By making timely concessions to its customers—no matter how forced—Gazprom had successfully adapted to Europe's new business model. At the same time, Gazprom managed to hang on to the traditional structure of long-term contracts, which is likely to endure into the next decade and beyond. In the new hybrid regime, as Mitrova and Boersma put it, "Volume is defined by long-term contracts; prices are defined by the spot market."[96]

But the revolution in the European gas market rolls on, and Gazprom's hybrid solution, which amounts to temporizing under pressure, is unlikely to survive under continuing challenge from European gas buyers. In the words of a recent report from IHS Markit, "By the early 2020s, most European gas customers will have a wider choice of potential suppliers and delivery routes, and thus an enhanced opportunity to choose among competitive supply options."[97] After twenty years of regulatory liberalization and antitrust enforcement, most gas consumers in Europe now enjoy highly liquid and competitive gas markets, with prices reflecting the balance between supply and demand linked to global gas markets. These conditions do not yet obtain throughout the European Union (southeastern Europe remains a mosaic of exceptions), but there is steady progress toward a uniformly functioning market—notably through the construction of bidirectional interconnector pipelines (including Ukraine, though it is not an EU member) and the wider implementation of network codes. As a result, though Europe's indigenous supplies continue to decline and its dependence on imports is growing, it has greater security of supply. It matters less today who supplies the gas than that market forces create checks and balances.

One consequence of the new order in Europe is that it transfers risk to suppliers, first and foremost to Russia. Long-term contracts are under constant renegotiation, and the process has grown increasingly litigious as buyers resort to arbitration. In what was traditionally an industry based on long-term relationships among players who worked together over their entire careers, new players have appeared who have less at stake and do not know their counterparts. Producer-customer relationships have weakened, which can be particularly dangerous at times of geopolitical tension. An urgent challenge for Gazprom today is to find innovative ways of redesigning long-term contracts and building new relationships.[98]

This chapter has been devoted to the crucial years from 2000 to 2020. Not surprisingly for such a large and complex organization, Gazprom has had a variety of different responses to the multiple challenges it faced, essentially showing three aspects of the company. The first, exemplified by the creation and growth of GM&T, is the progressive face of Gazprom—or, more properly, GazpromExport. The second, the dispute between the Yamal-firsters and the LNG-firsters, showed the power of the conservatives within Gazprom and the importance of internal divisions between different cultures within the company. The third, the long denial of the reality of the shale gale, shows how the company's unwillingness to react to a distant but revolutionary development delayed its response to shale gas, even despite strong urgings from the Kremlin.

But now Gazprom faces three new black swans. In 2016 the first US LNG reached Europe, opening a new chapter in the history of the Russian-European gas bridge. In the next decade and beyond, the future of the gas bridge will be determined above all by the competition between Russian pipeline gas and LNG, influenced by the growing threat of US sanctions. The second is the emergence of climate change as a major political issue, particularly in Europe, where natural gas has lost its aura as the virtuous bridge fuel and is increasingly under attack from environmentalists. The third is the continuing evolution of the Russian gas market, where domestic competitors are challenging Gazprom's long-established monopoly position. We return to these challenges in the Conclusion.

But first we turn to the crucial question of Ukraine, which more than any other has shaped the Russian-European gas relationship over the past three decades.

CHAPTER 11

Russia and Ukraine

Conflict and Collusion

This is the story of a prolonged and difficult divorce. For the past three decades, the Russian-Ukrainian gas relationship has been one of conflict and collusion.[1] Indeed, these were two faces of the same thing: a running battle over the division of Soviet-era gas rents between Russian and Ukrainian interests. But the root of the problem goes deeper. It was the result of the breakup of a previously integrated whole into two parts, in which one side controlled the gas and the other the transit. It is only since 2014, with the fall of the government of President Viktor Yanukovych, that Ukraine's gas relationship with Russia has begun to evolve toward something else. Though the precise terms are as yet far from clear—and will be affected by political and economic shocks still to come—its essence can be summed up this way: the Soviet gas legacy is inexorably fading.[2] Ukraine has already ceased importing Russian gas for its own domestic use. By the mid-2020s, if not before, Russian transit of gas through Ukraine may end. Thus the Russian-Ukrainian gas trade, which has been the most destabilizing element in Russian-European relations for the past thirty years, is gradually disappearing.[3] The long gas divorce is becoming final.

Russian Gas, Ukrainian Transit

The key to understanding the Russian-Ukrainian gas conflict is that both the Russian and Ukrainian gas sectors originated as a single whole. At the time of the breakup of the Soviet Union, the Soviet gas industry was the largest in the world and was highly integrated.[4] The dissolution of what had been a single state immediately changed the gas relationship between Russia and Ukraine. Though Ukraine had been the birthplace of the Soviet gas industry, by the last decade of the Soviet Union its own production was in decline as its fields ran down. Overnight, when the Soviet Union broke up, Ukraine became a transit country and a large importer. Russia had the gas; Ukraine had the pipes. Ukraine desperately needed Russia's gas, for which it could not pay. But it controlled the all-important Russian transit to Europe. At the time the Soviet Union broke up, over 80 percent of the gas that flowed from Russia to Europe passed through Ukraine. At the same time, in 1992, Ukraine consumed 110 billion cubic meters of gas (by far the most in Europe), with local production covering just 16 billion cubic meters (see Table 11.1).

The Soviet Union was responsible for the situation in which Ukraine found itself after the breakup, which in turn created the conditions for the seemingly endless negotiations and gas wars that followed. Soviet planners had exploited Ukrainian gas resources with the profligate consumption of a Soviet-type heavy industrial economy, expanding urban gas-fired infrastructure and industry throughout the 1970s and 1980s on the assumption that cheap gas would always be available. In the process, they exhausted Ukraine's gas fields, leaving it dependent on imported gas from West Siberia.

Given Ukraine's extreme need for gas (and its inability to pay at the same time), this was a situation made for conflict. Russia and Ukraine have been fighting over gas imports and gas transit ever since. To those on the sidelines it may seem like an ever-repeating cycle of quarrels and breakdowns. There have been three notable interruptions of supply—in 2006, 2009, and 2014—and several smaller ones before that.[5] When the elites on the two sides were not quarreling, they were often colluding, through a succession of dubious intermediaries and subterranean deals.[6] Gas has not only been the greatest source of conflict between Russia and Ukraine over the years, it has also been the greatest single source of illicit gains.[7]

Yet the central argument of this chapter is that the root causes of all this are fading. Ukrainian gas demand has declined substantially over the

Table 11.1 Ukrainian gas imports and transit volumes, 1990–2018 (billions of cubic meters)

	1990	1995	2000	2005	2010	2015	2018
Imports	94.5	65.5	59.2	60.2	36.5	18.1	16.3
Russia	60.8	52.8	58.8	25.1	36.5	7.8	2.7
of which payment for transit in kind	0.0	30.0	27.9	20.4	0.0	0.0	0.0
Turkmenistan	33.7	12.7	1.9	35.1	0.0	0.0	0.0
"Reverse" flow from Europe	0.0	0.0	0.0	0.0	0.0	10.3	10.6
Ukrainian gas transit to Europe (non-CIS) (reported)	102.1	110.2	109.3	121.5	95.4	64.2	83.8
Ukrainian transit as % of Russia's total pipeline shipments to Europe (non-CIS)	84.7	90.4	81.4	75.9	66.8	39.2	40.8

Data source: Matthew J. Sagers, *Eurasian Gas Export Outlook, May 2019* (IHS Markit Market Briefing, May 2019).

Notes: Numbers may not sum to totals due to rounding. The table uses the Russian measure of 8,850 kilocalories per cubic meter (in gross calorific value). CIS = Commonwealth of Independent States.

last thirty years, and direct imports from Russia have ceased. There is a growing potential for more Ukrainian indigenous gas production, which could eventually make Ukraine self-sufficient.[8] Meanwhile, Russia has built a series of bypass pipelines around Ukraine, and its dependence on Ukrainian transit has already shrunk from 80 percent of its exports to Europe to about 40 percent. The share will continue to decline, as Russia concludes its bypass policy with two more pipelines. Ukraine's own gas economy is gradually evolving toward a more liberal and commercial, if still fragile, regulated regime. The implications of these developments are clear: Ukraine already no longer needs Russian gas, and soon Russia will no longer need Ukrainian transit. In another decade or less, the gas factor in Russian-Ukrainian relations will be gone.

That is not to say that the way forward will be smooth. Russia's annexation of Crimea and its support for separatist regimes in two of Ukraine's eastern provinces have produced extreme hostility and distrust between the two countries. Russia and Ukraine remain engaged in an undeclared war, which shows no sign of ending soon. This has carried over to the gas business. The Ukrainian gas market now functions in two parts. One part is being supplied with a mixture of "reverse" gas from Europe (much of it consisting of Russian gas "recycled" by European trading companies from the West[9]) as well as indigenous gas, produced mostly in Kharkiv and Poltava Provinces. The separated eastern area, comprised of part of Donetsk and Lugansk Provinces, is being supplied by Gazprom from the east, essentially without compensation. Relations between the two countries' gas companies, Naftogaz and Gazprom, have sunk to a new low in the wake of an ongoing arbitration proceeding, discussed below. At the time of this writing (summer 2019), the prospects for a new supply and transit contract when the present one expires at the end of 2019 are clouded. And if there is no contract, another interruption of supply to Europe is a possibility.

Yet the larger significance of these trends is that the role of Ukraine as a gas bridge—indeed, as the very foundation of the Russian-European gas bridge over the decades, as we have seen it in this book—is rapidly disappearing as well, replaced by multiple bridges that bypass Ukraine, which no longer depends on Russian gas. Gas will continue to flow from Russia to Europe, but Ukraine's strategic role in that flow is already much diminished, and may soon vanish altogether.[10] The Soviet legacy will at long last be gone.

The Plan of This Chapter

Over the past three decades, the gas industry has evolved, both globally and in Europe. So have Russia and Ukraine. To make sense of this tumultuous and confusing stretch of time, in which all variables seem to be in motion simultaneously, I have divided this chapter into three periods. But throughout all three, the overall Russian-Ukrainian gas relationship is driven by the following trends:

- The first is the gradual normalization of economic relations within the Former Soviet Union, evolving from barter through private trader-intermediaries to more conventional monetary and commercial relations. Progress has been halting and punctuated by conflict at every stage, but the overall direction is clear, as the post-Soviet economies become marketized. This applies to both Russia and Ukraine.

- Second is the changing balance of Russian gas. Russia entered the 1990s long on legacy gas, and then in the 2000s it ran short as its legacy assets declined. In the 2010s it is long again, as the next generation of Russian gas comes on line, but at a higher cost. As a result, the legacy rents in the system have been shrinking, diminishing the scope for corruption.
- Third, the price terms offered to Ukraine have changed over time. The main event, by the mid-2000s, was growing Russian pressure on Ukraine for higher gas prices, which in turn produced a sharp decrease in Ukrainian gas consumption from 2006 on. The price pressure by Russia was also directly responsible for the supply interruptions of 2006 and 2009.
- The fourth driver is the steady deterioration of Russian-Ukrainian political relations, in response to Russia's growing ambition to expand its sphere of influence in the Former Soviet Union. Russia's repeated interventions in Ukrainian politics over the past three decades aggravated the chronic political instability of Ukraine, as Russia exploited the oscillation in political power within Ukraine from election to election. Russia's recent actions in Ukraine have made the overall relationship more difficult than ever.
- The fifth is the gradual spread of gas marketization and regulatory reforms from Western to Eastern Europe, together with the information technologies that support them. In the 2010s, these ultimately reached Ukraine, and are bringing change to Russia as well. Along with them came a growing role for the European Union in Russian-European gas trade and regulation, as well as mounting conflict over that role.
- The sixth is the steady determination of the Russian side, beginning in the late 1990s, to bypass Ukrainian transit by building alternative pipelines to Europe, and/or to gain control of the Ukrainian transit system. The latter having been unsuccessful, the Russian effort to bypass Ukraine continues unabated.

In the background is one further trend running through all three periods—namely, the recentralization of Russian politics and the strengthening of the Russian state from the near-chaos of the 1990s to the Putin era, contrasted with the perennial weakness and disarray of the Ukrainian state. Both countries are dominated by powerful elites that have been more concerned with wealth and power than with reforming the economy or the political system. Yet Russia possesses a stronger institutional base, which enabled Putin to rebuild central executive power quickly after his accession in 2000. In contrast, Ukraine as a state continues to struggle with a legacy of weakness (owing partly to its historic status as a secondary tier of the Soviet administrative system). In addition, Ukraine remains divided along multiple dimensions, both geographical and political. Thus it has been unable to build effective or stable political or economic institutions. It is ultimately this weakness that has prevented Ukraine from moving away from the Soviet era more decisively than it has and achieving greater autonomy vis-à-vis a stronger Russia.[11]

The good news is that since the fall of the Yanukovych government in 2014, Ukraine has started to address some of these weaknesses. The annexation of Crimea and Russian support for the insurgents in southeast Ukraine have had the effect of unifying Ukraine around its Ukrainian identity. Ukraine's previous strong economic ties to Russia have been weakened by the conflict, and this is gradually reorienting the Ukrainian economy toward Europe. Over the past half-decade Ukraine has begun to reform its economy and institutions, and while that process is still halting, Ukraine is unlikely to fall back under Russia's influence, as it could well have done prior to 2014. Progress has been the most striking in the gas sector, and reforms are continuing there, with an added strong push from Western donors. We return to this topic in the final section of this chapter.

Yet one should not underestimate the damage done by the Russian intervention and the de facto loss of portions of the eastern provinces. Russia in many respects has paralyzed the Ukrainian economy by saddling it with a heavy military burden. Ukraine appears condemned to remain in a geopolitical and economic limbo, suspended between a European Union that gives it half-hearted support, and a Russia with which it is in political and military conflict.

It is the interaction among these themes that makes the Russia-Ukrainian gas story so complex and yet so fascinating to follow. From one period to the next, as in the chapters of a novel, the characters shift and change roles, as they react to events and opportunities. Nevertheless, as we shall see, there is a logic to each period, as well as a progression from one to the next.

Decade One: Barter Nations, 1991–1999

During the 1990s, the Russian-Ukrainian gas relationship was in chaos, as Soviet institutions fell apart and both sides groped for a new basis for their relations. The gas trade was governed by intergovernmental agreements rather than contracts, but they were routinely violated. It was the era of the rise of the first Russian and Ukrainian private gas traders. This period ended in 1998–1999 with the collapse of commodities prices worldwide and the Russian government's default on its debts. During this decade, the import of gas to Ukraine and its distribution were dominated by the private traders, who parlayed political connections into chains of barter that enabled them, by hook or by crook, to extract value for the gas they brought into Ukraine.

How did the barter system work? At its core was the objective of turning one thing of value to the buyer (in this case, gas) into another thing of value, this time for the seller. The latter could be food, manufactured goods, clothing, IOUs, or state bonds. Typically this required constructing chains of barter, in which real money figured only at the end. For example, gas might be traded for electricity, which in turn might be traded for steel, which could be exported for dollars, the proceeds to be placed in an offshore bank in the account of the seller. The only limit was the ingenuity—and the connections—of the intermediaries.[12] In the case of gas, one especially important form of barter involved trading gas for IOUs from gas distributors, ultimately to be recycled into Ukrainian state debt. This explains the central importance of political protection at all levels of the chain.

It was during this early period that certain political themes first appeared that have been characteristic of Ukraine ever since.[13] Ukrainian politics was essentially founded on a compromise between pro-independence reformers and the elite of the Soviet-era *nomenklatura*. In 1994 one of the latter, Prime Minister Leonid Kuchma, formerly head of the Yuzhmash strategic missile works in the industrial city of Dnepropetrovsk, was elected president on a pro-Russian platform. He governed Ukraine for the next ten years. Kuchma's rise was typical of the role of Soviet-era technocrats in the late Soviet period, and it remained the pattern throughout his years in power. Many of Kuchma's protégés and allies came from the elite of Dnepropetrovsk, just as they had in Leonid Brezhnev's time—including, as we shall see below, the queen of the private gas traders, Yulia Tymoshenko.[14]

Ukraine was hit harder by the post-Soviet depression than Russia because of its manufacturing-based industrial economy. Between 1991 and 1997 real GDP fell by over two-thirds. By 1998, real per capita income was only half that of Russia (although in both cases one has to allow for the strong role of the informal, so-called shadow economy, which continues to play a strong part in both countries today). As in Russia, hyperinflation was made worse by soft credits and subsidies, as the central banks of the newly independent republics joined Russia in merrily printing money. Capital fled Ukraine as it did Russia. Ukraine, like Russia, did not begin to recover until the end of the 1990s, when commodity prices began to rise again worldwide.

For Ukraine, however, rising commodity prices were a two-edged sword, for while Ukraine's exports such as steel and chemicals did well when commodity prices were rising, Ukraine lacked sufficient indigenous energy resources and had to import hydrocarbons from Russia. In the Soviet command economy, gas had been supplied to Ukraine at low prices for industry and residential heat, and this policy continued through the middle of the following decade. Russia's inherited gas resources were initially considerable, as we have said: during the 1990s Russians enjoyed what amounted to a gas bubble, part of which was exported to Ukraine at discounted prices.[15] From an early date a substantial portion of the gas was paid for in barter, since both economies were severely demonetized.

By the mid-1990s the Ukrainian gas trade seemed to outsiders an exercise in chaos, yet there was an underlying logic to it. Ukraine consumed Russian gas, for which it mostly did not pay. Russia continued to ship gas to Ukraine, because it needed the transit to Europe and it had gas to spare;

when it did not, the Ukrainians resorted to "unsanctioned offtakes" (as gas professionals call them) of transit gas, but which the Russians called "theft" and against which they responded with threats and periodic cutoffs. As Gazprom accumulated Ukrainian gas debts, it used them to acquire Ukrainian industries as well as other concessions, such as control of an important naval base at Sevastopol. But the prize for the Russians was the Ukrainian gas pipeline system. Year after year, Gazprom sought to persuade the Ukrainians to trade its gas debts for control of the transit lines—in return for which it promised to keep gas prices low, as it did for Belarus. But the Ukrainian parliament consistently vetoed any surrender as a betrayal of national sovereignty, and the pipelines remained in Ukrainian hands.[16]

Such was the disorder that in 1996 Kuchma's government tried to replace it with a system of gas-trading concessions, under which selected traders were given exclusive rights to import and sell gas in territories assigned to them. It was a measure designed to carve up the market, and it worked only too well, ushering in a cartel of wholesale traders who enriched the business groups associated with them. By 1998 the trading-concession system was acknowledged to have been a failure. In 1998 Ukraine put its national oil and gas company, Naftogaz, in control of gas purchases from Russia and Central Asia. Yet gas relations were soon worse than ever. As Jonathan Stern comments, "The period between late 1998 and 2000 must count as one of the most acrimonious in Russian-Ukrainian gas relations," mainly because of the repercussions on both sides of the economic crash of 1998, which disrupted much of the barter system.[17] Yet for the first time a significant reform had been attempted. Russia and Ukraine made limited progress in developing a modus operandi to manage imports and transit. Gazprom agreed to pay for transit to Europe with volumes of gas that would be consumed in Ukraine. Transit tariffs and import prices were linked. The reform was strongly backed by the International Monetary Fund (IMF)—not the last of the its many attempts to reform the Ukrainian gas system.

Further complicating the gas trade during this period is the fact that it was actually a triangle, involving Turkmenistan (and other Central Asian producers) as well as Russia and Ukraine. The reasons for this were straightforward: during the period of the gas bubble, Gazprom wished to avoid competition from Central Asian suppliers in its European market.

Therefore, its policy was to arrange for traders to ship the Central Asian gas to the less profitable markets of the Former Soviet Union, where gas traded at lower prices and included a substantial share of barter. Ukraine was the most important of these.[18] The Turkmens had little choice but to acquiesce in what amounted to a rent extraction scheme, which benefited well-placed Russian and Ukrainian players with ties to the leadership of Gazprom during the Rem Viakhirev era. But it was not until the Chinese appeared in Ashkhabad, offering to build an export pipeline to China, that the Turkmens gained some control over their own gas and stopped shipping gas to Russia and Ukraine.[19]

Finally, it was during this period that Russia, exasperated by the frustrations of dealing with the Ukrainians, began building new export pipelines to bypass Ukraine, working together with European partners. The first bypass was the Yamal-Europe pipeline, which passed through Poland and Belarus and for which planning began as early as 1992. Second, in late 1997 Russia began negotiations with Turkey to build a pipeline under the Black Sea. This project, called Blue Stream, had three objectives: to bring more Russian gas to the Turkish market; to block Western plans for a pipeline under the Caspian; and above all, to provide another bypass around the Ukrainian transit system. Blue Stream began shipping gas in February 2003—a remarkably rapid achievement, considering the technical difficulties of laying pipe on the deep seabed of the Black Sea.[20] These first two pipelines underscore a key point in the whole Russian-Ukrainian story: that Russia's bypass policy is of more than twenty years' standing and has been pursued vigorously, virtually regardless of cost, throughout that time (see Map 11.1).

Yulia Tymoshenko: How the "Gas Princess" Got Her Start

Nothing could be more emblematic of the chaotic state of the Ukrainian economy and political system in the first decade of the post-Soviet era than the early career of Yulia Tymoshenko, a woman of iron will who rose from obscurity to become the symbol of Ukraine's quest for democracy and independence during the Orange Revolution in 2004. Tymoshenko has remained a political force in Ukrainian politics ever since,

Map 11.1 Westward gas export pipeline routes from the Russian Federation.

being prime minister twice and a perennial contender for the presidency. But for our purposes here the significant part of the story is the role she played in the barter gas trade in the 1990s, when she became one of the most important players as a protégée of Prime Minister Pavlo Lazarenko and then as a dependent of President Kuchma. Her company, Unified Energy Systems (UES), became one of the most successful gas distributors in the country. In those years, Tymoshenko was known as "the gas princess."

Much of the future political and business class of independent Ukraine came out of the Komsomol, the youth arm of the Communist Party, in the city of Dnepropetrovsk. Tymoshenko, who had grown up poor as an ordinary working girl, was an example. In 1988, at the height of Mikhail Gorbachev's reforms, Tymoshenko and her husband created a Komsomol cooperative called Terminal, the purpose of which was to trade in oil products. It was through this company that she gained her first exposure to fuels trading. She then cofounded a larger trading company called Ukrainskii Benzin (Ukrainian Gasoline, known as KUB). Thanks to the support of Lazarenko, then a local politician, KUB gained a monopoly on fuels distribution to the agricultural sector of Dnepropetrovsk Province.

In 1992 Lazarenko was named presidential representative in the province, and he soon became the de facto boss there. By this time, he had built a business empire throughout both the city and the province, which included KUB and Tymoshenko. Lazarenko put his political machine to work for Kuchma during the presidential election of 1994, and Kuchma rewarded him by naming him deputy prime minister in charge of energy. Then in May 1996, Kuchma named him prime minister.[21]

As Lazarenko rose, Tymoshenko rose with him. She was initially the commercial director of KUB; then she became the general director. Two months after Lazarenko's appointment as deputy prime minister, KUB morphed into a British-Ukrainian joint venture called United Energy Systems of Ukraine, with a charter capital of $10 million. Tymoshenko was soon the undisputed boss. By this time UES had moved on from oil products in Dnepropetrovsk to the much more profitable business of selling gas throughout the country. Suddenly Tymoshenko was in the big time.

President Kuchma's style as a political leader was to play all fields, rewarding different business clans with rents and perks but switching constantly among them. As a patron he was fickle and unpredictable—as Lazarenko soon discovered, when the president accused him of massive bribes and other crimes. Lazarenko fled to Switzerland. He was subsequently deported to the United States, where he served six years of a nine-year prison sentence for money laundering, wire fraud, and extortion. Without his support UES's gas business collapsed. From this point on, Tymoshenko's career shifted from business to politics.[22]

Tymoshenko's story and vivid personality understandably captured the imagination of the Ukrainian and Russian media, and since the mid-1990s she has never been out of the limelight as a politician, most recently as an unsuccessful contender for the presidency in 2019. As a gas trader, she briefly became the largest player in eastern Ukraine, but she was soon overshadowed by individuals with better connections to Kuchma and Ukraine's ultimate gas suppliers—Gazprom and Turkmenistan. Yet as one of the first new players in the post-Soviet fuels business, she illustrates better than anyone the connections between the Soviet era and the barter period of the 1990s. From the first, it was a business based on personality and political connections—and luck. Tymoshenko had all three.

Following the demise of UES, its role in the gas-trading business was taken over by a company called Itera, which had been founded by a Russian, Igor Makarov, a man with early ties to Turkmenistan. Makarov had been a champion bicycle racer on the Soviet Olympic team, and he parlayed his connections into an international gas-trading business, exporting Russian and Turkmen gas into Ukraine and other republics of the Former Soviet Union, trading it for food and other products to ship back into Russia and Central Asia. This required cooperation from Gazprom, and early on Makarov gained the support of Viakhirev, Gazprom's chairman, who saw in Itera a means of extracting value from the gas flow from Central Asia without the complications of dealing directly with the mess in Ukraine.[23]

With the sponsorship of Gazprom and the support of President Kuchma, Itera took over the task of supplying Ukraine. For the following eight years, until Aleksey Miller succeeded Viakhirev at the head of

Gazprom, Makarov's Itera was Gazprom's preferred trader in the republic. During the early years, Itera worked closely with Tymoshenko: Itera supplied the gas that she distributed.[24] In addition, Itera was able to get a piece of the Ukrainian market for itself.

Itera was more than a gas trader. With the support of Viakhirev and Gazprom, Makarov was ushered into the Russian upstream and soon became a major gas producer. But as with everything else during this period, it was an opportunity based on barter. Gazprom paid taxes to the government of Yamal-Nenets Okrug, but those taxes were paid in gas, leaving the regional government with the problem of turning the gas into food, supplies, and money. Enter Itera, with its connections to the Ukrainian market and beyond. This relationship, in which Itera played the useful role of monetizing gas, ultimately led to Itera's taking control of upstream gas licenses in West Siberia and becoming a gas producer itself. It was a prosperous chain of interests, which thrived through the end of the barter period. But with the advent of the Putin era and the replacement of Viakhirev by Miller at the head of Gazprom (described in Chapter 10), Makarov's star quickly faded.[25]

Decade Two: Russia Raises the Stakes, 2000–2013

The second period, from 2000 to 2013, was marked by four key events: the Orange Revolution[26] in 2004 and its subsequent failure; the growing pressure by the Russians from 2006 to increase gas prices to European levels; the "Great Recession" in the world economy in 2008–2009 and its economic impact on both Russia and Ukraine; and finally, the cutoff of Russian gas exports to Ukraine and Europe at the beginning of 2009, followed by the Russian-Ukrainian gas contract of the same year. During this decade there was a marked recentralization of politics in Russia around an increasingly powerful president, Putin—but not in Ukraine, where political power swung unpredictably between rival coalitions of politicians and oligarchs. Lastly, throughout this decade Russia continued its policy of bypassing Ukraine, with the construction of another export pipeline, Nord Stream. Putin and German chancellor Gerhard Schröder reached an agreement on the pipeline in 2005, which was followed by an investment decision in 2009.

After the economic crash of 1998, the era of barter came to an abrupt end in Russia. However, it continued in Ukraine, particularly in the Russian-Ukrainian gas trade, into the mid-2000s. In Russia, the rise of commodity prices and the arrival of a degree of prosperity were accompanied by important legal and financial reforms—notably, a bankruptcy law that limited the epidemic of unpaid bills that had bedeviled the country in the 1990s and supported the barter economy.[27] Ukraine followed the same path but more haltingly: the first meaningful economic reforms took place under the leadership of the liberal reformer Viktor Yushchenko, at that time the head of the National Bank of Ukraine (1993–1999) and subsequently prime minister (1999–2001) under Kuchma. The 2000s brought Ukraine a degree of prosperity, thanks to commodity exports, and money started to come back into the Ukrainian economy.

Yet the liberal reforms were overshadowed by Ukraine's chronic failure to improve the energy efficiency of its economy. There was marginal progress. The energy intensity of the Ukrainian economy stabilized, due to the expansion of the service economy and increases in energy prices to industry. Indeed, from 2000 to 2005, Ukraine's energy intensity actually fell by 40 percent, as investments were made in the most energy-intensive sectors such as steel. Even so, by the middle of the decade, Ukraine remained by a wide margin the most energy-intensive economy in the world—with an intensity almost four times that of Germany.[28] Basic heat and power systems were actually becoming less efficient, owing to systematic underinvestment in replacement and maintenance, not to mention low tariffs and the lack of metering.[29] The increase was especially large in the residential heating sector, where the share of gas had increased to 55 percent by 2004.[30]

The decade began peacefully enough. Putin, Russia's new president, attempted to build a personal relationship with President Kuchma and settle the unfinished business of the administration of Boris Yeltsin—notably Russia's borders and the status of the Black Sea fleet, but above all to bring order to the tangled gas relationship. During the first half of the 2000s Kuchma and Putin met repeatedly, sometimes as often as once a month, and the two men succeeded in alleviating some of the worst causes of conflict, in an agreement that Jonathan Stern called "a carefully structured attempt to break with a lawless past."[31] In particular, the two men negotiated at length about creating a consortium to manage the Ukrainian transit system, possibly with Western companies as partners.[32]

But the Orange Revolution blew Putin's strategy apart. He had attempted to persuade Kuchma to run for a third term, and when that failed Putin tried to block the election of a liberal pro-Western candidate, Yushchenko, in favor of an "Eastern" candidate, Viktor Yanukovych, who ran on a platform of close alliance with Russia. The brazen attempt by Putin's political technologists to manipulate the election in Yanukovych's favor and the subsequent explosion of popular anger in Maydan Square in Kiev were followed by the victory of Yushchenko and the Western-oriented Orange coalition.[33] It was a major defeat for Moscow, where it was perceived that the Orange Revolution was part of a wave of Western-sponsored "color revolutions" that would soon engulf Moscow.[34] Putin's bitter denunciation of the West at the Munich Security Conference in 2007 was a direct consequence of his conviction that he had been betrayed by the West.[35]

The second major event of the 2000s was the growing Russian pressure on Ukraine, beginning in 2005, to pay higher prices for gas. This was a major turning point. From $44 per thousand cubic meters at the beginning in 2005, the price of imported Russian gas surged to over $232 in 2009 and kept rising from there, as Gazprom sought to raise prices to the European level—which at this time still mainly meant the oil-linked netback price at the European border.[36] The Ukrainians, unable to pay, began importing less. From 60.2 billion cubic meters in 2005, Ukrainian gas imports from Russia dropped by over half, to only 26.8 billion cubic meters in 2009.[37] This was one of the major causes of the dispute of 2008 and the cutoff that followed in early 2009. But significantly, it also marked the beginning of a historic change—the gradual decline of Ukrainian dependence on Russian gas imports.

This had two main causes. The first was an immediate outgrowth of the Orange Revolution: faced with an unfriendly regime in Kiev, Russia no longer had any reason to offer gas at subsidized prices. But Gazprom's demand for higher prices had actually begun earlier and reflected the fact that by the end of the 1990s the Russian gas bubble had vanished. During this period Gazprom was increasingly short of developed gas, while for most of the decade demand in Europe was growing rapidly. Gazprom needed gas, and it needed capital to invest in the next generation of supply. These two factors—constrained supply in Russia and rising demand in Europe—help explain the increasing pressure on Ukraine to pay higher prices.[38]

Whereas in the 1990s Gazprom could afford to supply Ukraine and the western Former Soviet Union at low prices, by the first half of the 2000s this was no longer the case. For a time, Central Asian gas partially replaced Russian gas. But the Central Asians became increasingly reluctant to accept barter in place of cash, and as a result Central Asian gas became increasingly expensive. When the Turkmens refused to renegotiate the price, Gazprom cut them off. By this time the Chinese were making attractive proposals to build a pipeline from Central Asia to China. Even before Turkmen dictator Saparmurat Niyazov unexpectedly died in February 2007,[39] it was becoming clear that the future of Central Asia exports gas lay east to China, not north and west to Russia or Ukraine.

The immediate aftermath of the Orange Revolution was a resumption of the familiar conflicts in gas relations. The Russians and the Ukrainians quarreled over the structure of a consortium to manage the transit pipelines, the prices and terms for deliveries of Central Asian gas, the ownership of gas in Ukrainian storage, the meaning of the "European standard" for gas prices, debt settlement, and much more. In short, it was back to business as usual on the Russian-Ukrainian gas front, minus the good will that had initially prevailed between Putin and Kuchma in the first half of the decade. This period happened to coincide with a rapid rise of world oil prices—on which gas prices still tended to be based in Europe—with the result that by late 2005 European border prices were three to four times the level that Belarus and Ukraine were then paying.[40] From Gazprom's perspective, it was losing money on every molecule delivered to Ukraine. After the Orange Revolution the Kremlin had little reason to continue offering Ukraine "brotherly" low prices, especially after the Ukrainians' repeated refusal to grant Gazprom control of the Ukrainian transit system. From this time on, Gazprom began applying steadily increasingly pressure on Ukraine to pay higher prices, with the ultimate objective of reaching the European level.

These conflicts led to two interruptions, the first on New Year's Day 2006 and the second three years later. The 2006 cutoff originated with Ukraine, which offset shortfalls in Russian and Central Asian deliveries and higher prices with more offtakes of Russian transit gas to Europe, to which Russia responded with cutbacks of its own.[41] The standoff was brief—only three days—but severe, made more so by the fact that the Europeans were unprepared for such an event. Hungary lost 40 percent of its gas supplies, other Eastern European countries lost up to one-third, and France and Italy were short by over one-quarter.[42] Fortunately, the weather was mild for the time of year, and Gazprom quickly restored pressures in its export lines. By January 4 the situation was back to normal. Having crossed swords, Gazprom and Naftogaz signed a new five-year contract that addressed some—though not all—of the issues in dispute.[43]

But the resolution was only temporary. By late 2008, despite efforts at the presidential level to reach a new agreement, the two sides had fallen out again. On January 1, 2009, Gazprom cut off all supplies for Ukrainian consumption. Ukraine responded with offtakes, which Gazprom denounced as "theft."[44] On January 6 Gazprom reduced and then cut off supplies of gas to Europe via Ukraine. With flows at a standstill, international pressure mounted on Ukraine and Russia to come to an agreement. The two sides returned to the table and negotiated an eleven-year contract. Finally, twenty days after the beginning of the crisis, full deliveries resumed to both Ukraine and Europe.[45]

The 2009 cutoff and the contract that followed were watershed events. Since the 2009 cutoff, everything that has happened in Russian-Ukrainian relations, both commercial and political, has been measured in Europe by the question, "Do we dare rely on Russian gas? Could it happen again?" A decade later, the case for geopolitical risk connected to Russian gas exports rests very largely on the memory of the 2009 cutoff.

The irony is that from a strictly economic standpoint most of Europe was little affected by the cutoff: the impact was mainly felt in Bulgaria and the Balkans. By this time the bulk of the European gas system was so interconnected that gas could be swapped quickly across nearly the whole of Europe, and ample storage provided a cushion on which the Europeans could draw. The cutoff actually did Europe very little economic harm. Spot prices in northwestern Europe remained unchanged. Gas was redirected to the Balkans where possible (France's GDF supplied Croatia, Germany's Ruhrgas supplied Serbia, and British Gas and Total each brought a cargo of liquefied natural gas to Greece), and engineers worked hard to bring back into service long-unused pipelines (that linked Hungary to Slovakia, for example) or draw from Austrian storage. European gas companies actually saved money on the whole process, as their oil-linked contract prices were very high in the first months of 2009.[46] But against the backdrop of the overall deterioration in political relations since 2013–2014, these commercial and technical details are not the ones that matter. It is the cutoff that everyone remembers and that has shaped behavior ever since.

The cutoff of 2009 was, more than any other, the defining episode in the Russian-Ukrainian-European ménage à trois, yet it remains hard to explain, because both Moscow and Kyiv had so much to lose. For both countries (particularly for their main gas companies, Gazprom and Naftogaz) the cost was a lasting loss of commercial reputation. For Gazprom, the loss was particularly damaging. For four decades Gazprom had established a unique reputation for reliability. It would hardly have acted on its own. The only plausible account, then, is that the decision to cut off Ukraine—and then Europe—came directly from the top. But this raises further questions. Can Putin really have believed that Europe would take Russia's side against Ukraine's? One must remember that Putin had been personally involved in every stage of Russian gas policy from the moment he took office, and consequently he can surely have had no illusions about how Europe would react to a cutoff. Putin had taken an active part in every disappointing twist and turn of Russian-Ukrainian gas relations over the previous decade, from his nearly monthly meetings with Kuchma through the humiliation of the 2004 revolution, the dispute of 2006, and finally the showdown in late 2008. Yet at every point Ukraine had escaped his grasp, defied his efforts, and left him frustrated.[47]

In the 2000s interests allied with Russia were able to use profits from gas to buy up a wide range of gas-consuming industries as well regional gas-distribution companies. This created new roles for intermediaries, who found new opportunities no longer simply as incidental traders as in the 1990s, but as Gazprom's partners. The most notorious of these was Dmitro Firtash.

A Man for All Presidents: Dmitro Firtash

Firtash is unique among the Ukrainian oligarchs in having prospered under all of Ukraine's presidents. As he once put it. "I have simply been in the right place at the right time." But since 2014 he has been confined to his residence in Vienna by an international arrest warrant issued by the United States, so he is unable to travel to Ukraine.[48] Even so he remains a strong presence. He is the only significant gas player from the 1990s to have reinvented himself and his business model in each of the past three decades, which is a testimony to his considerable political and business skills.[49] But quite apart from his present legal problems in Vienna, there is a more fundamental question: has Firtash's time passed, and that of Ukraine's gas oligarchs generally, as the Ukrainian gas market begins to liberalize? We return to this theme in the third part of this chapter, "Progress amid Crisis," below.

Firtash is the very opposite of the flamboyant Tymoshenko. Firtash is reserved and cautious; he gives very few interviews, and when he does, he reveals little. He is one of the few entrepreneurs, one might say, about whom more is known about his first million than his last. The history of Firtash's career goes back to the beginnings of the barter era. He put together his first deal in 1988–1989, a trade of 4,000 tons of Ukrainian powdered milk against Uzbek cotton that was subsequently sold in Hong Kong. His profit, Firtash recalls, was $50,000. Shortly afterward he moved to Moscow. In those days, officials and aspiring businessmen from Central Asia gathered at the Rossiia Hotel, and it was there, in 1994, that Firtash made his first deals with the Turkmens—trading Ukrainian food products for Turkmen gas, which he then resold in Ukraine through an influential middleman, Ihor Bakay, an ally of President Leonid Kravchuk. Firtash began traveling to Ashgabat, the capital of Turkmenistan, where he met Makarov, the founder of Itera, and became his business partner. Over the next decade, throughout Kuchma's presidency and until the changing of the guard at Gazprom, Firtash and Makarov worked together.

Makarov was responsible for relations with Gazprom and Turkmenistan, while Firtash made the business work in Ukraine.[50]

Toward the end of the 1990s, however, Firtash's relationship with Makarov frayed, as Ukraine began to move from the era of regional traders to a more centralized management of gas imports under the newly created Naftogaz Ukrainy—headed by Bakay, Firtash's one-time sponsor. In addition, as Yushchenko emerged as the leader of a liberal opposition to President Kuchma, Firtash formed close ties with him, becoming his unofficial advisor on gas matters. But the most important break for Firtash was the advent of Aleksey Miller as the new head of Gazprom. In 2003, Miller transferred the franchise for the import of Turkmen gas to Firtash's company, Euraltransgas. Suddenly, Firtash controlled over half of the Ukrainian gas market. By the time of the Orange Revolution, in short, Firtash had executed his second well-timed transition. By 2003, when Gazprom ended its relationship with Itera as its preferred trader to Ukraine, Firtash was well placed to step in, with Gazprom's support.

At this point, Firtash's business model changed again. In 2004, Euraltransgas was replaced with a new company, RosUkrEnergo (RUE), in which Gazprom owned 50 percent and Firtash and his partners 45 percent.[51] In other words, Firtash was no longer simply a trader. He was now a partner with Gazprom, exporting to Ukraine's neighbors in Central Europe—notably through a Hungarian company called Emfesz, which Firtash controlled as part of an expanding conglomerate called Group DF.

Firtash's relationship with Yushchenko lasted only as long as fortune favored the Orange president. By 2006, as the Orange coalition fell apart, the influence of the eastern Party of Regions was rising, and Yanukovych, the defeated candidate of 2004, returned to the fore as prime minister. Before long, Firtash had transferred his allegiance to Yanukovych,[52] and his name became increasingly associated with the latter. Gazprom, and presumably the Kremlin, provided financial backing and low-priced gas,[53] which Firtash used to expand his holdings throughout Ukraine. By the end of 2008, on the eve of the 2009 cutoff, his Group DF owned three-quarters of the province-level gas distribution companies in Ukraine (the so-called *oblgazy*), to which Group DF sold gas. Firtash also invested widely in the chemicals and fertilizers business, which was even more profitable than retail gas. In one of his rare interviews, Firtash claimed to have lost money on every molecule of gas sold into Ukraine, because of *dotatsii* (cross-subsidies) in the Ukrainian market.[54] But now he increasingly invested his gas imports into "embodied gas"—that is, agricultural and industrial commodities made with gas in plants he owned, which he then exported to Eastern and Central Europe.[55]

But in 2009 Firtash's fortunes turned. The first reverse was the 2009 dispute with Russia, which led to a renegotiation of the entire Russian-Ukrainian gas trade through direct talks between Putin and Tymoshenko, both then prime ministers. Tymoshenko had long been an enemy of Firtash and an opponent of RUE. The result of her agreement with Putin was the new eleven-year gas contract; one of its features was that RUE was abolished. For Firtash, the timing could not have been worse. Ukraine had been hit hard by the Great Recession of 2008–2009, and Firtash's empire of fertilizers and chemicals was especially affected. It was at this point, in 2010, that Yanukovych was elected president of Ukraine. Firtash might have thought that he was well protected by his good relations with the new president,[56] and initially his businesses prospered as he acquired more plants and factories. But Yanukovych's fall, the annexation of Crimea, and the breaking away of the eastern provinces severely damaged Firtash's position as well as his holdings in the east.[57] Financial stringency soon followed. Firtash's Nadra Bank (the eleventh largest in the country) failed and was liquidated in 2015.[58] By the time of his de facto exile in Vienna, Firtash's fortune was badly depleted, although he remained in control of three-quarters of Ukraine's regional gas distribution companies.

The 2009 Contract

Since the 2009 contract has led to endless controversy and disputes, it is worth clarifying its significance.

The first noteworthy feature of the 2009 contract is that it was public. Gas contracts are normally closely guarded commercial secrets. But the 2009 contract was leaked to the newspaper *Ukrainska pravda* shortly after it was signed,[59] and the entire gas world was able to have a close look. What they discovered was that Tymoshenko had negotiated a singularly bad deal, with a base price that was far too high. People disagree over the reasons. Some, recalling her past as a freewheeling gas trader with close ties to Moscow, were suspicious. According to an explanation that was circulated by Wikileaks, Tymoshenko agreed to a high indexed price in exchange for an initial discount that applied through the end of 2009 and was supposed to help her win the presidency. No evidence has surfaced for this account, but many Ukrainians believe it to this day.[60] Others, including most of the people in the gas industry, saw it as a case of technical incompetence or excessive haste. But Yanukovych saw it as an opportunity to rid himself of a political rival: charging Tymoshenko with corruption, he clapped her in jail, where she remained until his overthrow four years later.

Another unfortunate result of the 2009 contract was that it locked the Ukrainians into a straitjacket, since it linked gas prices to oil prices—at the time still the universal practice in the gas trade. At first, all seemed to go well, since oil prices had dropped in the wake of the Great Recession and appeared to be headed even lower. But surprisingly, oil prices recovered in late 2009, and over the next five years they reached new heights. The Ukrainians were caught on the hook and were forced to pay increasingly high prices. It was only by making political concessions to the Russians that Yanukovych was able to negotiate a partial discount in 2010, in which he swapped an extension of the lease on the Black Sea naval base in exchange for lower prices on Russian gas. The deal was funded directly by the Russian government, instead of Gazprom, an indication of how far politics dominated commerce during the Yanukovych period.

Nevertheless, the 2009 contract marked an important transition in Russian-Ukrainian gas relations because it codified several key features of European gas contracts and provided the basis for some important reforms in Ukraine, as we shall see in the next section.

Decade Three: Progress amid Crisis, 2010–Present

As the third post-Soviet decade begins in 2010, the Orange Revolution has failed; the pro-Russian Yanukovych, in an astonishing political revival, has been elected president; and Putin is beginning a second major attempt to bind Ukraine to the Russian sphere of influence. The failure of that attempt, the collapse of the Yanukovych government, and the arrival in power of a pro-Western, pro-reform coalition under President Petro Poroshenko set the stage for the Russian-Ukrainian gas relationship from 2014 to the present.

This third decade is especially difficult to characterize in a single brief overview, because its major feature is a split between two opposite trends. On the one hand, with the collapse of the Yanukovych government, the long-standing tensions between Ukraine and Russia exploded into violence, with the annexation of Crimea, the takeover of portions of two eastern provinces by Russian-supported Ukrainian separatists, and jousting over access to the Sea of Azov.

Yet paradoxically, the very worsening of the political relationship accelerated the process of divorce on the gas front by catalyzing changes in two broad directions. First, both Russia and Ukraine have redoubled their efforts to end what remains of their mutual dependence in gas and transit. Ukraine has stopped importing Russian gas and now buys gas from the West. Meanwhile, Russia continues to build alternative export pipelines, the latest of which are Nord Stream 2 and Turkish Stream. Second, the advent of a Western-oriented, reform-minded government in Ukraine in 2014 opened the way for progress in marketizing and liberalizing the Ukrainian gas sector—a process that, however imperfect, goes far beyond anything achieved in the past.[61] These two trends together should logically lessen the danger of supply interruptions, as well as diminish the opportunities for the illicit arbitrage and subterranean deals that have been the curse of Russian-Ukrainian gas relations. Yet the hostile political situation sets limits on what can be achieved. It is ironic, to say the least, that at the very moment when Russian-Ukrainian gas relations stand the best chance of being normalized, they have become more tense and politicized than ever, and another transit interruption cannot be ruled out.

In the final section of this chapter we explore both of these trends. We look first at the domestic developments in Ukrainian gas policy, focusing on the initial steps in gas reform. We then turn to the evolution of the Russian-Ukrainian gas relationship.

The Beginning of Gas Reform in Ukraine

Following the fall of Yanukovych in 2014, the new government under Petro Poroshenko began wide-ranging reforms in the gas sector. It had little choice: the IMF insisted on reforms in the gas market as a condition for its continued support. The economy would have collapsed without external funding, and the government hastened to comply. In short order, from the most corrupt branch of the Ukrainian economy, the gas industry became a leading laboratory for reform—or, at least, for good intentions. The gas reforms remain contentious and face powerful opponents, and it is far from certain that they will succeed. But important changes have already taken place, and more will be coming under Ukraine's new president, Volodymyr Zelenskiy. The stakes cannot be overstated. If the gas reforms continue to be pursued in coming years, they will go a long way toward creating the conditions for Ukraine's self-sufficiency in gas.

Much of the Ukrainian gas reform plan is modeled on gas and power legislation and practices adopted by the European Union since the mid-1990s. Although Ukraine is not a member of the European Union, its adherence to the Energy Community Treaty obligates it to adopt the core features of EU gas and power legislation, notably the so-called Third Package on gas and power. The landmark launch of the Ukrainian gas reforms took place in October 2015, when the parliament passed a framework law titled "On the Natural Gas Market." As is typical of legislation in both Russia and Ukraine, a framework law only specifies general objectives, whose actual implementation requires the passage of subacts—that is, regulations that spell out the details. And that is where the real politics begins. For the past four years Ukraine has been locked in battle over the subacts.[62] There are four main reforms under way: in domestic prices, import policy, the structure of the industry, and the conditions for upstream investment.[63] We review each one briefly below.

Deregulation of Prices

The most important reform by far has been the deregulation of domestic prices and the cutback (although not yet the elimination) of cross-subsidies. This happened in two steps. On October 1, 2015, when the law "On the Natural Gas Market" went into effect, prices for industrial consumers and so-called budget-financed consumers *(byudzhetniki),* such as schools and hospitals, were decontrolled immediately. Whereas formerly the state energy regulator set industrial prices monthly, now suppliers can charge industrial consumers and *byudzhetniki* whatever the market will bear. Seven months later, on April 1, 2016, tariffs for residential consumers were quadrupled, while prices for heat distributors were more than doubled, both in dollar terms. A price increase schedule agreed with the IMF had envisaged introduction of full market prices by mid-2017, but the Ukrainian government, led by Prime Minister Vladimir Groysman, decided to jump to market prices in one step. Thus, the most egregious sources of market distortion and inefficiency were greatly diminished overnight.

Note that the latest price reform differs fundamentally from the periodic price increases that were decreed from time to time in the past in Ukraine. Those were one-time regulated increases, the effects of which were quickly wiped out by inflation, and as a result domestic gas prices typically remained as heavily subsidized as before. Now, under the new system, the benchmark for domestic prices in Ukraine is explicitly linked to the south German hub price (NetConnect Germany) as well as to the hryvnia-to-dollar exchange rate, the dollar-to-euro exchange rate, and the estimated cost of transportation from the German hub to Slovakia and then to the western border of Ukraine. It is supposed to be adjusted quarterly to reflect changes in these parameters as market conditions change. To offset the impact of higher gas prices on the disadvantaged, the government instituted direct payments to the poorest households.

The increase in residential prices has been fiercely resented by ordinary consumers and has become a political issue. As a result, the price reform has not been fully implemented in practice. At the urging of the IMF and World Bank, the government agreed to raise the prices for residential customers and district heating companies step by step until they reached parity with import prices, and then to abolish the two-tier price system. There was a point toward the end of 2016 when residential prices were indeed getting close to the level of import prices, but then the government backed off.

Still, the consequences of the partial reform were felt immediately. The most spectacular was that the state company Naftogaz, which had run chronic deficits since its creation in 1998—the most recent, in 2014, was a record $10 billion—suddenly became modestly profitable. This in turn eliminated a major burden on the Ukrainian budget: indeed, practically overnight Naftogaz became the largest single contributor to the budget.[64] Another consequence is that with the near-disappearance of subsidies some of the incentive for cheating, notably through misreporting of residential consumption, has diminished. There is still room for corruption, however. For example, the *oblgazy* frequently fail to pay Naftogaz, on the ground that consumers have supposedly not paid their gas bills. The *oblgazy* are also accused of inventing fake residential customers so that they are allocated more gas than they actually need.[65]

As the new Ukrainian president, Volodymyr Zelenskiy, took office in the spring of 2019, he received a welcome honeymoon present. Gas prices in Europe plummeted in the months before the election, making residential prices in Ukraine suddenly higher than market prices in Europe. This gave Ukraine an unexpected win-win opportunity to reduce prices to ordinary consumers while continuing to satisfy the pricing formula agreed on with international lenders, particularly the IMF. Thus, the sensitive pricing issue may be defused for the time being. If Europe is entering a period of low gas prices for the next few years, as some studies predict may be the case, then the result will be a break for Ukrainian consumers—and the president.

Import Policy: Reverse Flows

Direct gas imports from Russia have been replaced by private-sector imports from the West, commonly referred to as "reverse flows." These actually began under Yanukovych, with a small-scale agreement in 2012 with the German utility RWE to import gas from Hungary via a looping pipeline. The second contract came in January 2014 via Slovakia.[66] This time the volumes were much larger, taking advantage of an unused pipeline between Slovakia and Ukraine, which was reversed to run east instead of west.[67] In 2015, the first full year of reverse flows, Ukraine imported 10.4 billion cubic meters from the West, of which 9.7 billion cubic meters came via Slovakia.[68]

Since then, the reverse gas trading business has grown and diversified. In 2017, reverse flows amounted to nearly 13 billion cubic meters,[69] supplied by over eighty private companies of all kinds. They include a wide range of players and business models: some sell to the state company, Naftogaz, and some directly to distribution companies; some, like Engie and Equinor, are European giants, while some are Ukrainian traders. In short, a market has suddenly come alive where none existed before. As a result, Ukraine has not directly purchased Russian gas since November 2015.

There has been a lively debate over exactly where the reverse molecules come from, and whether the reverse flow could be vulnerable to Russian intervention. Most of the molecules actually consist of Russian gas, which is sold to Ukraine out of the European buyers' portfolios. Could it then be vulnerable to a Russian interruption? Legally no: European buyers are perfectly entitled to reroute and resell gas once they have purchased it, regardless of source. But physically? In 2015 Russia briefly attempted to stop the reverse flow via Slovakia by cutting back its own transit exports by an equivalent amount. The attempt failed and was soon abandoned, apparently because the weather was mild and there was ample surplus transmission capacity available. The threats have not been repeated since. Nevertheless, the Ukrainians have shown concern, calling for an expansion of transmission capacity—notably via a new Poland-Ukraine interconnector.[70]

Restructuring: The Unbundling Issue

A fundamental part of the EU's Third Package is a restructuring of the traditional integrated gas company, of which the centerpiece is unbundling—that is, the separation of other parts of the business, such as gas sales, from transportation, resulting in two independent entities.[71] In Ukraine, this means breaking up Naftogaz and moving its pipeline subsidiary, Ukrtransgaz, into a newly created company, Gas Pipelines of Ukraine (Mahistral'ni Gazoprovodi Ukraini, or MGU).

A government decree was issued in 2016, making unbundling official policy. But Naftogaz dragged its feet, insisting on keeping control of Ukrtransgaz in its own hands.[72] Naftogaz needed the revenues from transit, which offset its losses from what it discreetly called "public service obligations"—that is, its remaining sales at below-market prices to residential consumers. If it lost control of Ukrtransgaz, Naftogaz would go back into the red.[73] The government, meanwhile, was under growing pressure from international lenders to show progress on the reform. In October 2018 an unusual joint letter "highlighting the urgency of unbundling" was sent to the Ukrainian government by the European Commission, World Bank, European Bank for Recovery and Development, the Energy Community Secretariat, and the US State Department. The letter demanded that Ukraine act speedily to fulfill its commitment to unbundling as a member of the EU-sponsored Energy Community.

Thanks to this constant prodding, the last two years have brought progress. In 2018 MGU was established as an independent company, with an independent supervisory board chaired by the former head of the European Agency for the Cooperation of Energy Regulators. In 2019 a joint working group was created with five European transit system operators, to enable MGU to assume its responsibilities as a Ukrainian transit system operator. MGU is supposed to take title to the transmission assets from Ukrtransgaz in 2020 and begin operation as a freestanding entity.

This change—if it happens—could mark the beginning of a new era in the Ukrainian gas sector.[74] Until now, the pipeline system has been a cash cow, supporting multiple interests both inside and outside of the government. It has been one of the chief sources of rents in the legacy system. Under the reform, MGU would be a regulated utility. Transportation tariffs, instead of being set by negotiation, would be governed by standard Europe-wide methodologies. Any remaining transit of Russian gas through Ukraine would be conducted under more transparent—and, it is hoped, more stable—rules.

Upstream Investment

Ukraine has substantial gas resources that, if developed, could go a long way toward assuring Ukraine's gas independence. At current production rates Ukraine has up to fifty years of reserves already identified.[75] The main drawback of Ukraine's gas portfolio is that it is highly mature, and much of it is scattered among hundreds of small fields, generally located at great depths. Nevertheless, given the right investment climate, Ukrainian indigenous production could reach 40 billion cubic meters per year, possibly within a decade.[76]

But this assumes that present obstacles to upstream investment can be overcome. The effort to reform the upstream is just beginning, having been relatively neglected by international donors and Ukrainian reformers, compared to midstream and downstream issues. Yet there has already been encouraging progress, which has already had an impact on production.

The first breakthrough is actually a by-product of price reform. For the first time in its history, the main state-owned production company, Ukrgazvydobuvannya (UGV), is no longer obliged to sell its output at the low transfer prices paid by its parent company, Naftogaz (most of the time, these prices were so low that UGV was not even able to cover its operating costs). As a result, UGV is able to finance a more ambitious upstream exploration and development program. This in turn augurs well for UGV's future production.

Yet Ukraine so far has failed to attract significant foreign investment or domestic entrepreneurship in the gas upstream. Large international oil and gas companies showed some initial interest several years ago, but they have all left Ukraine. The future of gas production in Ukraine lies with the kind of smaller companies that have been the chief drivers of the shale gas revolution in the United States. But upstream investment has been choked by a forbidding thicket of burdensome regulations, punitive tax rates, uncertain policies, and perceptions of high political risk. Nevertheless, there has been some encouraging early movement. A new fiscal regime for new gas wells went into effect in January 2018, featuring a low 6–12 percent royalty rate (depending on well depth). The challenge for the Ukrainian government will be to keep this lower rate fixed so as to give would-be producers stable expectations on which to venture their capital. So far, such stability has been missing.

In addition, in early 2018 Ukraine passed a special law aiming to deregulate the upstream. This law, combined with a number of other regulatory adjustments in 2017–2018, significantly reduced the number of bureaucratic approvals required to develop a hydrocarbon deposit. It also introduced more modern field development rules and made the licensing process more fair and transparent.

The next step will be to offer foreign investors something of value. A new, transparent licensing system has been adopted, and licenses are being offered through auctions. However, the first auction under the new rules netted only a modest $5 million and drew no foreign players. To attract outside players Ukraine will need to make upstream data available and accessible for analysis, a prospect that brings no enthusiasm from Ukrainian bureaucrats. As one well-informed observer concludes, "It all takes time and capital."[77]

Finally, a number of foreign investors have been in discussions with state companies, which still control around 80 percent of production and reserves, about possible ways to invest in their licenses. But to date no acceptable mechanism has been found to invest foreign money directly into upstream projects operated by Naftogaz's upstream subsidiaries which include UGV and Ukrnafta, the latter of which is controlled by a powerful Ukrainian oligarch, Ihor Kolomoiskyi.

Despite the progress achieved to date, the balance between reform and reaction remains fragile. Tymoshenko ran as a populist candidate in the 2019 presidential election, promising to roll back the gas price increase, but she was eliminated in the first round—thus removing the most immediate threat to the reform. The positions of the new president, Volodymyr Zelenskiy, and the new parliament are as yet unknown at the time of writing (summer 2019), but so long as Ukraine depends on external funding and international lenders continue to exert pressure, an outright retreat seems unlikely.

The Evolving Russian-Ukrainian Gas Relationship

Meanwhile, Russian-Ukrainian gas relations remain as contentious as ever. As the 2010s came to an end, Gazprom and Naftogaz were engaged as usual in multiple quarrels over debts, penalties, and tariffs. In December 2017 and February 2018, two decisions by the Arbitration Institute of the Stockholm Chamber of Commerce brought a new exchange of venom between Kiev and Moscow, reminding European bystanders how vulnerable the Russian-European gas trade remains to the bad political relations between the two countries.[78]

The Arbitration Institute is a remarkable institution.[79] Founded in 1917 as part of the Stockholm Chamber of Commerce (although independent of it), it has established an unparalleled reputation as the world's leading body for the neutral resolution of East-West business disputes. Although it is not a court, it is frequently referred to as such, indicating the legitimacy it has acquired over the decades as a highly competent body, and governments are required under the New York Convention to enforce its decisions.

The first decision concerned the treatment of exports from Russia to Ukraine under their 2009 contract. It was in several respects a victory for Ukraine, inasmuch as the Arbitration Institute threw out a number of Gazprom's claims. In particular, it dismissed the Russians' demand that the Ukrainians pay $56 billion for failing to observe the take or pay clause in the contract. But the Arbitration Institute also found in Gazprom's favor over Naftogaz's underpayment or nonpayment for gas delivered in December 2013 and the second quarter of 2014, and ordered Naftogaz to pay $2.02 billion, plus interest. Gazprom chose to interpret the decision as a victory. "Naftogaz owes us over $2 billion," said Aleksandr Medvedev (then still Gazprom's deputy chairman) exultantly, "and the clock is ticking."[80]

But the second decision, two months later, brought an explosion from Moscow. This time the Arbitration Institute found that Gazprom had failed to meet gas transit obligations under the 2009 contract and ordered Gazprom to pay $4.63 billion in transit fees for having made insufficient transit shipments. Here too the arbitrators' rulings were mixed: the award fell far short of the $16 billion that Naftogaz had demanded, and the Arbitration Institute rejected a claim by Naftogaz for a review of transit rates in the 2009 contract. Each side could boast a measure of victory. But the combination of the two judgments left Gazprom with a net $2.63 billion to pay (now $2.8 billion at this writing, assuming interest—and Moscow reacted in fury.

The affair quickly escalated over the following days. "The Stockholm Arbitration Court made an asymmetrical ruling," declared Miller.[81] Gazprom vowed it would not pay and began proceedings to terminate the 2009 contract. Despite the exchange of hostilities, however, transit flows to Europe—in the middle of an exceedingly cold wave—were not affected.[82] But the entire quarrel cast a pall over Russian-European gas relations, as Europe wondered what would happen next.

The conflict will not be settled soon. Gazprom is appealing the processes of the arbitration body to a Swedish appeals court, a procedure that will take a year or more. During this time Ukraine and Russia will be locked in litigation. Naftogaz, meanwhile, has threatened to attach Gazprom's assets throughout Europe and may start seizing and auctioning them. Because of the New York Convention, which makes the decisions of the Stockholm tribunal binding, there are strong pressures on Gazprom to yield, but at this writing Gazprom is continuing to refuse to pay.[83] In addition, Gazprom for the present still needs Ukrainian transit, especially if the Nord Stream 2 pipeline is delayed. Yet Gazprom and Naftogaz are refusing to talk to one another, and as a result the 2009 contract may expire without any new agreement. Beyond that lies unknown territory.

Thirty Years of History: Summing Up

Despite the latest controversy, one must not lose sight of the long-term trends. Over the thirty years since the end of the Soviet Union, the Russian-Ukrainian gas relationship has changed radically. The changes can be summed up under the following five main points.

First, gas rents have shrunk. As Russia's surplus of legacy Soviet-era gas declined, it went from being long on gas (1992–1998) to short (1999–2008). Thanks to investments in new supply from the mid-2000s, it is now long again (since 2009), but there has been a steady increase in the marginal cost of Russian gas. The costs of Soviet-era gas are no longer sunk. The result is a virtual disappearance of gas rents available for distribution to favored interests. As time goes by, gas is less and less a source of corruption.

Second, barter has vanished as the gas trade has moved to money. Since 1998 (in Russia) and 2001 (in Ukraine), the economies of the two countries have been remonetized, and barter in all its forms—most of them shadowy—has mostly been replaced with cash payments, a more transparent mode of exchange. The largest remaining in-kind transaction, payment for transit in gas, was finally eliminated in 2006. As a result, intermediaries have faded, and the role of politically favored traders has declined.

Third, since 2008 Turkmen gas has disappeared as an alternative source of gas for Ukraine. This has turned what was once a complex and ever-shifting triangle, with many intermediaries, into a more straightforward two-player game between Ukraine and Russia alone. Instead, most Turkmen gas now goes to China.[84]

Fourth, Russia's interest in acquiring control of the Ukrainian export pipeline system has disappeared. On the contrary, Russia's policy for the past twenty years has been to bypass the Ukrainian system, a goal that it has nearly achieved with the completion of Nord Stream 2 and Turkish Stream.

Lastly, the introduction of gas reforms in Ukraine may ultimately change the basis of the Russian-Ukrainian gas relationship. However, politics has for the present won out over business. Only time will tell whether reforms in Ukrainian gas, and particularly the creation of an independent European-style transit system operator, will survive the present hostilities and ultimately lead to a new commercial relationship.

Is Ukraine Headed for Gas Independence?

A favored narrative in the West, especially in Washington, is that the key to understanding the Russian-Ukrainian gas relationship is that gas gives Russia a powerful geopolitical lever, which it has repeatedly used to manipulate Ukrainian politics and to bind Ukraine to its sphere of influence. There is no question but that under President Putin Russia has indeed pursued these objectives. But there is a problem with the gas weapon part of the narrative. Whenever Russia has really sought to influence Ukrainian politics or policy, it has for the most part not used gas as a weapon. Rather, it has deployed an array of classic means of coercion, ranging from political technologists to covert military support, and including disinformation, computer hacking, financial pressures, trade sanctions, weapons transfers, and simple blackmail and bribery. In short, Russia has no lack of weapons, without resorting to gas.

Instead, the handful of instances in which Russia has deployed the gas weapon have involved specific issues related to the gas trade, mainly gas debts and the control of transit. The prime example has been Russia's repeated attempts to gain mastery of the Ukrainian gas transit system, which were clearly aimed at curtailing Ukraine's leverage over Russia's gas exports to Europe. The same point could be made about the gas cutoffs of 2006 and 2009, which were prompted mainly by disputes over gas debts and the application of contract terms.

Indeed, the gas weapon as a geopolitical instrument has been ineffective in Ukraine. In the first decade and a half of the post-Soviet period, the Russians underpriced gas exports to Ukraine to win goodwill. But arguably they actually gained little tangible geopolitical benefit in exchange. Arguably, such tactics worked in Belarus and in the smaller republics of the Former Soviet Union, where gas has been combined with other instruments of leverage.[85] But the fact that Russia has embarked instead on a program of pipeline diversification to bypass Ukraine testifies to the failure of the gas weapon to procure significant geopolitical—and, for that matter, commercial—advantages there.

Thus, the view of gas as a weapon fails to explain much of the history of the Russian-Ukrainian gas relationship, and especially the two landmark events, the gas cutoffs of 2006 and 2009. As Simon Pirani writes, "In the Russo-Ukrainian 'gas wars' of 2006 and 2009, the dynamics of post-Soviet transition, economic events, and shifting commercial relationships were more significant than political factors."[86]

There is, to be sure, a sense in which gas has been used indirectly to build a collusive relationship from which several generations of Ukrainian rulers and their backers have directly benefited, helping keep Ukraine corrupt and dysfunctional and giving Russia leverage it could always threaten to use. But this is very different from a policy instrument as it is conventionally understood, for the simple reason that it compromises the wielder as much as the wielded. A full release of the *kompromat* behind the Russian-Ukrainian gas relationship over the last thirty years would likely be embarrassing for both sides.

This chapter has attempted to show how factors such as these have dominated the story. Thus, it argues that the main cause of the endless conflicts in Russian-Ukrainian gas relations is the Soviet inheritance—the tie between Ukrainian dependence on Russian gas, on the one hand, and Russian dependence on Ukrainian transit to Europe, on the other. In the background was Russia's initial surplus of legacy gas and Ukraine's extreme reliance on gas for its inefficient economy. These factors, inherited from Soviet history and the Soviet breakup, are now fading or are gone altogether.

As with everything else about Russia and Ukraine, the last paragraph is subject to multiple caveats. The first is Ukrainian domestic consumption. The change since 1992 has been breathtaking: gas demand dropped from 110 billion cubic meters in 1992 to under 30 billion cubic meters in 2016 (excluding the areas controlled by separatists). The causes include the decline in the economy as a whole, the rise of the service sector, and gains in the efficiency of industrial consumption.[87] But the greatest breakthroughs are the decontrolling of domestic prices and the reductions in cross-subsidies. Even so, Ukraine remains one of the least efficient energy consumers in Europe, and there is clearly scope for additional efficiency gains. The easiest gains have now been made. Further decreases in consumption will require much more investment in industry and district heating systems. Given the acute scarcity of foreign capital available to Ukraine, this investment will remain modest, and domestic consumption is likely to remain fairly close to the current levels of 32–33 billion cubic meters per year well into the next decade. The same point applies to the prospects for increased indigenous production.[88] Nevertheless, the net result has already been the disappearance of direct Ukrainian dependence on Russian gas, and its replacement by reverse imports.

The second caveat is that further progress toward independence depends on continuation of the gas reforms. With the defeat of Tymoshenko and the victory of Zelenskiy in the 2019 presidential election, the threat that the gas reforms might be interrupted or reversed by an overtly populist president has been averted. The new president is broadly favorable to close association with the European Union and a continuation of the reform agenda. But his views on gas and gas reform are unknown. If Ukraine slides back to subsidized domestic prices and cross-subsidies, if Naftogaz postpones its unbundling and other reforms, and above all if Kiev agrees to a new transit contract that is perceived as excessively favorable to Moscow, the new president's reputation with foreign lenders and investors would be tarnished and Ukraine's progress toward energy independence would be impeded.

Yet ultimately the final dissolution of the Soviet gas legacy, with its inheritance of collusion and conflict, is inexorable. Seen in a long-term perspective, this will be a positive—both for Ukraine and Europe, and ultimately for Russia.

CHAPTER 12

Russian-German Gas Relations

Any discussion of Russian-German gas relations since 2000 must begin with two puzzles. The first relates to the immediate present: on the one hand, Russian-German political relations have never been as difficult and uncertain since the end of the Soviet era as they are today; yet on the other hand, the Russian-German gas relationship is booming, causing difficulties in US-German relations. Russian gas exports to Germany have broken all records, and Germany now relies on Russian gas for nearly 38 percent of its gas imports, up from 35 percent in 2015.[1] Two German companies are among the five corporate backers of a new Russian pipeline to Europe, the Nord Stream 2 pipeline, with the support of the German business community—despite the opposition of the EU Commission and several East European EU members, as well as that of the US government. Germany continues to export pipe and equipment to Russia for its gas pipelines and infrastructure.

At the same time, Russian-German political relations are, as a recent analysis by the German Institute for International and Security Affairs (Stiftung Wissenschaft und Politik, or SWP) puts it, "a special relationship in troubled waters."[2] This has been especially true since the overthrow of President Viktor Yanukovych's regime in Ukraine in 2014, the annexation of Crimea, the Russian-supported separatist takeover in the Donbas, and the shooting down of the MH17 Malaysian airliner by a Russian rocket over eastern Ukraine. The sanctions imposed by the United States and the European Union have been supported by Chancellor Angela Merkel, the German government, and most business groups.[3] Yet on the gas front business remains vigorously active. How can we explain this apparent contradiction?

The second puzzle stems from the revolutionary changes that have taken place in the regulatory and commercial setting for the gas trade, resulting from the liberalization of the German gas market discussed in Chapter 8. This spelled the end of the Groningen era of long-term oil-indexed contracts and marked the rise of hubs and spot markets, both in Europe as a whole and in Germany. These were accompanied by a decade of conflict between Gazprom and its European customers over issues such as contracts and prices. Yet as intense as these disputes have been, they have not led to an interruption of the gas trade or even the threat of one. On the contrary, the major conflicts in the gas trade have been resolved by negotiation or arbitration within the framework of existing legal contracts and commercial dialogue, despite the overall cooling of political relations between Germany and Russia. Why has the overall frostiness of political relations not had more of an impact on these controversies over gas?

In the background looms a third puzzle. Until recently the policy debate over Russian gas was dominated by issues of economic efficiency versus security. Yet both have been overtaken by a third set of issues, connected to the environment and the German energy transition, called the *Energiewende.* This is ultimately a more radical challenge than either of the other two, because it appears to call into question the very function and future of gas in the German and European economies. Already one likely result can be glimpsed in the distance, but it is a paradoxical one. Although gas is regarded with increasing disfavor by environmental advocates, the medium-term result of the *Energiewende* in the 2020s is likely to be increased German demand for gas, particularly Russian gas. There are three major reasons for this. First, in December 2022 the last remaining nuclear power plants will close. Second, the construction of a new north-to-south ultra-high-voltage DC transmission line, essential to the further penetration of renewables into the power sector, is likely to be delayed to the mid- or late 2020s. Lastly, mounting public opposition to coal (as well as German determination to improve its performance in limiting greenhouse gas emissions), combined with pressure from Brussels, will accelerate the decommissioning of German's coal-fired power plants. These three factors combined will leave a hole in Germany's energy supply, which can be filled only by gas. By the end of the 2020s, Russian exports of gas to Germany may well leave today's records behind, which will increase the need for transit capacity. We return to this question in the Conclusion.

Needless to say, this prospect offers provocative food for thought for both the efficiency and security communities. And little wonder that in the SWP report mentioned above, the German energy analyst Kirsten Westphal and her coauthors Aurélie Bros and Tatiana Mitrova conclude, "The major change on the economic and commercial sides stems from the loss of a long-term vision."[4] Yet no long-term vision is possible if the very terms of the debate are about to be upset.

The plan of this chapter is to attempt to unravel the first two puzzles, while leaving the third for the Conclusion. The basic explanation argued here is that German gas policy toward Russia is the result of two different sets of values—security versus efficiency—and the different communities that defend each. Where the balance is struck between the two depends on which issue and which period one is talking about, but on the whole efficiency has been the dominant value in practice.

Efficiency versus Security: Two Communities of Opinion

In Europe as a whole, and Germany in particular, there are two communities of opinion—effectively, two camps—on the subject of Russian gas. For the first community, economics and efficiency are the paramount explanations of events and the surest guides to policy. In this view, the real issues that divide the European Union from the Russian government in gas policy, and Gazprom from its European customers, are above all commercial and regulatory. Prices and volumes are matters of business strategy, not government policy. Pipelines are either economic or they are not. Conflicts, when they occur, are above all the result of the changes in the European gas market. For those who hold these views, an interruption of supply would be commercial suicide for the party that attempted it; and the impact of any interruptions can easily be mitigated with fixes such as interconnections and storage, which make good business sense in any case. This is broadly the view of the German gas industry and analysts specializing in the economics of gas, as well as the more business-oriented parts of the German government, such as the Federal Economics Ministry.

For the other community, the Russian-German gas trade is fundamentally about geopolitics and security, particularly the threat of gas as a weapon for Russian political and strategic ends. In this view, Gazprom's pricing practices in Eastern Europe are part of a systematic policy of granting or withholding political favor. Its reluctance to embrace European liberalization simply reflects Russian discrimination against its former satellites and possessions. In this reading, a pipeline is a means of political leverage, part of the arsenal of an energy superpower, which will naturally attempt to use it as a geopolitical resource, and an export strategy is a bid to dominate. As for supply interruptions, the point is that they could happen, and to disregard the possibility is naïve. This is broadly the view of the foreign affairs and security communities, both in Germany and throughout Western Europe.

The split in the literature and in the policy world between these two groups is striking. Very few scholars or institutions have attempted to bridge the divide. It is a partly a matter of differences in professional background and employment: energy companies and economics institutes not surprisingly defend efficiency; policy think tanks stress security; and governments are divided across ministries. But the gap also divides Europe into two different intellectual cultures in two geographic zones.[5] In Western Europe, where liberal ideology has spread over the past generation, the efficiency view tends to predominate, particularly in the business world. In contrast, in Eastern Europe, where this revolution is only now beginning to penetrate, the second view holds sway. One's view, of course, is also a function of one's proximity to Russia. Poland, for example, is squarely in the security camp.

Germany sits in a special place, at the center of both communities. Consequently, not surprisingly, its stand is ambivalent and complex. As we have seen, market liberalization has won the day in Germany, which has tended to reinforce a traditional bias in favor of the efficiency argument in gas policy toward Russia. Indeed, until recently Germany's overall post-Soviet policy toward Russia has been predominantly economic, a continuation of the historic *Ostpolitik* that created the gas bridge in the first place. Writing on the eve of the Second Maydan demonstrations and the overthrow of the Yanukovych government in Ukraine in 2014, and the ensuing crisis in East-West relations, Stephen Szabo, an American expert on Germany and the author of a well-regarded book on German-Russian relations, wrote, "Economics is the driving factor in the German-Russian relationship." He went on, "Almost all of Germany's Russia-watchers see this as the constant factor and one that favors a geo-economic approach over a [security]-oriented one. Whatever the ups and downs in the broader relationship, the economic one remains a success story from the German perspective and remains its anchor."[6]

Is the same still true today, despite the recent traumatic developments in Russian-German political relations? Where gas is concerned, there is an underlying community of interest that continues to drive the gas relationship in an economic direction. Germany needs gas, while Russia has the cheapest and most abundant gas available. Germany is Europe's largest importer of Russian gas, and it is Russia's second largest source of imports—consisting chiefly of machinery and manufactured goods, which, in effect, are paid for largely through Russian exports of oil and gas.[7] Gazprom's gas sales to Germany continue to be covered by long-term contracts (even though the prices within them are increasingly set through spot markets).[8] German industry remains as always committed to active trade with Russia, despite US and EU sanctions aimed at punishing Russia for its intrusions into Ukraine. Certain themes have an almost eerie familiarity, such as the resentment of parts of the German business community over the United States' toughening of sanctions against Russia and Washington's recent attempts to block a Russian gas pipeline (see below), just as in the early days of the gas bridge back in the early 1980s. Yet despite the overall political cooling, the basis for the economic relationship remains as complementary today as it was in past decades.

Thus, on one level, Germany perceives no contradiction in the present gas situation. As Angela Stent writes in a recent analysis, "As a geo-economic power, where trade is seen as a vital aspect of national security,

Germany traditionally has defined its interests largely in terms of commercial realpolitik, viewing the pursuit of its economic interests as the ultimate test of the success of its foreign policy."[9] This has not changed.

In contrast, for the other camp, the liberalization of German gas has only increased potential German insecurity, especially at times when political relations worsen. Thus, for Kirstem Westphal of the SWP, the German gas market has changed from "a bilateral monopoly to an imperfect market." She goes on, "Gazprom has enhanced its position in the German market, whether for better or for worse in terms of supply security depends on transparency, effective control, and the implementation of rules and regulations, but certainly requires close monitoring."[10] But these conditions are as yet inadequately strong in Germany, Westphal argues, because there has not been sufficient attention to long-term institutional protections, which can be provided only by the state, and adequate laws and policies. The weak point of the present gas relationship, according to this view, is security. Hence the temptation, for the security-minded, is to attempt to take greater control of the regulatory issues, particularly in Brussels, using litigation as a weapon (see the discussion of the Ostsee-Pipeline-Anbindungsleitung [OPAL] affair below).

It is only recently, chiefly as a result of events in Ukraine, that the concerns of the security community have begun to undermine the long-held German consensus favoring economics first. Until about 2011–2012, a common thread ran through German policy. The gas trade was underpinned by a complementary political relationship, the *Ostpolitik.* But since then, and for much of the present decade, Germany's political policies toward Russia have been increasingly distrustful, verging on hostility. Yet even in this latest decade Germans and Russians have attempted to insulate the gas relationship from East-West geopolitics. The gas trade's present prospering demonstrates that this remains the case today. The fundamentals of the gas bridge still operate as they traditionally have. But how will the growing tension be resolved?

We shall consider three main phases. During the first period, from the 1970s to the German reunification in 1990 and the collapse of the Soviet Union, the gas trade and its political framework operated much as it had during *Ostpolitik.* During the second, from the period of liberal Russian reforms in the 1990s to about 2005, Gazprom entered Germany as an investor and operator, and German-Russian political relations reached a high point under Vladimir Putin and Chancellor Gerhard Schröder. However, in about 2005, Germany and Russia entered a third phase, a period of growing uncertainty and discomfort, during which both Germans and Russians attempted to maintain a balanced political and economic course in the face of deteriorating US-Russian relations and complications arising from Ukraine. Since about 2012, the uncertainty and discomfort have grown. The balance still favors economics, but will that last?

After reviewing the history we shall turn to a case study that illustrates the tension between politics and economics in Russian-German gas relations, the dispute over the Nord Stream 2 pipeline.

The Way It Was: The Inherited Structure, 1970s–1990

By the middle of the 2000s, the two gas industries, Russian and German, had been in partnership for four decades, on both personal and institutional levels. They were accustomed to dealing with one another on the basis of a system that balanced profit and risk—over a substantial period of time—in a blend of arrangements that had proven their worth over time. As Aurélie Bros and her colleagues describe the traditional system:

> The contractual relationship was designed for the long term and was based on a bilateral political and commercial consensus. The market structures matched perfectly—a centrally planned economy with a regionally monopolized gas market. The business models of both sides were based on long-term, oil-indexed delivery contracts, with terms of 20, 25, or 30 years, including a minimum take-or-pay obligation to purchase at least 75 to 85 percent of the named quantity. . . . Demarcation at the border was clear: gas was delivered to the "flange" at the border.[11]

In the well-known formula, "The producer bore the price risk, whereas the importer bore the risk of failing to sell the full quantity of the contracted gas."[12] This principle governed the gas trade for decades. Russian and German gas professionals came to know one another over the course of their entire careers, and the legalistic rigor of the gas contracts was in practice softened by informal understandings and compromises.

At the center of the relationship was the fact that "throughout the Cold War, political and economic interests converged on both sides."[13] Yet what made such smooth convergence feasible was the structure of the long-term contracts, which operated automatically to manage risk in the absence of traded commodity markets for gas. The indexation to oil meant that the contract price could not be manipulated by either side or by either side's host government through its tax or regulatory regime. The presence of take or pay clauses in long-term contracts underwrote the burden of investments in the upstream and the pipeline system. Informally, it was understood that neither side would make profits while the other suffered losses, and there was broad agreement on what was considered fair. If all else failed, the contracts contained renegotiation clauses that enabled the two sides to come to the table and rework the terms in response to changing market conditions. Thus, the structure of long-term contracts, and the relations of acquaintance and trust that they fostered, enabled both sides to protect themselves from commercial as well as political risk. It is a remarkable fact that this system survived both market fluctuations and the geopolitical thaws and frosts of the Cold War from the late 1960s until 1991.[14]

The Era of Partnership and High Hopes, 1990s–2005

The eventful 1990s and early 2000s were marked at the beginning by the German reunification; in the middle by the Putin-Schröder friendship and the era of good feelings between the two leaders; and at the end by the tense relations between Putin and German chancellor Angela Merkel against the backdrop of deteriorating Russian-Ukrainian relations after the Orange Revolution, as described in Chapter 11.[15] In energy relations the high point was the entry of Gazprom into the midstream of the German gas market through its stake in Wingas and its growing partnership with the two leading gas companies in Germany, BASF / Wintershall and E.ON / Ruhrgas, as described in Chapter 8. This was followed by a second phase, in which the German companies sought to invest in the Russian upstream. During this period, German-Russian political and energy relations were aligned more closely than at any time before or since.

This was a time of hopefulness about the future of Russia, East-West relations, and the potential for liberal change in the world. The European Union placed high hopes on the power of treaties, joint projects, and institution-building of all sorts to bring Russia into what the Russian reformers called the "civilized world" of European norms and values. Symbols of this effort were the European Energy Charter of 1991 and the European Energy Charter Treaty of 1994, the Partnership and Cooperation Agreement of 1997, and the "four common spaces" of the Saint Petersburg summit in 2003.[16]

The drive to deepen and diversify ties was particularly strong in Germany and took the form of multiple agreements in practically all spheres of the energy sector, including energy efficiency. However, the German government was skeptical about the multilateral, EU-based approach to Russian energy, and as a result largely went its own way on a bilateral basis. As Westphal sums up the attitude of the Federal Economics Ministry and the Ministry of Foreign Affairs, "the EU-Russia Dialogue was widely viewed as the Commission's 'cup of tea.'"[17] In contrast, the formula that summed up Germany's approach at the time was "rapprochement through linkage" *(Annäherung durch Verflechtung)*.[18]

Of these and many similar initiatives little remains today except acronyms in the footnotes of history books. For all the calls for constructive interdependence, it soon became apparent that the two sides, Europe and Russia, remained far apart in their understanding of its meaning. As Bros, Mitrova, and Westphal note about this period, "Whereas Germany and its EU partners had perceived regime-building to be the basis for economic cooperation to minimize transaction costs and create level playing fields, Russia increasingly resisted supporting this rule- and norm-based understanding of interdependence. Moreover, national sovereignty over natural resources remained paramount in Russia, given their role for the Russian economy and the Russian budget."[19]

During this period Germany was clearly the dominant partner. There had been a fundamental shift in the political relationship following reunification, "from a German dependence on Russia for inter-German relations to a Russian dependence on a unified Germany for its post-communist transition."[20] The economic bridge between Germany and Russia remained essentially unchanged in kind, with Russia continuing to export energy and import manufactured goods as the Soviet Union had done. But the overall volume of exchange shrank, as Russian GDP fell by nearly half following the Soviet collapse. The low point came in 1998–1999, as a result of the Russian default on its state debt in August 1998, which caused severe losses to German investors who had put money into Russian short-term state bonds.[21] (Beginning in 1998 Ruhrgas took minority stakes in Gazprom acquiring 6 percent of the company for $1 billion as a means of providing capital inflow into Russia to help its balance of payments.) Yet once oil prices began to recover in 1999–2000 soon followed by gas prices, which were keyed to oil, the economic relationship began to prosper as never before, and it continued to blossom as oil and gas prices reached record heights both before and after the recession of 2008–2009.[22] Hans-Joachim Spanger, a German authority on Russian-German relations, captured the optimism of this decade and a half: "Russia was transformed," he wrote, "from an unreliable border to a market of unlimited possibilities."[23]

Relations remained warm throughout the chancellorship of Helmut Kohl, reaching a high point under his successor, Gerhard Schröder (1998–2005). The key was the close personal friendship between Schröder and Putin, the new Russian president. As Spanger observes, "Schröder became the major advocate for German investment in Russia and for an energy policy dialogue, arguing in 2004 that the confidence of Western investors in Russia had been fundamentally renewed and re-established."[24]

Schröder was not initially disposed to invest personal or political capital in Germany's relationship with Russia. Germany had economic difficulties of its own in the 1990s, and Schröder's top priority was domestic policy, especially a reform of the sclerotic employment and welfare systems. Moreover, Schröder was skeptical of what he viewed as his predecessor Kohl's excessively friendly ties with President Boris Yeltsin. As Stent writes, in Schröder's view "Kohl had developed too cozy a relationship with the erratic Russian president Boris Yeltsin—including sharing the sauna with him—and Schröder vowed to take a more critical stance toward Russia."[25]

Yet almost as soon as they met, Schröder and Putin developed an exceptionally close personal rapport, which endures to the present day. In January 2001, the two presidents and their wives celebrated the Orthodox Christmas together by touring the sixteenth-century Kolomenskoye estate in Moscow in a horse-drawn troika. As Stent describes the scene,

> Together they visited the fourteenth-century Sergiev Posad monastery, which is regarded as the spiritual center of Russian Orthodoxy, and were greeted by women in traditional folk dresses and a choir chanting solemn Russian liturgy. There they met with Patriarch Alexy II, the head of the Russian Orthodox Church. The sleigh ride captured not only the spirit of Christmas but also carried the spirit of the new relationship between Russia and Germany.[26]

This was but the first of many meetings, both formal and informal, between the two men over the following five years. By the time of Schröder's departure from office in 2005, he and Putin had become close personal friends. The high point in the relationship came that year, at the time of the parliamentary elections in which Schröder lost the chancellorship. In September 2005, ten days before the German election (which it had been widely predicted that Schroder's Social-Democratic Party SPD would lose), Putin came to Berlin, and the two leaders looked on as Gazprom, E.ON, and BASF signed a contract to build a pipeline under the Baltic Sea, called the North European Gas Pipeline, which subsequently became known as the Nord Stream 1 pipeline.[27] Predictably, Eastern European states—by this time newly admitted members of the European Union—were indignant. President Alexander Kwasniewski of Poland denounced the contract as the "Putin-Schröder Pact." His defense minister, Radoslaw Sikorski, was even more direct, denouncing the pipeline as comparable to the 1939 Nazi-Soviet Molotov-Ribbentrop Pact that divided Poland and the Baltic Republics.[28] But the reaction in Western Europe was largely positive. After the signing in Berlin, Putin went on to London, where "he spoke about energy security and Russia's willingness to provide Europe with fuel. His audience listened approvingly."[29]

German public opinion was taken aback, however, when shortly after his resignation as chancellor, Schröder accepted the position of chairman of the shareholders' committee of Nord Stream, although there was little opposition to the pipeline itself. The security community today, not surprisingly, believes that Schröder went too far in emphasizing economic ties over political issues, at the expense of security. As Westphal writes, "[Schröder's] interest-based policy turned a blind eye to the political shifts in Russia."[30] Even at the time, Schröder's policy was controversial, as many accused him of keeping silent about human rights in exchange for Russian gas. As he left office in 2005, it was widely expected that his "schmooze-policy" *(Schmusekurs)* would give way to something more balanced.[31] But over the following fifteen years, Schröder's personal relationship with Russian business has become, if anything, steadily more intimate.

We have seen in Chapter 8 how Gazprom came into the German gas market in 1989–1990 and came to play an active role in the midstream in partnership with two leading German companies, Ruhrgas (now E.ON) and BASF / Wintershall. After a few years of successful relations the two German companies began to seek positions in Russia itself. Wintershall and E.ON Ruhrgas would invest in upstream properties in West Siberia where German technology and capital would provide a boost. But in exchange, Gazprom wanted the German companies to sell it stakes in the German downstream. This approach was in theory highly complementary. The Russians would invest downstream, the Germans upstream, and both would share in the transportation. Thus, each side would balance the risks: each could hold the other hostage in case of difficulty. The model came to be known as a "barbell."

Yet both sides were reluctant to give the other side what it wanted. Gazprom was resistant to granting a foreign company access to its fields. Negotiations between Wintershall and Gazprom over the Yuzhno-Russkoe field in the promising Achimov formation of the giant Urengoy field went on for nearly a decade from the time the first strategic memorandum was signed to the moment Wintershall actually began work. Logically, the Achimov formation should have been an important part of Gazprom's interim strategy. With an estimated 700 billion cubic meters of reserves, Yuzhno-Russkoe was intended to supply gas to Germany through the Nord Stream 1 pipeline. Although it lay deep beneath the ground in a complex geological formation and was correspondingly even more of a challenge than conventional dry gas, the prize was attractive, because the Achimov formation was rich in gas liquids that could be marketed separately, making the entire project economic. Yet Gazprom was reluctant to allow the Germans to market their own share of the liquids, agreeing only to buy them from Wintershall at a "fair price." As for the gas, Gazprom proposed to reserve two-thirds of it for the domestic market, an arrangement the Germans found unappealing. Yet the Germans were being asked to provide the bulk of the financing. So the talks dragged on.[32]

Similar talks involving Ruhrgas took nearly as long, despite the fact that Ruhrgas had taken an equity stake in Gazprom and had a seat on the Gazprom board.[33] Even the intervention of President Putin failed to budge Gazprom from its positions.[34] Clearly, at that rate, the harder-to-reach deep formations of the West Siberian fields, which the Germans proposed to develop, would not be part of Gazprom's supply portfolio any time soon. It was not until the second half of the 2000s that the strategic concept of balancing upstream against downstream positions was finally realized in actual projects.

The key to the arrangement was the Russian companies' interest in the lower end of the German gas market. This was an outgrowth of the Russians' long-standing conviction that the lion's share of the value of their exports was being siphoned off by middlemen—including the Russians' own German customers. Not everyone agreed with this judgment, which appeared to owe more to Soviet-era ideology than to market reality. As European gas professionals knew, most of the value in exports lies in the upstream—indeed, that was why they wanted to be there. Burckhard Bergmann, a member of the Gazprom board, spoke disapprovingly of Gazprom's aim to take positions in Germany's midstream:

> They should look at it from an economic, not an ideological point of view. The world has changed, and you can sell to final consumers indirectly, through the distribution systems of third parties. The tariffs are regulated . . . so that today there is no danger or risk of not gaining access to the final consumer. One must ask oneself, why haven't other

> large producers gone down that path? Statoil, Exxon, Shell aren't acquiring stakes in distribution systems, but on the contrary are selling them off. For Gazprom it would make more sense to invest in upstream production than in sales or in acquiring stakes in the final consumers. As a member of the board of directors of Gazprom, I think the company should be looking first and foremost to its own cost-effectiveness. It's a question of what comes first. I think that for Gazprom that means production and the upgrading of its pipeline system. That's why I firmly recommend that Gazprom should invest its capital inside Russia.[35]

But old business models died hard, and Gazprom persisted in its faith in vertical integration. (Aleksey Miller argued in addition that vertical integration was a guarantee of security of supply for European customers.) It pressed E.ON hard to concede stakes in Ruhrgas,[36] in exchange for a 24.5 percent stake in the Yuzhno-Russkoe field in West Siberia.[37] But E.ON's CEO, Wulf Bernotat, resisted granting access to the domestic German market, on the ground that it would give Gazprom a key position within E.ON and make its relations with other gas suppliers more complicated. Bernotat offered Gazprom stakes in the Hungarian gas sector instead, which caused protests from the Hungarian government. The resulting standoff caused negotiations to be dragged out for over two more years, until finally a deal was struck in 2006.[38] (In the end, Gazprom accepted the Hungarian assets.)

The interesting point about the negotiations over Yuzhno-Russkoe is that a major commercial negotiation took place over a period of years with no observable intervention by the German government. Whereas on the Russian side Putin was active in support of the plan (although apparently to little avail), for the German companies Yuzhno-Russkoe was a private-sector affair. The issues were primarily commercial, not geopolitical.

Clouds on the Horizon, 2005–2012

By this time, however, Russian-German political relations had grown more difficult. There were three fundamental reasons. The first was a deterioration in Russian-Western relations between 2000 and 2007, culminating in Putin's historic Munich Speech of 2007, which is widely viewed as a turning point in Russian-Western relations.[39] The second was increasing East-West tensions surrounding Ukraine and the Baltic Republics, catalyzed by the Orange Revolution of 2004. The third was the growing impact of EU regulatory reforms on the structure and operations of German utilities, especially in gas and electricity. Despite the efforts of the German political and business elites to keep German-Russian relations on an even keel, by the end of the decade they had become uncertain and troubled.[40]

Symbolizing the increasingly fraught relationship was the difficult personal relationship between President Putin and Chancellor Merkel. All accounts agree that relations got off badly when they first met. Today, after a decade and a half of many meetings and phone calls, their relationship has settled into a routine that allows essential business to get done, but without warmth or trust. Analysts differ over just when a tipping point occurred in Merkel's view of Putin—for John Lough of Chatham House, for example, it was the Russian-Georgian War in 2008,[41] whereas for Stent it was 2011, when Putin announced he would return to the Kremlin as president—but all observers would agree with Merkel's biographer, the journalist Stefan Kornelius of the *Süddeutsche Zeitung,* when he writes:

> They have followed similar paths in life, almost as if they were mirror images. . . . Whenever Putin and Merkel meet, two world views collide. For Merkel, the fall of the Berlin Wall was a liberating experience, whereas for Putin . . . it was a deeply traumatic event. He sees the collapse of the Soviet Union as a historic defeat.[42]

These two radically different views of the world, and the different approaches to it that the two leaders epitomize, provide the context for understanding the bitterness and sense of personal betrayal that both felt at the time of the fall of the Yanukovych government in Ukraine in 2014 and the subsequent takeover of the Donbas by Russian-supported separatists. Merkel's feelings run especially deep, because they go to the heart of her beliefs about the modern world. Her words in her report to the

German parliament still resound today as the most eloquent statement by any Western leader on the retrograde nature of Russian policy in Ukraine:

> For—I cannot say it often enough or with enough emphasis—the clock cannot be turned back. Conflicts of interest in the middle of Europe in the 21st century can only be successfully overcome when we do not resort to the examples of the 19th and 20th centuries. They can only be overcome when we act with the principles and means of our time, the 21st century.[43]

Yet Merkel remains above all a pragmatist, and she understands that German public and political opinion are sharply divided over Germany's Russia policies.[44] This is particularly the case with gas, as we shall see below when we come to the Nord Stream 2 pipeline. The Russia sanctions adopted by the United States have created a new source of pressure on the chancellor. She has been able to hold the line in support of sanctions and has received some support from German business: revealingly, while the German Ostausschuss (the influential German Committee on Eastern European Economic Relations) has criticized German support of the Russia sanctions, the even more influential Bundesverband der Deutschen Industrie (Federation of German Industry), has supported the chancellor's policy. Merkel maintains regular contact with Putin and the Kremlin, but without warmth or accommodation.

This brings us back to the contradictions with which we began the chapter. Given the split in German public and political opinion, combined with Merkel's determination to hold the line on sanctions while maintaining the pragmatic middle, how has Germany handled gas politics? The main answer, from the chancellor's office to the business community, is that gas and gas-related issues are wherever possible a matter for the private sector. The state gets involved through its regulators, but one should note that the influential German Bundesnetzagentur (BNA), the state network agency that is responsible for pipelines, is an independent agency. This role is recognized and valued by the top leadership. On the twentieth anniversary of the founding of the BNA, celebrated in May 2018, Merkel paid tribute to the agency's unique regulatory role as "deserving a high degree of credit for a functioning economy in Germany."[45]

Meanwhile, Commercial Relations Go Their Own Way

In gas relations as in geopolitics, 2005 was a high point in the Russian-German era of good feelings. But a mere three years later, economic conditions changed sharply, in the wake of the recession of 2008–2009. Gas demand in Europe plummeted, and European utilities found themselves in a perfect storm of commercial problems: loss of monopoly, accompanied by surplus supply and falling demand. Meanwhile, oil prices resumed their sharp rise (owing to increased demand from China) and grew steadily through mid-2014.[46] This put the European gas utilities, especially the major German players, in a bind, because their supply contracts with Gazprom were still linked to oil prices, while their customers now had the option of buying gas at low prices on the emerging trading hubs. "In 2009," Jonathan Stern relates, "European hub prices fell significantly, to levels as much as 50 percent below oil-linked contract prices . . . and averaged 25–33 percent below oil-linked prices for the next five years."[47] E.ON and RWE, two of the largest buyers of Russian gas, were particularly affected.

Here was an issue that went to the heart of the debate between economics and efficiency versus geopolitics and security. How was it resolved, and by whom—the companies or state players?

In public, Gazprom stood firm behind the traditional oil-linked long-term contracts and insisted it would not budge on the oil linkage as the basis for pricing. One result was a furious exchange of broadsides between Gazprom Export's Sergei Komlev and European gas analysts, in particular the leading academic experts in the field, from the Oxford Institute for Energy Studies. Komlev argued that gas was special and therefore could not be priced like an ordinary commodity.[48] In effect, long-term contracts were sacred. Not at all, retorted the Oxford Institute, which denounced Komlev's position as contrary to economic logic.[49] But Komlev's position was backed by Putin, who declared in July 2013 that Gazprom would "continue to support gas pricing based on oil / oil products indexation to ensure fair prices and stable development of natural gas resources."[50]

In this debate there were few public declarations by German leaders or ministries on the supposedly proprietary issues of pricing policy or contracts. There is hardly any literature from German institutes and think tanks similar to that from the Oxford Institute. But in international conferences and private working groups, the Germans made their position clear. At the Offshore Northern Seas conference of 2010, for example, the newly named CEO of E.ON Ruhrgas, Klaus Schäfer, declared: "Hubs are the reference point when customers talk to us. . . . Long-term contracts [that is, with oil-linked prices] in their current form no longer reflect the market."[51] And there are many public statements by senior German utility executives to the same effect. In other words, in Germany the issue was handled as a matter for the private sector.

In contrast, with the Kremlin's support, Gazprom stuck to its guns—at least in public. But in actual contract negotiations its behavior was considerably different and amounted to a tacit, if reluctant, acceptance of the emerging market-oriented rules of the game. In reality, it had no choice: its most important customers in northwestern Europe were rebelling. Beginning in 2012, in a series of difficult price negotiations, they pressed Gazprom for three key concessions: a lower base price; a more relaxed take or pay point (70 percent of the annual contract quantity instead of the 85 percent that had been the norm); and most significantly, an agreement that any significant excess over the hub price would be repaid to the buyer in the form of rebates. (The latter, in particular, were sizable sums, amounting to billions of euros.) Whenever Gazprom resisted, one European company after another turned to the Arbitration Institute of the Stockholm Chamber of Commerce. Faced with a barrage of cases, as well as growing evidence that its partners were in acute financial difficulty, Gazprom had little option but to give ground. The German companies were part of the general movement. Thus, in July 2013 the Stockholm Arbitration Institute ordered Gazprom to pay 1.5 billion euros in rebates to RWE. Three years later, in April 2016, Gazprom and E.ON settled a similar case out of court, but the result was the same, as Gazprom made further concessions in pricing. In other cases just the threat of arbitration was often sufficient. The details are not publicly known, since gas prices and contracts are closely held proprietary secrets, but the general pattern is quite clear.

This episode is of special interest for three reasons. First, it underscores the point that despite worsening Russian-German geopolitical relations, the commercial relationship, though difficult, followed the overall pattern of northwestern Europe—that is, tough negotiations accompanied by legal proceedings with billions of dollars at stake. But at no point did this produce a crisis between the two states or the hint of a possible interruption of supply. Second, the de facto accommodation by Gazprom took place despite the fact that (indeed, because) the European gas sector, and the German one in particular, were going through the greatest upheaval in the history of the industry. Yet the community of interest between buyers and sellers was sufficiently strong (despite public rhetoric to the contrary) to enable businesses to adapt and carry on, even as interstate relations continued to worsen. Gazprom, in effect, behaved like a private company rather than an arm of the Kremlin. Third, the gas buyers won, thanks to the increasingly dominant structure of hubs and markets.

Thus, until the post-2013 conflict in Ukraine, German businesses and regulators largely succeeded in insulating even the most controversial gas issues from the larger geopolitics of Russian-German relations. Now, however, economic and efficiency issues and geopolitical and security issues threaten to fuse. In the rest of this chapter, we shall attempt to illustrate this point through two case studies, the OPAL affair and the Nord Stream 2 controversy.

Nord Stream 1 and the OPAL Affair

The controversy over Nord Stream 1 followed lines that have since become familiar from the debate over its successor, Nord Stream 2.[52] The Eastern European governments were vehemently opposed, particularly that of Poland. Yet in retrospect the debate over Nord Stream 1 was less intense than the one over Nord Stream 2.[53] The European Commission designated Nord Stream 1, then called the North European Gas Pipeline, a project of European interest, on the ground that it would strengthen the European energy market and reinforce the security of supply (which has an ironic ring in retrospect). At the time Nord Stream 1 was agreed on, Russian-German relations were still at their high point. The only controversial aspect of Nord Stream 1 in Germany, as noted above, was the appointment of former Chancellor Schröder as chairman of its shareholders' committee.[54] But by the time Nord Stream 1 was inaugurated in 2011, relations between Moscow on the one hand and Berlin and Brussels on the other had deteriorated. The controversy that then broke out was not over whether Nord Stream 1 could be stopped, but whether it would be allowed to function at more than a fraction of its capacity. The controversy centered on a pipeline called OPAL.

The OPAL Dispute

The first major confrontation over the new rules concerned the application of third-party access (TPA) regulations to OPAL.[55] The dispute centered on the two onshore pipelines that transmit throughout Germany the Russian gas that arrives via Nord Stream 1, known as OPAL and NEL (Nordeuropäische Erdgasleitung). OPAL is a transnational interconnector with a capacity of 36 billion cubic meters, which runs 470 kilometers south from Greifswald, on Germany's Baltic coast, to the Czech border. In the Czech Republic OPAL links with the Gazelle pipeline, which delivers gas to southern Germany and eastern France. NEL has a capacity of 22 billion cubic meters and runs 440 km westward from Greifswald. It terminates at Rehden, a major storage facility in western Germany, from which the gas can be shipped to the Netherlands or Belgium.

Nord Stream 1 could not function until NEL and OPAL were approved. Gazprom, as the operator, argued that since OPAL crossed the Czech-German border it qualified as an international project that boosted the regional security of supply and should therefore be exempt from the TPA requirements of the Third Package. In addition, Gazprom argued, 100 percent of OPAL's gas came via Nord Stream 1. Hence, there could be no shippers other than Gazprom and Wintershall (Gazprom's partner on the project) that could wish to book OPAL capacity, so TPA requirements did not apply. The European Commission disagreed, and in 2009 it issued a statement ruling that 50 percent of OPAL's capacity must be made available to third parties. This effectively barred Gazprom from using more than 50 percent of OPAL's capacity and placed a limit on flow rates via Nord Stream.

The Russian reaction was vigorous and indignant, and it came straight from the top. Putin, who had stepped down as president between 2008 and 2012 but remained very much in charge as prime minister, denounced the Third Package and its application to OPAL in heated terms. In a speech to German business leaders in November 2010, Putin called the Commission's action "robbery" *(razboy):*

> Companies have invested hundreds of millions of dollars and euros into this project. And everything was done by the rules. But now a different decision is being taken retroactively, and [the investors] are being denied the use of their own property. How is that possible?
>
> Up to today no one has been able to explain to us how the new rules are supposed to work. . . . Gas reaches OPAL through Nord Stream. How will it be transported onward through the pipelines built at the expense of ourselves and our German partners? It's our property. What, then—we're not allowed to transport the gas to final consumers?

But Putin was only warming up. He went on:

> You know, we've got lots of problems in our country, and sometimes we don't act correctly, I'm prepared to admit that. But we constantly hear from our partners in North America and Europe the same thing. . . . If you want to be members of the civilized family, then behave in a civilized way. And what's this? Our colleagues have forgotten some elementary, fundamental rules. One of them is that "a law shall not have retroactive effect." It was formulated as far back as the time of the Great French Revolution. So how is this possible?[56]

Over the following months Putin kept up a drumbeat of criticism. In February 2011, as the deadline for the official entry into force of the European Union's Third Package approached, Putin led a delegation of Russian officials to Brussels to meet with European Commission Chairman José Manuel Barroso. There Putin denounced the Third Package as a "confiscation of property."[57] Over the following months, he remained personally involved in the progress of Nord Stream 1. In September 2011, when the first gas flowed through the pipeline, Putin was the one who announced the news; and later that month, he was present at the official inauguration.

The counterargument of the Russians and their German partners consisted of two points: first, that OPAL and NEL had been completed before the Third Package went into effect. The second and more fundamental argument was that cross-border pipelines such as Nord Stream 1 (and its extensions) should be granted a special exempt status on the ground that they contributed to European energy security, a point that Brussels had appeared to endorse when it designated Nord Stream 1 a project of "European interest." Since OPAL and NEL were simply extensions of Nord Stream 1, the Russians argued, they deserved to be exempt as well. Indeed, the Russians added, such megaprojects as the combination of Nord Stream 1, OPAL, and NEL were not subject to the Third Package in the first place, since they ran partly outside the European Union's jurisdiction. The European Commission should first fix this "gray zone" in its own legislation, the Russians insisted, before attempting to apply it to pipelines that were already built.

But the Russians' arguments were rejected by the Commission. "Russia and Europe have different understandings of the meaning of competition," Barroso had told Putin when they met in February 2011.[58] Two years later, as the deadline approached for the transposition of the Third Package into the legislation of the member-states, the Commission's position had only hardened. As far as the Commission was concerned, the Third Package was now law.

OPAL had been planned to begin commercial operation on November 1, 2012, but with no third parties having access and without a TPA exemption it could not be operated at more than 50 percent capacity. In December 2012 the Russians took their case to the EU-Russia summit in Brussels. It was a difficult conference, and there was no meeting of minds. But it was at least agreed to begin a new round of negotiations. The Russians asked for a temporary exemption to allow OPAL to operate at full capacity while negotiations proceeded, but this request too was turned down[59] (although in March 2013, because of cold weather in

Europe, the Commission granted a one-month reprieve, which was subsequently extended through the spring—an irony that did not go unnoticed by the Russians).[60] However, at this stage the two sides were so far apart that it was not until April 2013 that a four-man negotiating working group was named and talks resumed.

By now Putin had returned to the presidency and Dmitri Medvedev, back as prime minister, took over the dialogue with the Commission with a fresh team headed by a new energy minister, Alexander Novak. In the negotiations there was a new pragmatic tone, and although Putin continued his public campaign against the Third Package in such venues as the Gas Exporters' Forum in July 2013, inside the four-man working group the Russians were quietly modifying their opposition, while the Commission showed greater flexibility. By early June 2013, Novak told a correspondent that the Commission had proposed "five options, of which one was the closest to a compromise."[61]

Gazprom and Wintershall's central contention with the European Commission's TPA demand was that since all gas entering OPAL was delivered via Nord Stream 1, there could be no other potential shippers claiming title to gas at Greifswald, and hence there would be no third party seeking access to OPAL. As proof, the Russians pointed to the fact that when an open season had previously been held for OPAL capacity, no parties had shown interest.

In late November 2013 a solution to this problem began to emerge, based on the notion of virtual reverse flows through NEL, the east-west pipeline connected to Greifswald. Under the proposal, third parties with title to gas in western Germany and the Netherlands could in effect swap their volumes for gas delivered to Greifswald via Nord Stream 1. These volumes could then be shipped south via OPAL into the Czech Republic and on to the Net Connect Germany (NCG) exchange (southern Germany's gas trading hub). In effect, anyone with gas on the Dutch Title Transfer Facility could book reverse-flow capacity via NEL and take ownership of gas delivered via Nord Stream, then ship this on to the NCG exchange via OPAL.[62]

This rather elegant solution allowed both parties to save face: Gazprom could use most if not all of OPAL's capacity while adhering to the spirit of the Third Package, and the Commission could take comfort from the calculation that the prospect of new entrants accessing the capacity would bring competitive pressure to bear on Gazprom's pricing strategy. The BNA, the German regulator, duly endorsed the arrangement in a settlement agreement in November 2016. All that was missing for the compromise to be implemented was the official endorsement of the European Commission.

But the Ukrainian crisis derailed the arrangement. In late 2016 and early 2017 the Polish state oil and gas company filed a series of complaints with the European Court of Justice (ECJ), while in parallel another Polish entity challenged the Bundesnetzagentur (BNA)'s settlement agreement before the German Higher Regional Court in Düsseldorf. The Poles demanded a suspension of the OPAL arrangement of the previous year. But both the ECJ and the BNA rejected their demand and allowed auctions to proceed. This was a temporary decision, however, which did not rule on the legal merits. At this writing, the OPAL affair still goes on.[63] A final ruling by the ECJ was issued in September 2019. The future of OPAL is now uncertain.

The OPAL affair is significant for two reasons. First, despite the conflict in Ukraine and the cooling of German-Russian gas relations, the German government kept a low profile and allowed the case to proceed through the courts and the independent regulator. In contrast, as we have seen, Putin became the chief negotiator on behalf of Gazprom, thus politicizing the whole affair on the Russian side. Yet behind the scenes, the behavior of Novak, the energy minister, and of Gazprom Export was resolutely pragmatic. The case might have become a litmus test of East-West tensions, as the Poles clearly wished it to become. But the Germans and the Russians, in this case, adhered to the legal process and insulated the regulatory issues from the political process, as did the European Commission.[64]

Second, the OPAL affair stands at the gateway to the Nord Stream 2 dispute. Having lost the battle over Nord Stream 1, its opponents were determined not to be denied a second time. We turn now to Nord Stream 2.

The Nord Stream 2 Dispute

Nord Stream 2, like its predecessor, is a two-string pipeline originating from the Saint Petersburg area in the Gulf of Finland and passing under the Baltic Sea to a landing point at Greifswald, in northern Germany. The intended 1,200-kilometer route is nearly identical to that of Nord Stream 1 (see Map 11.1). Nord Stream 1, with two strings in service since 2011 and 2013, respectively, has a design capacity of 55 billion cubic meters a year. Nord Stream 2, will have the same capacity. Thus the Nord Stream pipeline system will have a combined capacity of 110 billion cubic meters, enough to deliver all the Russian gas currently shipped to Central and Western Europe.

Nord Stream 2 has ignited a bitter controversy among friends and foes of the project, touching on virtually every aspect—economic, political, geostrategic, and legal. It is backed by five major European energy companies (Shell, Engie, OMV, Uniper, and Wintershall), which have promoted it as a more direct and lower-cost route than the existing route via Ukraine to bring gas from the Yamal Peninsula in West Siberia to Europe. Nord Stream 2 will connect to a transit pipeline called the European Connector Pipeline (Europäische Anbindungsleitung, or EUGAL), which will take most of the gas through Germany to the Czech Republic and ultimately to the Baumgarten hub in Austria, destined mainly for southern and southeastern Europe.

Nord Stream 2 has been condemned by its opponents as a purely geopolitical project. But there is a strong economic case to be made for it as the lowest-cost route from Siberia to Europe. If one looks at a globe instead of a flat Mercator projection, it becomes apparent that the route from Yamal to Greifswald lies along a great-circle route, making it the shortest route to northern Europe and indeed (if one continues the circle by way of OPAL) to Slovakia as well.

Nord Stream 2 is actually one of a pair; the other part is Turkish Stream, which consists of two pipelines with a total capacity of 31.5 billion cubic meters per year. The first string of Turkish Stream, which has already been laid, is destined to supply the Turkish market. The second string will pass through Bulgaria, Serbia, and Hungary, likewise to Baumgarten. Curiously, while Eastern European opposition to Nord Stream 2 has been unrelenting, Turkish Stream actually enjoys support from some of the countries it will pass through. Both cases involve gas from Russia; both bypass Ukraine; the gas from both will supply mainly Western and Central Europe; and some gas from both, no doubt, will end up in Ukraine as part of its reverse gas imports from the West. Indeed, the entire European gas system is now so interconnected and marketized that the gas from these two pipelines can be traded almost anywhere in Europe.

Nevertheless, Nord Stream 2, as the larger project, has attracted most of the attention. Moreover, its symbolic importance is crucial: even more than Turkish Stream, it will mark the completion of Russia's long-standing policy, adopted over twenty years ago, of diversifying its gas exports by bypassing Ukraine. After the construction of Blue Stream, Yamal-Europe, and Nord Stream 1, the share of Russian gas exports transiting through Ukraine dropped from a high of 91 percent in 1994 and 1996 to just under 41 percent in 2018. Nord Stream 2 and Turkish Stream are the capstone. Once they begin operation, they will reduce Russian dependence on Ukrainian transit to 30 billion cubic meters per year or less.[65]

But will Nord Stream 2 be allowed to operate? As of this writing that is still unclear. Nord Stream 2 remains highly controversial.[66] Supporters argue that it will offset Europe's declining indigenous production. Opponents argue that it will increase Europe's dependence on Russian gas. In addition, Nord Stream 2 is inextricably tied to the ongoing dispute between Russia and Ukraine over gas transit. Once the two new pipelines begin operation, revenues earned by Ukraine for transiting Russian gas to Europe will be diminished and perhaps ultimately eliminated. For these reasons, it has been strongly opposed by the US government and the Eastern European members of the European Union. But will their opposition succeed?

We will not attempt to review here the complex history of the project, on which there is already a large literature.[67] Rather, we will focus on the part that deals with the subject of this chapter, Russian-German energy relations. The central argument of the following section will be that the key to the operation of Nord Stream 2 ultimately lies in German hands, but who will decide, and on what basis, is complex and still uncertain.

Cautious Support from the German Government

In the German parliamentary system there is a standard procedure under which opposition deputies can raise written questions with the government's ministries on current policy issues and expect thorough (if not necessarily prompt) answers. These are called *Kleine Fragen* (small questions). In November 2015 a group of deputies from the Green Party submitted a series of twenty-four such questions to the Federal Economics Ministry about Nord Stream 2. The ministry's answers (which took six months to come back) summed up the government's position: Nord Stream was a purely private-sector venture and as such did not concern the ministry or the government.

The ministry appeared equally relaxed about the security implications of additional Russian gas exports to Germany, arguing that the question turned on competitiveness: "Gazprom's position in the European internal market depends primarily on the competitiveness of Russian gas deliveries in competition with other providers." Yet the ministry took no position on whether the Nord Stream pipeline was actually economic, stating in its memorandum, "That is entirely a matter for the companies involved."[68]

Chancellor Merkel's position has been consistently more cautious. On the whole she has supported the Nord Stream 2 project, on essentially the same ground as the Federal Economics Ministry. "It is first and foremost an economic project," she stated in December 2015, at the conclusion of the EU summit that year. "There are private investors for it." But she has also stressed from the beginning that Ukrainian interests had to be protected. As she stated at the same summit, "In any solution Ukraine must continue to play a role as a transit country."[69]

The question of Ukrainian transit is the major unknown in the Nord Stream 2 equation. Although Merkel's words sounded straightforward, it has been unclear from the beginning how far the German government was prepared to go to defend Ukrainian interests. Having linked its approval of Nord Stream 2 to the defense of Ukrainian transit, might it seek to oppose the project if Ukrainian interests were not protected? Would it have the power to do so? And how would this question be perceived in Moscow?

The strongest and most visible backer of Nord Stream 2 in the previous German government was Sigmar Gabriel, then Germany's deputy chancellor and minister for economics and energy (and subsequently foreign minister) and one of the leading figures in the SPD. However, his role landed him in controversy. In October 2015, Gabriel traveled to Moscow and was received at President Putin's official dacha at Novo-Ogaryovo, outside Moscow. The two men reviewed the state of Russian-German economic relations, and especially the status of Nord Stream 2. The importance of the meeting stemmed from the cooling of relations between the two governments since the cancellation of the annual Petersburg Dialogue the previous year, when many German delegates refused to travel to Sochi in protest against Russia's annexation of Crimea and Russian-sponsored actions in eastern Ukraine. German public opinion was especially outraged over the shooting down of the MH17 Malaysian airliner over the Donbas by a Russian antiaircraft missile the previous July.

But by October 2015 a year had passed, and new problems claimed the world's attention. Meanwhile, German businesses were beginning to show signs of impatience over the sanctions and countersanctions imposed in the wake of the Ukrainian crisis, and the public mood was beginning to shift, if not in the direction of warmth, at least toward normalization. Foreign Minister Frank-Walter Steinmeier (also from the SPD) even proposed a step-by-step loosening of the Russia sanctions in exchange for Russian concessions on Ukraine. Gabriel's visit to Moscow coincided with fresh efforts to create a replacement for the defunct Petersburg venue, and he was accompanied by a large retinue of German business leaders.

Yet the visit turned into an embarrassment for Gabriel. During a supposedly confidential exchange à deux with Putin in the informal setting of the presidential dacha, Gabriel assured the Russian leader of his own strong support and that of his government for the Nord Stream 2 project and expressed confidence that approval for the project lay with Germany alone. Unbeknown to Gabriel, the conversation was being recorded, and the next day it was posted verbatim on the Russian presidential website. Anyone who cared to look (particularly in Brussels and the East European capitals) could read Gabriel's words:

> What's most important as far as legal issues are concerned is that we strive to ensure that all this remains under the competence of the German authorities, if possible. So if we can do this, then opportunities

> for external meddling will be limited. And we are in a good negotiating position on this matter.
>
> And in order to limit political meddling in these issues—you are, of course, aware, this is not just a formality—we need to settle the issue of Ukraine's role as a transit nation after 2019. There are technical reasons for this: you know that Ukraine's gas transportation system is not in very good state. And, of course, the financial and political role it will play for Ukraine, as will the reverse flow of gas.
>
> As regards everything else, I believe we can handle it. What's most important is for German agencies to maintain authority over settling these issues. And then, we will limit the possibility of political interference in this project.[70]

Gabriel was criticized in the German media for having appeared to play the role of lobbyist for Nord Stream 2. His lukewarm defense of Ukrainian transit may indeed have given Putin the impression that the German government would continue to put commercial interests first, even though Gabriel's key point was that the German government would play the main role.

The change of government in Berlin in 2017–2018, together with expanded sanctions from Washington, led the German government to use tougher language toward Russia. In the newly rebuilt Christian Democratic Union–Christian Social Union (CDU-CSU)–SPD coalition, Gabriel was replaced at the foreign ministry by Heiko Maas, a man who had no previous experience in foreign affairs but who immediately took a sterner line. In his first interview, with *Der Spiegel,* he declared flatly, "Russia has defined itself more and more as separate from and in partial opposition to the West. . . . Unfortunately Russia is increasingly hostile."[71]

Chancellor Merkel likewise toughened her stance on Nord Stream 2. In a visit to Putin in Sochi in May 2018, she stressed that the support of the German government for the project was contingent on credible guarantees by Russia that the transit of gas through Ukraine would continue. Putin was noncommittal in reply, agreeing only that Russia would continue Ukrainian transit as long as it made economic sense.[72] Nevertheless, German public opinion chose to interpret this as a promise.

The European Union Enters the Picture

In 2017 the European Union entered the picture in a much more active way. The European Commission and the Parliament have been consistently opposed to Nord Stream 2. In early 2017 the Commission put teeth into its opposition by promoting a modification of the Third Gas Directive, the so-called Third Package. The main innovation was that it would now apply to pipelines connecting non-EU member-states with the European Union. Though the measure did not mention Nord Stream by name, that was clearly its target. However, the measure was held up for a year, chiefly by German opposition. Finally, in February 2019 a breakthrough compromise was reached, under which the authority to enforce the new gas directive rules was entrusted to the regulator of the country where the pipeline landed. In addition, the national regulator was given the power to issue any exemptions. This appeared to put the Germans in the driver's seat, and the outcome was widely interpreted to mean that Nord Stream 2 would go forward.

But matters are not so simple. Since the adoption of the compromise amendment to the EU Gas Directive, once Nord Stream 2 enters the territorial waters of Germany, a stretch of about 53 kilometers to the shore, it will require a permit from the German government before it can begin operation. Specifically, it is the BNA, that must rule on whether Nord Stream 2 has satisfied the requirements of EU law.[73]

The main relevant fact here is that the BNA is an independent agency. Independent agencies are unusual in German administration. Although the BNA is formally subordinated to the Ministry of Economic Affairs and Energy, its decisions, which are made by quasi-judicial ruling chambers, cannot be revoked by political authority. It has a strong legal mandate (the Energy Industry Act), and its independence is reinforced by the fact that its decision making takes place in public proceedings. The BNA can issue a permit or grant a temporary exemption—or it can delay a decision, thus effectively preventing Nord Stream 2 from operating. Whatever it does, however, its decision is supposed to turn on whether Nord Stream 2 has met the European Union's legal requirements, not on the pros and cons of the project itself or its larger implications for energy security.

According to the amended Gas Directive, there are four conditions that must be met: Nord Stream 2 must be under the control of an independent operator. There must be a regulated tariff (this applies only to the approximately 53 kilometers that lie in Germany's territorial waters). There must be third-party access to the pipeline. And finally, ownership must be unbundled (at present Gazprom is the sole owner of both the gas and the pipe); this too applies only to German territorial waters.

Who will decide whether the conditions have been met and thus control the decision to grant or deny permission to operate? The Federal Economics Ministry, headed by a close Merkel ally, Peter Altmaier, stated in January 2019, shortly before the EU compromise was reached: "Germany is a law-ruled state where private investments, as in the case of Nord Stream 2, are made in line with legal criteria. The federal government will not interfere into such a process as it has no legal grounds for that."[74] Yet shortly following the compromise, Altmaier told the *Financial Times,* "Unbundling or third-party access will be key. We will have to discuss that with the investors, with Gazprom, but I'm optimistic that a good, sustainable solution can be found." That statement suggested that Altmaier at least believed that the ultimate decision was negotiable, and that the government itself, not just the BNA, might ultimately make the key decisions.

Despite the independent status of the BNA, it does not make its decisions in a vacuum. It would be surprising indeed if it were not influenced, directly or indirectly, by the prevailing mood in the German government. (It is perhaps relevant to note that Jochen Homann, the president of the BNA since 2012, was formerly a state secretary in the Federal Economics Ministry. He is an economist by training, not a lawyer.)

Here the striking fact is that although the German leadership, media, and public opinion all are on record as supporting Nord Stream 2, they have lately become more ambivalent in response to perceived Russian intransigence in dealing with Ukrainian gas and gas transit. The most interesting, and possibly revealing, player to watch is the rising star of German politics, Annegret Kramp-Karrenbauer, a protégée of Merkel who is presently the chairman of the CDU and thus Merkel's successor as chancellor. In an unusually candid interview in February 2019, Kramp-Karrenbauer (who is widely known in Germany simply as AKK) conceded that she is not enthusiastic about Nord Stream 2, even though she continues to back it: "It is not a project that I support with all my heart. . . . But fundamental decisions were made earlier." AKK is no unconditional friend of Russia. As she said in the same interview, "Russia's agenda includes the destabilization of Europe and of Germany."[75] The German media are also ambivalent. As Alan Posener, an editorialist for *Die Welt,* put it recently, echoing AKK's formula: "It is especially the interests of our Eastern European partners that should be our 'heart's project.' There are alternatives to Russian gas, but none to European solidarity."[76] German public opinion may have cooled slightly, too. By late February 2019 only 56 percent of those polled supported Nord Stream 2 (with 16 percent opposed), in contrast to 73 percent in favor the month before.[77]

Lately there has been growing German annoyance over what has been perceived as the Russians backing away from Putin's 2018 commitment. In an interview with the Russian newspaper *Gazeta* in late February 2019, Energy Minister Novak spelled out the reasons why, in the Russian view, the Ukrainian gas transmission system is not economic compared to Nord Stream 2:

> The Ukrainian route today is 2 to 2.5 times more costly than existing pipelines that are being used today for the transit of Russian gas, more than, for example, Yamal-Europe and Nord Stream 1. . . . That is, first, because the Ukrainian transport system was built about 50 years ago, there were then old technologies; [next], the infrastructure is relatively worn out; and there are large losses during the transport of gas. The realization of the new transport systems which are being built now, such as Nord Stream 2 and Turkish Stream, is based on the latest up-to-date technologies that are being used in this field. These are large-diameter pipelines, with high pressures, high transmission speeds, and low losses.[78]

German politicians and the media reacted angrily. The deputy chairman of the CDU-CSU faction in parliament, Johann Wadephul, blasted Novak's remarks in an interview in *Handelsblatt,* Germany's leading business newspaper, stating, "If Novak's declaration is really the last word, this would be a substantial breach of trust vis-à-vis Germany, and where Nord Stream 2 is concerned, it would call into question the reliability of Russian promises."[79]

Seeking to clarify the Russian position, Prime Minister Medvedev then gave a lengthy interview in Luxemburg in early March. On Ukrainian transit, Medvedev declared: "We are ready to preserve gas transit through the Ukrainian pipeline system even beyond 2019. Naturally, provided certain conditions are met. We have spoken about this many times already. To put them briefly: these are a settling of relations among the interested companies, profitable economic and commercial parameters for the deal, and a stable political situation."[80] Needless to say, such vague conditions, particularly the last one, are hardly likely to calm German concerns. At this moment, discussions between Gazprom and the Ukrainian gas company Naftogaz are at a standstill, despite the efforts of the European Commission to act as an intermediary. Gazprom insists that it will not agree to a new supply and transit contract so long as Naftogaz demands payment of the $2.6 billion arbitration award made to the Ukrainian company last year by the Arbitration Institute. To collect the Russian debt, Naftogaz has begun to attach Gazprom assets in third countries, further complicating relations.

The Russians' position on the transit negotiations, however, is plausibly due less to intransigence than to caution over the fluid political situation in Ukraine. In one of the major political surprises of recent years, in April 2019 a complete newcomer to politics, a television personality named Volodymyr Zelenskiy, defeated the incumbent president, Petro Poroshenko, and is now president of Ukraine. In the parliamentary elections that followed in July, Zelenskiy's newly formed party, Servant of the People, won an unprecedented absolute majority. Yet Ukrainian policy, particularly on energy, is as yet undefined.

Meanwhile, EU politics are fluid as well. The compromise amendment to the Gas Directive was approved by the outgoing EU Parliament before it adjourned in April. But in May, elections for a new EU Parliament took place, followed by the selection of a new president of the Commission and new presidents of the European Council and the European Central Bank. As one analyst observes, "This is the first time in EU history that posts and seats in so many key institutions have been filled in the same year."[81] Thus it is unclear, to say the least, whether the February 2019 compromise amendment to the Gas Directive will actually hold up.

The last unknown in the equation is the behavior of the United States. In the past year, Congress has effectively taken control of US sanctions policy.[82] Anger is mounting in Congress over what it regards as the dilatoriness of the Trump administration in the implementation of sanctions policy, and there is growing sentiment in favor of punishing Russia by whatever means available. New sanctions legislation is currently making its way through the House and the Senate, backed by strong bipartisan support. Once Congress realizes that the recent compromise amendment to the EU gas directive does not actually close the door definitively on Nord Stream 2, it may be added to the proposed new laws, although much will turn on whether there are new geopolitical incidents during the balance of 2019.

Previous informal efforts by the US government and its diplomats to warn the Europeans away from Nord Stream 2 have been rejected by the German leaders as attempts to interfere in Germany's business. In February 2019 the US ambassadors to Germany, Denmark, and the European Union wrote an open letter urging the Europeans to reject Nord Stream 2. This brought a strong reaction from German politicians and business leaders. "It is not the best dealings between friends and partners to threaten one another with sanctions," AKK said. "Washington must listen to the German replies: we are presently diversifying, we have other sources of supply."[83] In an attempt to placate Washington, the German government offered to support the construction of two liquefied natural gas terminals in Germany (despite the fact that there is already a large surplus of regasification capacity in Europe), but this concession only underscored Germany's determination to make its own decisions. Indeed, US legislation imposing sanctions on Nord Stream 2 might backfire, by consolidating German opinion in favor of the project and overcoming the government's ambivalence over allowing it to operate.

To complicate matters further, in March 2019 the Danish government refused to give its approval for Nord Stream 2 to pass through its territorial waters and demanded that the builders undertake an environmental assessment of an alternative route, running through Denmark's exclusive economic zone farther south. This will delay Nord Stream 2. The Danish position thus ties progress on Nord Stream 2 to a settlement of Ukrainian transit. In theory it would be in the Russians' interest to make concessions on Ukrainian transit as quickly as possible. The combined capacity of Nord Stream 2 and Turkish Stream is not enough to meet daily and seasonal peaks in European demand, and thus for the Russians some continued access to Ukrainian transit is essential. But it could be well into 2020 before a deal can be reached. The second obstacle may be that if Naftogaz presses its policy of attaching Gazprom assets to recover its Stockholm arbitration award, the Russians and Ukrainians may be even less inclined to talk than they are today.[84] At this writing, in short, there is no telling how long Nord Stream 2 may be delayed.

This chapter has traced the evolution of Russian-German gas relations from the 1990s to the present. Until the second half of the 2000s, the political and business aspects of the energy relationship were conducted under the sign of *Ostpolitik* and were mutually reinforcing. In foreign policy, Germany viewed Russia as a strategic partner. Strategic partnership was also the hallmark of the business relationship in gas, symbolized by the active role of Gazprom in the German gas midstream and the investments of German companies in the Russian energy upstream. The gas relationship still operated under Groningen rules, which fit the traditional structure of the German gas industry. Both politics and business, in short, pointed in the same direction and reinforced one another.

By the second half of the 2000s, however, the evolution of Russian politics, relations within the Former Soviet Union, and the expansion of the European Union, together with the rise of the neoliberal market model in Western Europe, brought upheaval to the gas world on both the geopolitical and business fronts. A commercial and political regime that had grown out of *Ostpolitik* and provided security and predictability in European-Russian relations became within a decade a factor of mistrust and potential insecurity. Much of the decision-making authority that once resided in Berlin moved to Brussels, where Germany, although the leading player, must take careful account of other European voices.

Yet the present frostiness of political relations between Berlin and Moscow has not led to any significant curtailment in the scope or volume of Russian-German gas relations. This chapter finds that one explanation for this puzzle is that the political and business sides of the relationship have been in the hands of separate communities and have been handled by different rules. Gas disputes over issues of pricing and contracts have been settled on pragmatic grounds, as both the Germans and the Russians have made tacit concessions and adapted to the requirements of the Third Package. The same pragmatic spirit has been evident at the level of the European Commission, as the OPAL case demonstrates.

The great exception has been the Nord Stream 2 pipeline, in which geopolitical and economic arguments have fused into a single emotion-filled debate. Security has returned to the fore over the Nord Stream 2 affair. Yet even as geopolitics swirls around the pipeline, Germany has sought to keep business and politics distinct by insisting on the private-sector nature of the project and on Germany's prerogative to make key business decisions on a bilateral basis. Based on past form, this position is likely to continue in the future, and Nord Stream is likely to flow, if with some delay.

Yet as one looks ahead into the 2020s, the tension between security and energy in Russian-German gas relations is likely to increase. The German energy transition, the *Energiewende,* makes the long-term position of gas uncertain. In the near term, imbalances in the design of the *Energiewende* are likely to increase Germany's need for gas, particularly Russian gas, but the longer-term outlook is uncertain. At the same time, it is difficult to see any fundamental improvement ahead in Russian-German geopolitical relations. Which shall prevail, the commercial or the geopolitical?

CHAPTER 13

Battle Joined, War Averted

In late September 2011, following complaints by Lithuania of anticompetitive behavior by Gazprom, the European Commission's Directorate-General for Competition (DG-COMP) launched a series of "dawn raids" against nearly two dozen subsidiaries and affiliates of the Russian company, to obtain documents and computer files. The raids were reported in dramatic tones in the *Financial Times,* based on interviews with DG-COMP's team:

> First came the stake-out. On foot and by car they took the measure of the buildings, the entrances and the layout, plotting the fastest way to seal off corridors, shut servers and seize computers. With the morning came the synchronised raids, launched against almost two dozen offices spread across 10 countries, a series of commercial outposts stretching from Bulgaria in the south to Estonia in the north, tracing an arc of Gazprom's affiliates and customers across Europe's eastern flank.[1]

Even if they did not take place exactly at dawn (DG-COMP usually carries out its raids at a more civilized hour), the raids were dramatic enough, underscoring DG-COMP's sweeping powers, as recounted in Chapter 7. Even more dramatic, however, was what followed: a three-and-a-half-year investigation by DG-COMP based on the material it had seized, followed by three more years of negotiation between DG-COMP and Gazprom. In spring 2018, the DG-COMP affair finally came to a close, with an amicable settlement that did not conceal the essential outcome: a sweeping victory by DG-COMP and a wholesale series of concessions by Gazprom. As a result, the marketization revolution has finally spread to Eastern Europe and—after stiff resistance—has been accepted by Gazprom. Thus the DG-COMP affair, on both symbolic and practical levels, marks the victory of the neoliberal wave that has been the central subject of this book.

At the outset, it appeared that the DG-COMP affair might lead to another crisis in Russian-European gas relations, including the possibility of a third cutoff. In the middle of the affair came the collapse of the Ukrainian government and the events surrounding the occupation of Crimea and the separatists' takeover of eastern Ukraine. Yet the most remarkable feature of the investigation and its aftermath was that throughout the geopolitical turmoil of 2013–2014 and after, the two sides managed to keep talking productively and ultimately reached a settlement. How did that happen? That is the subject of this chapter.[2]

The Initial Russian Reaction

The initial Russian reaction was one of shock and anger. "I hope nobody has been arrested or jailed in Europe for contracts with Gazprom," Vladimir Putin was reported to have said sarcastically. "Not yet," replied Aleksey Miller.[3] Gazprom proclaimed its innocence, insisting that its European contracts were in "full compliance" with international law.[4] Putin promptly issued a decree forbidding Gazprom to release corporate data to investigators, although by this time the horse was already out of the barn.

The Russians immediately claimed that the raids were politically motivated. As Vladimir Chizhov, the Russian ambassador to the European Union, put it, the Lithuanian government had complained to the Commission in retaliation for a lawsuit brought by Gazprom before the Arbitration Institute of the Stockholm Chamber of Commerce over what Gazprom claimed was the "forced nationalization" of a pipeline by the Lithuanians. "This coincidence leads one to believe that our respected

Lithuanian colleagues are trying to use the European Commission to influence these legal proceedings, whose results may be not quite in Lithuania's favor," Chizhov said.[5] Commentary on the DG-COMP investigation in the Russian media was not only indignant but disbelieving. Aleksey Grivach, a well-known gas analyst, wrote, "How can you pick a quarrel and attempt to impose multi-billion-euro fines . . . on practically the only source of additional volumes?"[6]

The investigation then went quiet for nearly two years, as DG-COMP pored over the materials it had seized and prepared its objections. Given the initial Russian reaction, there was much speculation in the media about the likelihood of confrontation, but as DG-COMP neared the conclusion of its investigation, the Russians appeared to be increasingly inclined to compromise. In November 2013, Prime Minister Dmitry Medvedev sent a letter to European Commission Chairman José Manuel Barroso and dispatched Deputy Energy Minister Anatoliy Yanovskiy to join the discussions. In early December, after Joaquin Almunia, the European Commission's director-general for competition, Gazprom Export's general director, Alexander Medvedev, and Yanovskiy had held meetings, it appeared as though a settlement might be at hand.

In an interview with the *Financial Times,* Alexander Medvedev stated, "We are trying to work together to find a solution before a statement of objections is issued." He expressed optimism that a compromise with DG-COMP might be found "in a relatively short time, in three or four months."[7]

But at this stage the Ukrainian crisis intervened. The chilling effect of the crisis could already be felt at the EU-Russia summit on January 28, 2014. Instead of the customary two days of meetings, capped by a formal dinner, the summit was pared down to the bare minimum—four hours of talks and a working lunch, with no dinner. Yet even so, significantly, the summit was held on schedule—it was the thirty-second biannual meeting since 1998—and the participants maintained a tone of businesslike civility. In their final press conference, the three main participants—Putin, Barroso, and European Council President Herman Van Rompuy—did their best to emphasize the positive while not concealing their disagreements. By late February 2014, however, with the breakdown of the political accord brokered in Kiev among representatives of the European

Union, Russia, and the Ukrainians, followed by the flight of President Viktor Yanukovych to Russia, geopolitical events overshadowed all other business. All further talks on regulatory and commercial issues were suspended. DG-COMP insisted that its investigation was continuing as planned, without being affected by the Ukrainian crisis. Yet it was clear that all sides had shifted their focus to Ukraine, and the DG-COMP investigation had effectively been placed on hold.

The Statement of Objections

A year and a half later, geopolitical relations between Russia and Europe were no smoother, but some of the immediacy of the Ukrainian crisis had passed, and the Commission felt able to proceed. In April 2015 it issued its long-awaited statement of objections to Gazprom's business practices in Eastern Europe. Three and a half years had passed since the initial raids.

By this time a new team had taken charge of the Commission, with the appointment of Jean-Claude Juncker as president and his selection of a new team of commissioners—which notably included a highly dynamic figure, Margrethe Vestager, as his commissioner for competition. At the same time, the new EU leadership contained a record number of key players from Eastern and Central Europe, beginning with Donald Tusk, a former Polish premier, but also including such figures as the new Commission's vice president for energy, the Slovak Maroš Šefčovič. The new team lost no time in signaling its determination to reenergize the role of the Commission as a driving force in EU energy policy, particularly in gas and power. But now, instead of going after the Western European energy companies, as had been the case under the preceding team in Brussels, the new Commission targeted Eastern Europe—and Gazprom.

The key figure in the Gazprom case has been Vestager, the competition commissioner until 2019, and now the executive vice-president of the Commission. The daughter of two Lutheran pastors, Vestager entered politics at the age of twenty-one and rose to become deputy prime minister of Denmark, her last job before moving to Brussels. In her first term she quickly gained a reputation for steely resolve—as some put it, "out-Kroesing Neelie Kroes"—as she took on some of the most powerful corporate giants in the world, including Google and Apple as well as Gazprom. Vestager's approach to competition law, like Kroes's, is shaped by a strong moral sense. She once declared: "What is at stake is as old as Adam and Eve. For all the economic theories and business models, it all comes down to greed."[8] With that kind of outlook, one might have expected Vestager to bring the same prosecutorial zeal to the Gazprom case as she had to cases against Google and Apple, which resulted in hefty fines and penalties. It is all more interesting, then, that the final result, after long negotiation, was a settlement and no penalty.

DG-COMP's Charges

DG-COMP's charges against Gazprom were contained in a 600-page document that was delivered to Gazprom on a confidential basis in April 2015.[9] However, the document could be summed up in just one word: "partition." As DG-COMP spelled out in its preliminary fact sheet, "The Commission's preliminary view is that Gazprom is breaking EU antitrust rules by pursuing an overall strategy to partition Central and Eastern European gas markets with the aim of maintaining an unfair pricing policy in several of those Member States."[10]

According to DG-COMP's case, Gazprom implemented its strategy in three different ways: first, by hindering cross-border gas sales through so-called destination clauses; second, by charging unfair prices; and third, by making gas supplies conditional on obtaining unrelated commitments from wholesalers (chiefly concerning support for gas transportation infrastructure).

As sweeping as the charges were, several of their features conveyed the impression that from the first the Commission was making a conscious effort to focus its charges and limit the risks of a geopolitical confrontation. DG-COMP's fact sheet was at pains to show that the target was not Russia or Russian gas per se. The document listed several recent antitrust actions taken against other energy companies, notably against Gaz de France (GDF) (now Engie) in 2004 and against Electricité de France (EDF) and E.ON in 2009. Equally significant, the statement of DG-COMP's objections came just one week after the announcement of its case against Google; this was presumably not a coincidence. The message to Gazprom (and the Kremlin), in other words, was that it was not being singled out, and that DG-COMP was pursuing with equal vigor all suspected violators.

Thus, from the beginning the tone and content of DG-COMP's public statements on its charges were far from the more vivid language sometimes used by officials of the previous Commission—as when in 2014, for example, Günther Öttinger (then energy commissioner) spoke to an audience in Poland about a "game of divide and rule proposed by Moscow [which] cannot be and will not be accepted by EU member states."[11] The new Commission's approach appeared to be to speak more softly even as it wielded a bigger stick, focusing on the strictly antitrust legal issues instead of broader geopolitical ones.

Why Eastern Europe?

The key feature of DG-COMP's charges, however, was that they were quite specifically targeted. Two countries (Poland and Bulgaria) were named in the case of unrelated commitments; five (the three Baltic Republics, Poland, and Bulgaria) in the case of unfair pricing; and eight (the Baltic Republics, Poland, Bulgaria, the Czech Republic, Slovakia, and Hungary) in the case of hindrances on cross-border sales. In other words, in two countries all three charges applied; in three countries only two did; and in another three countries only one did. This reinforced the general impression that DG-COMP's case was not a blanket indictment across the board but was carefully limited, presumably to the cases that the evidence in its possession would best support.

What these eight countries had in common was that despite their accession to the European Union beginning in 2004, they had previously belonged to the Soviet Union and / or its Eastern European sphere of influence, and they remained connected to Russia and Russian gas through essentially the same network of bilateral relations—often described as the "spoke-and-wheel" principle—that had previously characterized the Soviet Union's relations with its satellites in the eastern trade bloc called the Council for Mutual Economic Assistance (colloquially referred to as the Comecon system), and of course in the case of the Baltic Republics within the Soviet Union itself.

Nearly twenty-five years after the end of the Soviet Union, the countries that had been Soviet satellites (with the exception of eastern Germany, the former German Democratic Republic) were still gas islands, largely cut off not only from the rest of the European gas system but also from each other. As a result, they remained heavily dependent on Gazprom for their supply, and therefore on its prices and terms. It was only recently that some Eastern European countries—such as Lithuania, which opened a liquefied natural gas import facility in 2014—had begun to take steps to reduce their gas dependence (and their vulnerability to Russian pressure).[12]

In other words, the violations of which the Commission was now accusing Gazprom were not new; they were largely a continuation of existing structures and practices. It was, in fact, the eastward advance of European antitrust law that was the new element. Ten years after the accession of the Eastern European members, the European Union's legal *acquis* was finally becoming the law of the land.

The Possible Outcomes—Prohibition or Settlement?

Following the announcement of DG-COMP's statement of objections in April 2015, the case followed a well-traveled route. Gazprom had twelve weeks to respond with proposed remedies—that is, policy changes that would remedy the violations of antitrust law of which it stood accused. This opened a period of negotiation, during which the two sides attempted to resolve their differences. If they succeeded, then Gazprom's proposed remedies would be accepted in what is known as a "commitment decision." If they did not, Gazprom would be subject to a "prohibition," which would include the imposition of penalties and fines that in theory could run up to 10 percent of Gazprom's annual turnover. But this would not be the last word: Gazprom could then challenge the Commission's action before the European Court and after that, if necessary, before the European Court of Justice.[13]

One could have concluded from this road map that the final resolution of DG-COMP's charges would take a long time, perhaps several years, before all legal appeals were exhausted. But in reality DG-COMP had a powerful additional weapon at its disposal. At the point when it imposed a prohibition, the entire 600-page document, containing the full statement of objections and all the supporting evidence, would become public. Since DG-COMP was the sole judge of whether to impose a prohibition, it had, in effect, the power to drop what would amount to a legal cluster bomb on Gazprom. The publication of the full statement of objections would have immediate and far-reaching effects, setting off a cascade of arbitration cases and damage claims.

In short, it was not the threat of large fines at some distant point in time that was the Commission's main source of leverage in this case, but its near-term power to inflict major damage on Gazprom's finances and reputation. Given this situation, there was clearly a strong incentive for Gazprom to make the concessions needed to avoid a prohibition.

One can imagine the sharp exchanges that must have taken place between Gazprom and the Kremlin. To judge from the Kremlin's initial sharp reaction to the "dawn raids," the Kremlin appeared inclined to defy the Commission and all its works. Gazprom's reaction, in contrast, was likely to have been much more pragmatic, especially since on the face of it, it appeared that compromises on the three specific charges at the core of the statement of objections would be possible. Indeed, possible actions on the first two charges were reasonably straightforward, inasmuch as there was already precedent for them:

- *Restrictions on cross-border sales:* The illegality of such restrictions under EU law was already well established. In 2004 and 2009 the Commission had imposed fines on GDF and E.ON over destination clauses in their contracts with Eni and EDF, respectively. There was also a strong precedent involving reexport restrictions by Gazprom. In 2003, Gazprom and the Commission had reached an informal settlement regarding Gazprom's inclusion of destination clauses in its contracts with Western European buyers, as a result of which Gazprom stopped the practice. There was thus no reason in principle why a similar arrangement could not easily be reached regarding Gazprom's Eastern European buyers.
- *Unrelated commitments imposed on wholesalers:* This battle appeared to be already on the way to resolution on the Commission's terms, after Gazprom abandoned its South Stream project earlier in 2015,

> in the face of determined opposition by the Commission. The conflict had come to a head in December 2013, in a lively public exchange between Prime Minister Dmitry Medvedev and a Commission official before the European Parliament. To Medvedev's assertion that "Nothing can prevent the construction of South Stream," the Commission official replied, "What the Commission would hardly accept is that you put to us a pipeline that is built . . . and then hand over the baby to us."[14] Since that time, the Commission's determination to prevent South Stream from operating was repeatedly restated and ultimately carried the day.

In the media commentary that followed DG-COMP's statement of objections, a number of experts argued that the third charge, that of "unfair pricing, would be the most difficult to prove in a court of law.[15] But as noted above, DG-COMP did not actually have to prove anything, because it was the sole judge of whether to impose a prohibition, with all the damage that the legal cluster bomb could produce. Therefore, as a practical matter the definition of a "fair price" was whatever DG-COMP chose to accept. In that situation, any pragmatic benchmark would do.[16] One rough-and-ready solution, for example, would simply be to take the average prices of gas exports across Europe (or a selected European market such as Germany) as published in various national customs statistics and to define as "fair" any price that did not depart from these averages by more than a certain percentage. All that would be required would be some adjustments in the internal pricing formulas in the existing contracts.

Alternatively, DG-COMP and Gazprom could agree to use market-based pricing as the most appropriate criterion, given that the spot market, based on liquid and transparent hubs, had become the dominant mechanism for pricing gas in Western Europe. Market-based pricing is gradually spreading east and south and has even reached Ukraine, at least as a basis for pricing reverse flow exports into western Ukraine. Over time, market-based pricing is becoming the norm. Gazprom had increasingly accepted that fact, as evidenced by several recent contracts concluded with Central European buyers, such as Slovakia.

In short, as DG-COMP and Gazprom settled down over the summer of 2015 to negotiate the terms of a possible final settlement, there appeared to be room for compromise on the three specific complaints at the center of DG-COMP's case. And so there proved to be, but it took another year and a half, until the spring of 2017, to work out the details.

Spring 2017: A Preliminary Settlement Is Announced

In March 2017 a preliminary commitment decision was announced, which already made clear that the essence of a deal had already been struck: Gazprom would make key concessions on all points in exchange for a settlement in which no penalty would be imposed. Two items were of special note:[17]

- *"Swaps" to reach inaccessible areas:* As noted above, the inherited "spoke and wheel" pattern of gas pipelines in Eastern Europe left certain countries, such as Bulgaria and the Baltic Republics, supplied only from Russia and isolated from connections with the rest of Europe. To overcome this problem, Gazprom and DG-COMP agreed that European suppliers could "swap" gas into those countries. In other words, Gazprom would ship the actual gas while taking the European supplier's gas in exchange elsewhere in Europe. The importance of this concession was that it obligated Gazprom to go beyond the mere removal of obstacles and actively promote better access.
- *European hubs as a key benchmark:* The preliminary commitment decision went a long way toward establishing the spot prices prevailing on the major European exchanges as the key criterion for judging the "fairness" of prices, rather than using oil-indexed contracts. This was a major concession by Gazprom, in that it extended to Central and Eastern Europe the pricing benchmark that Gazprom had already accepted in Western Europe. Although oil-indexed prices were not explicitly banned, it was clear that Gazprom had finally abandoned one of its most closely held positions.

Following the publication of the preliminary agreement, DG-COMP conducted a wide-ranging market test, during which all interested parties could give their reactions and submit proposed changes. There were complaints, for example, that there were no limits on the fees that Gazprom could charge for swaps. Another reaction was that the commitment to hubs was not sufficiently explicit and needed to be strengthened. There followed further negotiations, during which Gazprom appears to have made additional concessions.

Spring 2018: The Final Settlement

The announcement of the final settlement of the DG-COMP case was a major event. In both symbolic and practical terms it stood as the culmination of the drive that had begun with Jacques Delors in the 1980s, to realize a true internal market for energy in Europe. Vestager's words, in announcing the decision, gave a subtle hint of the tactical flexibility that had characterized both sides during the previous three years of negotiations, when she referred to the decision as a "tailor-made rulebook for Gazprom's future conduct." Vestager was at pains, as she had been throughout the process, to emphasize that the decision, while specific to Gazprom, was not aimed at Russia or motivated by anti-Russian sentiment. "As always," she emphasized, "this case is not about the flag of the company." Yet this reassuring message was coupled with a warning to Gazprom: "The case doesn't stop with today's decision. Rather, it is the enforcement of the Gazprom obligations that starts today."[18]

In short, the settlement of the DG-COMP investigation, which was achieved in the middle of one of the most serious crises in East-West relations since the end of the Cold War, showed both the power and persistence of DG-COMP as well as the ability of the Russian side to adapt to the regulatory and commercial framework created by the three Gas and Power Directives over the previous thirty years. The case is not over, however. Gazprom remains, in effect, on probation for the next eight years—that is, for the duration of the obligations in the settlement. During that time, DG-COMP is firmly ensconced in the role of enforcer and overseer. Whether both sides will play their roles remains to be seen, but the DG-COMP case nevertheless marks the most important historical precedent since the end of the Cold War, one that may improve the overall tone of the Russian-European gas dialogue. The gas bridge, once again, has held.

Conclusion

The Future of the Gas Bridge

This book has been a story in three parts. The first recounted the rise of a new industry, initially in the Netherlands and then the North Sea, together with the creation of a vast network of pipelines that wove Europe into a single integrated gas system. In size, reach, physical interconnectedness, and complexity, the European natural gas industry is exceeded only by electricity and automotive transport. Natural gas was seen as a convenient fuel that was politically more secure than oil, cleaner than coal, and safer than nuclear power and that had almost no visual impact on the landscape. These qualities were essential to its growth and remain important today (see Table C.1).

This first story also took us east, to the origins of the Soviet and Russian gas industry, its beginnings in Ukraine and West Siberia, and its survival through the chaos of the Soviet collapse. This part of the story focused particularly on the origins of Gazprom and sought to explain the mixture of commercial motives, political constraints, and corporate culture that governed the gas industry's relations with the Soviet state in a closed centrally planned economy. The main comparison was to Norway, which represented a very different model of company-state relations, open to the outside world and aimed at developing cooperative relations with international companies offshore, under the benevolent but watchful eye of the Norwegian government. The book then went on to describe the

Table C.1 Natural gas demand in Europe by sector, 1990–2018 (standard billions of cubic meters, gross calorific value)

	1990	1995	2000	2005	2010	2015	2018
Residential	92.5	114.2	131.6	145.5	149.9	127.4	136.3
Electric power	50.5	64.1	103.9	139.6	163.3	107.0	135.1
Industrial	136.1	128.4	140.2	138.6	121.4	117.6	121.1
Commercial	40.0	48.5	49.1	56.8	61.8	57.4	60.4
Transportation	0.4	0.5	1.0	3.2	3.6	4.7	6.8
Hydrogen generation	0.0	0.0	1.8	4.4	4.7	4.7	4.7
Agricultural	6.0	5.3	5.2	5.3	4.9	4.0	4.4
Other	31.8	43.1	51.5	60.5	66.1	57.3	58.6
Total	357.0	404.1	484.4	553.8	575.8	480.1	527.4

Data sources: Rick Vidal, *2019 Updates to the Rivalry Macro and Energy Data Sets* (IHS Markit Global Scenarios Data, June 2019); historical data from the International Energy Agency and the US Energy Information Administration.

Notes: Numbers may not sum to totals due to rounding. Europe includes Turkey, Cyprus, Estonia, Latvia, Lithuania, and Malta. The "industrial" category includes feedstocks. The "other" category includes energy sector uses, distribution losses, and statistical differences.

origins of the gas bridge between East and West in the 1960s and its evolution through the subsequent decades.

The second story of this book was very different, consisting of ideas as opposed to things. It told of the wave of neoliberal thinking that arose in Great Britain and the United States in the 1970s and stimulated the market-oriented, supply-side political and economic thought that drove the Margaret Thatcher revolution in Britain. In the 1980s the tide moved to Brussels, becoming the basis for the Single European Market initiatives under Jacques Delors. These led to the succession of gas and power directives drafted by the European Commission and pressed on Europe's frequently reluctant national governments. The watchwords were open access, transparency, and competition, which were poles apart from the business models that had governed the European gas industry up to that time.

The new regulatory doctrines from Brussels, when applied to the gas and power sector, met with strong resistance from most of the European member-states and the gas and power utilities. This was particularly the case in Germany, where the battle between Brussels's single-market doctrine and Germany's traditional model of overlapping and adjacent monopolies for gas and power lasted for over a decade, until finally, in the second half of the 2000s, the resistance of the German gas monopolies abruptly collapsed. As a result, the structure of the German gas industry was transformed, with once-powerful players disappearing or being absorbed, and the traditional allocation of rents in the gas value chain was disrupted.

All of this came as a surprise to Moscow. As the Russians began to look at their European customers with new eyes in the 1990s, freed from the constraints of Soviet planning and from the norms of the Soviet-European gas trade, they hardly anticipated that the rules of the game were about to change radically in Europe as a consequence of a movement in which they had played no part. Starting in the mid-2000s they were caught up in the revolution that was unfolding around them, especially in the northern half of Europe.

The victory of the market-oriented model would not have been so complete if it had not been for simultaneous trends in the technology of the gas industry. The rapid expansion of liquefied natural gas (LNG) began turning what was historically a regional industry into a global one, by making gas tradable across world markets like oil or any other commodity. LNG has already increased the diversity of sources available to Western Europe, and access to LNG is spreading to Eastern Europe as well. But that is not all. In the 1990s information technology transformed the way gas was traded, enabling the creation of thriving spot markets in an industry that had formerly operated exclusively through long-term contracts. As a result, gas sellers and buyers now increasingly trade their product on computerized hubs, doing business with people whom they may have never met, who may or may not own gas fields or pipelines or other physical assets, and for whom gas contracts may be measured in days or hours. While the tools of trading allow for the sophisticated management of some financial risks, suppliers perceive that there has been a shift of overall risk onto their shoulders.

The trend toward liberalization and marketization in gas markets in Europe can be visualized as a moving line on a map, starting from the northwest quadrant—initially Britain and then Brussels—and then moving east and south, first into the Netherlands, Germany, France, and

Italy; and then to Central and Southern Europe; until it finally reaches the Balkans, the Baltic Republics, and Ukraine, where moves toward market liberalization in gas have become a leading part of the broad movement of political and economic reform in Kiev. It is even reaching Russia, although these are early days, and the path of possible restructuring there is still highly uncertain.

Starting in the mid-2000s a third story began, as Russia reverted to a state-centered and strongly nationalistic system under President Vladimir Putin. As part of this, the Russian state regained control of Gazprom. The company, under a new management tightly subordinated to the Kremlin, undertook an ambitious program of investment in upstream gas, both in eastern Russia and in the far north, accompanied by a continued expansion and diversification of its export pipeline system, much of which was designed to bypass Ukraine. As a result, the Russian gas industry has secured its next generation of supply. As it moves toward midcentury, it will enjoy a comfortable surplus of developed gas and transportation capacity that already makes it the lowest-cost exporter in the European market. However, Gazprom is increasingly under pressure from domestic competition, following the rise of powerfully connected independent gas producers, who, having already established themselves in the domestic market, are now challenging Gazprom's traditional monopoly of exports.

A central theme in the third story is the slow and reluctant adaptation of the Russians to the new business environment in the European gas market. The neoliberal tide in gas initially reached Russia only weakly, except for a handful of reformers in Moscow. When the Russians began selling gas in Europe, it was according to the commercial practices first developed in the 1960s in the Netherlands. During the long conflict between the German gas industry and the European Union, the Russians were mostly passive observers, absorbed as they were by economic and political issues at home. The changes taking place in the gas industry in Europe in 2004 and after therefore came as an unwelcome surprise to the Russians. The stage was set for a series of conflicts, as Gazprom was compelled, by a combination of pressure from regulators and customers, to adapt to the new European model. In the third part of the book we looked at three of the epic battles of the past decade that grew out of these pressures: the price concessions forced on Gazprom by its European customers, the investigation of Gazprom by the European Commission's Directorate-General for Competition (DG-Comp), and Nord Stream and the Ostsee-Pipeline-Anbindungsleitung.

The East-West gas bridge has so far survived all challenges. Frosty geopolitical relations coexist with mutual accommodation in business dealings, as Russia approaches the third decade of the twenty-first century exporting gas in record volumes. Its export specialists have learned to trade gas on a spot basis on the new gas hubs of Europe, operating under the new corporate and regulatory structure of the European gas industry. But the personal foundations and contractual bonds of long-term gas relationships, once a defining characteristic of the industry, have been weakened or, in some cases, severed altogether. Unlike the situation in the Cold War, when the gas bridge served a stabilizing and confidence-building function based on mutual economic interest and long acquaintance, today's gas relationships, despite the gas trade's present prosperity, are vulnerable to growing East-West tensions. Yet so far the gas bridge endures.

But will it continue to do so? In this Conclusion we shall explore two possible futures. The first is what one might call an evolutionary scenario, in which present trends are extrapolated to midcentury. Economics and technology coexist with politics, although not without conflict. It is an optimistic scenario, but one in which gas professionals believe strongly. The second is a very different story, in which environmentalism—the politics of climate change—becomes a major driving force. We discuss the possible consequences, using the example of Germany, where the environmental movement has already penetrated deeply, with far-reaching effects on policy.

The Golden Age of Gas

It is the golden age of gas in the global economy. Over the next thirty years, natural gas will be the world's bridge fuel, the only fossil fuel that will grow globally. By midcentury, the share of gas in global energy demand will rise from approximately one-fifth today to over one-quarter. By that time, gas will rival oil as the world's largest supplier of primary energy, as oil demand reaches a plateau and then declines slowly. Strong growth in gas supply will be driven by an expanding array of sources—especially US shale gas—supported by abundant reserves. Natural gas, once a regional fuel, will increasingly be traded worldwide as LNG, the share of which will surpass that of exports by pipeline. The combination of plentiful supplies, coupled with the expansion of LNG and strong competition among suppliers and shippers, will keep the world amply supplied at moderate prices.

But in Europe, the picture looks considerably different. Renewables, chiefly solar and wind, are driving a revolution in the power sector, which will be further accelerated by the advent of electric vehicles. By midcentury, solar and wind may account for half of the total European power capacity, which means in practice that they will supply a quarter to a third of European electricity, despite the intermittency of supply from renewables, which rely on the sun shining and the wind blowing. If there is no technological breakthrough in power storage, there will be an increased need for gas to offset this variability.[1] However, demand for gas will be constrained by the spread of heat pumps and condensing boilers for residential heat, hitherto the prime market for natural gas in Europe. These opposing forces will largely cancel one another out. Thus, on balance total European demand for natural gas will remain stable at roughly today's level, while oil, coal, and nuclear power will all decline.

Europe's own supplies of natural gas (so-called indigenous gas), mainly from the Netherlands, will continue to decline. This means that a growing share of Europe's gas supplies will have to be imported. The gas market at midcentury will be divided between pipeline gas (chiefly from Russia and Norway) and LNG (available from a wide array of suppliers, including the United States and Qatar as well as Russia). Russia's development of a new generation of West Siberian reserves and a diversified array of new export pipelines will give it formidable trumps in the competition between LNG and pipeline gas that is shaping up as one of the chief dramas in the European gas market in coming decades.

Delors's age-old vision of a Single European Market—a central theme of this book—will have been realized in gas. The European gas market will be so densely interconnected that gas can be easily traded across regions, and it will matter little what delivery route is used. As a result, regional price differentials between pricing hubs will be sharply reduced. Movements in prices will be driven mainly by world supply and demand and by competition between pipeline imports and LNG.

Despite the decline in indigenous gas and the need for more imports, there will be ample transportation capacity to bring gas to Europe, thanks to two generations of massive investment in pipeline building by Russia and by Western companies. This will keep transportation tariffs low and offset the risks whenever high global prices pull LNG away from Europe.

In sum, as the combined result of an efficient gas market, ample gas supplies from multiple sources, and an abundance of diversified transportation capacity, gas buyers will have more options. To quote a study by IHS Markit, "The issue of who supplies gas will be far less significant than the fact that market forces create checks and balances that bring cost-effective gas supply to consumers."[2] This will have two consequences. First, in contrast to the traditional formula, according to which the seller bears the price risk, while the buyer bears the volume risk, both price and volume risk will shift to the seller. Second, the potential geopolitical leverage that control of the pipeline system once gave to the seller will disappear, although not short-term market power.[3]

This picture summarizes trends that are already strong and can plausibly be extrapolated into the future. It assumes a world in which economic and technological forces are dominant, and in which business pragmatism prevails over geopolitics. In other words, it assumes that the gas bridge will survive, at least for another generation.

But like all scenarios, this one is vulnerable to multiple uncertainties. Three are especially salient: geopolitical conflict, Russia's export strategy, and Ukrainian transit.

Geopolitical Conflict

The gas bridge in its day underpinned the *Ostpolitik* and helped stabilize the Cold War in Europe. It subsequently became a symbol of the Russian-European business relations that sprang up after the end of the Soviet Union. When it was first launched in the 1960s and 1970s, it was an example, unique during the Cold War, of an alliance of politicians and businesspeople in the West, together with state functionaries and technocrats in the East, to exchange gas for pipe, technology, and finance across the divide of the Iron Curtain.[4] For nearly a half-century, the gas continued to flow, surviving both Cold War tensions and the collapse of the Soviet Union.

However, the hopes of the 1990s and early 2000s have given way to the bitterness and hostility of the late 2000s and 2010s. Russia's annexation of Crimea and sponsorship of separatist forces in eastern Ukraine, military pressures on the Baltic Republics, and cyberinterference in Western politics have created what is in many respects a new Cold War, as Russia seeks by means new and old to reassert its sphere of influence in the Former Soviet Union and in Eastern Europe, while the United States and the European Union push back with sanctions. At the time of this writing, there is little prospect of an easing of tensions.[5]

What do these political developments imply for the future of the gas bridge? As the share of Russian gas in Europe's gas supply reaches record levels, and as Russia completes a new generation of export pipelines, does Russia not have unprecedented leverage over Europe?

The revolutionary changes in the European gas market suggest that the answer is no. For all the reasons discussed above—the increasing interconnectedness of the European transportation system, the diversification of import sources thanks to LNG, and the availability of storage—the European gas system is strongly resilient today and will become even more so in the future, despite the decline of Europe's indigenous sources. Behind this is a simple fact: because of changes in gas technology and market structure in Europe and around the world, a pipeline shipper has less and less leverage compared to the past. This is true not only in Western Europe, but increasingly also in Eastern Europe and the Former Soviet Union. The ultimate illustration is Ukraine, which is now able to substitute "reverse" imports from the West for Russian imports, even though the actual molecules may be Russian.

Events over the past decade that have been described in this book show the strength of these propositions. As we have seen, Gazprom and the Kremlin have adjusted to the multiple changes in the European gas market, even while denouncing them. Pragmatism in practice has prevailed. As we look to the future, to a world in which oil demand has peaked and Russia's oil revenues may be declining, Russia will attach even more importance to its gas exports and its commercial reputation. Russia's interest, in short, will be not only to maintain the gas bridge but to keep it stable.

Russia's Export Strategy

The Soviet Union's gas export strategy consisted of two parts. The first was exports to Eastern Europe, where countries were supplied at low prices in exchange for political loyalty. The second was exports to Western Europe, which were conducted on a strictly commercial basis for hard currency. For two decades after the collapse of the Soviet Union, Russia attempted to continue much the same policy. Yet in the 2000s the eastward advance of marketization and EU regulation, together with Gazprom's need for capital, caused Russia to change its traditional approach to gas exports. The great turning point was the DG-Comp affair, in which Gazprom was sued by former Soviet satellites and forced to abandon several of its Soviet-era policies. The story of the past decade has been Russia's adaptation, however reluctant, to normal European commercial practices, notably including the growing share of Russian exports priced on spot markets. Russia increasingly plays by a single set of European rules throughout the European gas space, in the East as well as the West.

Russia will enjoy a strong competitive position in Europe in coming decades because of the sunk costs of its newly developed reserves in the Yamal Peninsula and its expanded pipeline system. As a result, it will be able to be the lowest-cost seller, and it will be able to compete successfully with imported LNG. But Russia faces choices. The first is whether to fight for market share by aggressively cutting prices or to keep to higher prices, even if that means a lower market share. So far Russia appears to be opting for the former.

Gazprom's export strategy is not a simple matter of economics. Its managers and the Kremlin understand very well that any all-out offensive to conquer additional market share in the European market by aggressively lowering prices to squeeze out competitors is likely to bring a vigorous political reaction—from both the European Commission and a number of EU member-states. Note that the perception of risk cuts both ways: Europe fears the risk of an excessive Russian market share, and Russia fears the risk of rejection of Russian gas by Europe. Gazprom understands this; the big question is whether the Kremlin does as well.

In coming decades Russia will increasingly direct its gas investments to expanding its exports to the East, particularly in the form of LNG. Urged on by the Kremlin, which sees LNG as the most progressive way of producing and exporting gas, Russia will expand exports of LNG to the potentially huge Chinese market. Climate change will make the northern sea route via the Arctic Ocean more competitive, and a program of generous state support for infrastructure will help Russian LNG overcome its higher cost compared to that of international rivals. With the passing of the era of the great diversification pipelines to Europe, most of Russia's new investment in gas will go to LNG for the eastern market, as well as expansion of the Power of Siberia pipeline system to China, supported by East Siberian fields.

Nevertheless, over the next decades, Europe will remain the main export outlet for Russian gas. What will this mean for Ukrainian transit, and ultimately for Ukraine's gas independence?

Ukrainian Transit

In Chapter 11 it was argued that Ukraine's strategic role as a transit country for Russian exports, which has been dwindling for the past two decades, will soon disappear, but the end of the road may not be smooth. Indeed, at the time of this writing it is far from clear that Russia and Ukraine will succeed in negotiating a new transit contract after the eleven-year deal signed in 2009 expires at the end of 2019. The interests of both sides lie in their coming to an arrangement. Nevertheless, the beginning of the 2020s is highly uncertain.

But that is not the end of the story. By the later 2020s, as indigenous production continues to decline, Europe will need more imported gas. Most of it will come by pipeline, as we have said, primarily although not exclusively from Russia, by way of the bypass pipelines. Yet the total capacity of all the new pipelines may not be enough to accommodate all of the European demand. Under that scenario Ukrainian transit may revive.

By that time, however, Ukraine will likely have implemented the EU gas directives. This will have far-reaching consequences. One of these will be the creation of a transmission system operator with authority to set transit tariffs. These will no longer be set by negotiation between the parties but according to accepted methodologies, which will include such things as calculations of the cost of capital. The practical result may well be lower transit tariffs, which may help make the Ukrainian route more competitive but will sharply lower the revenues accruing to Ukraine. The Ukrainian route may revive, but it will no longer be a cash cow for the Ukrainian government and other interests, as it has traditionally been. That in itself will be a major change.

In sum, none of the three uncertainties just discussed appear likely to disrupt the scenario that opened this chapter. But as we shall see below, other uncertainties await—particularly the rise of environmentalism and the politics of climate change.

An Environmental Scenario: Expectations for Gas Confounded

After the first decade of the twenty-first century, one might have expected that natural gas would be the great beneficiary of the trend toward decarbonization. Natural gas has many advantages: it is cleaner than coal and far more convenient, readily available from an expanding number of sources, and delivered through an extensive and increasingly interconnected transportation and distribution system. It offers an easy route to improved energy efficiency, and promoting gas has proved in some parts of the world—notably the United States and Great Britain—to be a quick win as a means of reducing emissions of greenhouse gases.

Yet for a full decade, as the gas expert James Henderson put it, "these expectations have been confounded" in Europe:

> The argument that gas is the cleanest fossil fuel and should at least displace coal in the energy mix, especially in the power sector, has had little traction with policy-makers. . . . In fact, over the past decade

> gas's share of the European energy mix has declined sharply in the face of a rise in renewable energy, while coal demand has remained remarkably robust.[6]

Why did this happen? There are a number of causes: competition from low-priced coal (resulting in part from the displacement of coal by gas in North America and its export to Europe, as well as exports from Russia), low carbon prices (under the weak EU Emission Trading System, at least until its recent reform—the consequences of which are as yet unknown), and growing environmental concerns about gas (over such issues as hydrofracturing and methane leakage as contributors to greenhouse gas emissions). Above all, gas's place in the power market has been held back by the plunging costs of solar and wind, coupled with the requirement that power from renewable sources be delivered first.[7] Natural gas is regarded with growing suspicion by the public and policymakers alike; it has, in effect, lost its aura as the virtuous fuel. An energy source that logically should be the bridge fuel between the fossil past and the renewable future may not be able to play that role.

Nowhere are these issues more salient than in Germany. The German energy transition today is unbalanced.[8] Its heavy emphasis on the role of renewables in the power sector, together with its reluctance to cut back on the use of coal, continues to shut gas out of the power sector. Gas has been the unloved child of the German *Energiewende*. The government's long-range climate plan, presented in Marrakesh in November 2016, calls for an exit from gas as the centerpiece of its decarbonization strategy, which aims to reduce carbon dioxide emissions by 80–95 percent from the power, heat, and transportation sectors by 2050.[9] A policy document issued in August 2016, the "Greenbook for Energy Efficiency," which represents a major shift in emphasis in Germany's energy policy toward energy saving, hardly mentions natural gas.[10] There is little indication that this approach to policy will change soon. A high-level governmental commission, charged with setting a policy for coal, debated for nearly all of 2018 as it struggled to balance the urgency of cutting emissions against the need to protect the economies of regions that mine brown coal. In the end, it produced a compromise schedule for an exit from coal by 2038, which will require extensive investment.[11] But significantly, it had little to say about what will replace coal once it exits from the mix, especially once Germany shuts down its last nuclear power plant in 2022.

As Germany becomes increasingly reliant on renewable sources of power, chiefly solar and wind, it will face with increasing urgency the problem of intermittency—how to keep the grid in balance when the sun and wind are not available, a condition that Germans call *Dunkelflaute,* or being "becalmed in the darkness." The only solution now available is to have additional power available from conventional sources, whether nuclear, coal, or gas. But if Germany has no nuclear, and the exit from coal proceeds on the schedule adopted by the commission, then what alternative is there to gas?

Technology will undoubtedly provide a partial answer in the form of improved batteries, which will enable a portion of the power produced by renewables to be stored. A more ambitious vision even sees electric vehicles as mobile storage, which can take power from the grid when they are moving and store it when they are parked. Over the longer term new forms of energy will become available. The greater long-term challenge to the gas bridge may come not from LNG but from synthetic gas, using energy from renewable power. The plummeting costs of solar and wind create the possibility of synthesizing gas (in a process called "power to gas") in a way that is carbon-neutral and therefore attractive to environmentalists and policymakers alike. The synthesis of hydrogen using surplus renewable power represents another long-term pathway that may ultimately offer substitutes for natural gas. Thus, technological innovation, which until now has favored the growth of the gas bridge, may begin to work against it, as synthetic gas and hydrogen begin to take market share from natural gas, regardless of its source.

But these long-term trends are unlikely to materialize on a large, transformative scale before midcentury. In the interim they leave space for natural gas, and even by 2050 there will continue to be demand for it, particularly in the residential and industrial sectors. And if there is demand for gas, Russia will be available to provide it. In short, even under an environmental scenario, the Russian-European gas bridge would survive for another several decades.

Yet as the political momentum for decarbonization grows in Europe, new legislation will impose tighter limits on CO_2 emissions, and these in turn will gradually constrain the use of natural gas. This will sharpen competition among suppliers, and especially between LNG and pipeline gas. There will be no new transcontinental pipelines to Europe. Marginal transportation capacity will go unused. Ukrainian transit, as the highest-cost transit route, will finally disappear. In sum, under an environmental scenario, the gas bridge survives, but it stagnates.

For the next decade, the "golden age" scenario seems the likelier of the two. By the 2030s, however, the environmental scenario gains in plausibility, until by mid-century it may well become the dominant one.

The central question of this book has been: How, despite changing circumstances and aggravated geopolitical conflict, has the Russian-European gas trade endured? The answer is that over the years the gas bridge has served a shared economic interest that has stood the test of time. It has been successfully adapted to changing technologies, regulatory and commercial regimes, and ideologies. For much of that time, the gas bridge has served a useful political purpose as well, as the foundation of the European *Ostpolitik* and, more recently, as a new business model has become established, as a moderating influence at a time of mounting geopolitical conflict. Yet the underlying driver is simplicity itself: since Russia has gas and Europe needs gas, the gas trade serves both sides profitably. That was the foundation of the gas bridge in the middle of the Cold War, and it remains true in the redivided Europe of today.

Throughout this book the metaphor of the bridge has been used with multiple meanings. It refers first to the physical gas bridges that connect the periphery of Europe to its core (via pipelines from Russia, North Africa, the Caspian, and Norway), as well as the many regional interconnections that make the gas bridge robust. The physical bridge also includes the invisible structure of technological innovation that has made the bridge possible—large-diameter pipelines and compressors, high-quality steels, computerized control systems, combined-cycle gas turbines, deepwater construction, and the like. Natural gas, despite its seeming simplicity, is a high-technology fuel, the product of advanced engineering, and one of the aims of this book has been to show the importance of technology for the origins and growth of the East-West natural gas bridge. The steady progress of technology over the decades has made the physical bridge today more robust than ever.

The latest transformation in the gas industry is the rapid growth of LNG. At first glance this might appear to pose a threat to the established trade by pipeline, but on reflection one can see that it actually adds to the flexibility and robustness of the gas bridge. First, LNG has much of the same distribution infrastructure as pipeline gas; it merely represents an additional way of bringing supply into the gas system. LNG adds to the diversity of supply routes into Europe. It is now being used by Russian gas producers as well as more distant suppliers. (Indeed, from the Russian perspective, LNG can be seen as simply another Ukrainian bypass.) Thus, LNG illustrates how continuing technological advances can reinforce the existing gas bridge, at least for a time.

The second meaning of the bridge metaphor in this book—and the second answer to the core question with which we began this section—has been the bridge of ideas formed by neoliberal thinking and marketization, which initially crossed from the United States and Great Britain to Brussels and subsequently brought upheaval to the established European gas industry in the form of new regulations and commercial structures. The bridge of new thinking took over two decades to become established in continental Europe, but after facing much opposition it is being extended irresistibly eastward and southward and is becoming the established way of doing business in the gas industry throughout the European Union and beyond.

The spot market—the physical expression of the new regulatory and commercial order—has become the dominant means of selling gas in Europe. The volumes traded and priced on European gas hubs have reached about two-thirds of total sales and are still growing rapidly.[12] Two hubs, the Dutch Title Transfer Facility and the British National Balancing Point, still account for the bulk of total spot sales in Europe, but other hubs in Eastern and Southern Europe, particularly the two German gas hubs, GASPOOL and Net Connect Germany, are developing rapidly. The hubs are interconnected and ensure a degree of transparency and a sufficient total turnover to support liquid markets that provide confidence for the participants. Now they are being extended to Eastern Europe, and the result is a reinforcement of the gas bridge, as a common platform of business models and regulation is created throughout Europe. The Single European Market is becoming a reality for gas in Europe, including Russian gas imports and transit.

The possible future threats to today's liberalized gas market are likely to arise not from new ideologies, but from the very success of marketization. One major question for the industry as a whole, including the Russians, is how and whether the traditional virtual bridge of the industry—long-term contracts and relationships—can be refashioned to provide more stability and a more even sharing of risk between buyer and seller.

A third meaning of the bridge metaphor is the informal ties between people. Natural gas from the beginning has been the product of enthusiasm and vision. Throughout the industry's history it has been a striking fact that people who work in it have been true believers in gas. This in turn helps account for the key role of leading individuals throughout its history and the role they have played in building the East-West community of interest, from Aleksey Kortunov and Nikolai Baibakov down to Alexander Medvedev and Leonid Mikhel'son (the founder of Russia's LNG producer, Novatek), on the Russian side, and from Burckhard Bergmann at Ruhrgas to Herbert Deterding of Wintershall in Germany and their peers in France, Italy, and Austria on the Western side. This book has attempted to show the critical contributions they have made over the past half-century.

But there is now a gnawing doubt within the European industry over its leadership and its future. It is still taken as an article of faith among gas people that natural gas is the bridge fuel to a cleaner environment, but that is proving to be a difficult message to convey. The liberalization of the gas industry and the breakup of traditional utilities have atomized what was previously a tight-knit community, and it no longer speaks with a single voice. To compete in the decades ahead, the natural gas industry will need a new generation of leaders who can persuade policymakers and the public that natural gas (whatever its source) has a vital role to play in the coming energy transition. This is a dilemma faced by the entire gas industry, east and west.

The fourth and final meaning of the bridge metaphor is the geopolitical one. In historical perspective, much of the geopolitical conflict surrounding the gas bridge is a long-term legacy of the Soviet era and the long battle over the liberalization of the European gas market: the mixture of conflict and collusion in Ukraine and the Baltic Republics, the separation of Eastern Europe into gas islands, the efforts of the Russians to bypass Ukraine, the long resistance to the single European energy market, the slow spread of marketization to Central and Eastern Europe—all these are the distant echoes of the gas system put in place under the Soviet system and the traditional European industry. But the Soviet past is gradually fading, the new European gas market is increasingly an established fact, and the gas world is moving on. But how long will the gas bridge survive the new challenges ahead? If shared economic interest prevails in the face of political divisions, the gas bridge may continue to provide a stabilizing force between Russia and Europe for another generation and perhaps beyond. But will geopolitics allow? Will the politics of climate change cause it to be downgraded? How long before technology makes it obsolete? Those are the big questions ahead, for both Europe and Russia.

Notes

Introduction

1. The oil market went through a similar revolution decades ago. Russian oil is sold into a wide-open global market, and although Europe relies on Russian oil as much as on Russian gas, oil exports are viewed as less of a security threat. Furthermore, no one even mentions Russian coal exports to Europe as a security threat, least of all in Brussels.

1. Two Worlds of Gas

1. Daniel Yergin, *The Prize: The Epic Quest for Oil, Money, and Power* (New York: Simon and Schuster, 1991), p. 92.
2. *Historical Statistics of the United States (Millennial Edition Online),* http://hsus.cambridge.org.proxy.library.georgetown.edu/HSUSWeb/toc/tableToc.do?id=Db155-163 (accessed May 7, 2019). This source is only available online via the Georgetown University library to members of the Georgetown University community.
3. On the continuing legacy of the Soviet way of development in the former Soviet economies, see the insightful chapter by Clifford Gaddy, "Room for Error: The Economic Legacy of Soviet Spatial Misallocation," in Mark R. Beissinger and Stephen Kotkin, eds., *Historical Legacies of Communism in Russia and Eastern Europe* (Cambridge: Cambridge University Press, 2014), pp. 52–67.
4. On this aspect of the Soviet planning system, see Peter Rutland, *The Myth of the Plan: Lessons of Soviet Planning Experience* (La Salle, IL: Open Court, 1985).
5. Quoted in Wolf Kielich, *Subterranean Commonwealth: 25 Years of Gasunie and Natural Gas* (Amsterdam: Uniepers B.V., 1988), p. 18.

6. Alain Beltran and Jean-Pierre Williot, *Les routes du gaz: Histoire du transport de gaz naturel en France* (Paris: Le Cherche Midi, 2012), p. 24.
7. Ibid., pp. 17–21.
8. Ibid.
9. Nadja Daniela Klag, *Die Liberalisierung des Gasmarktes in Deutschland* (Marburg, Germany: Tectum Verlag, 2003), p. 159.
10. Kielich, *Subterranean Commonwealth,* p. 15.
11. Beltran and Williot, *Les routes du gaz,* p. 21.
12. Kenneth Hutchinson, *High Speed Gas: An Autobiography* (London: Duckworth, 1987), p. 152.
13. Ibid., p. 178.
14. Dieter Helm, *Energy, the State, and the Market: British Energy Policy since 1979* (Oxford: Oxford University Press, 2003), p. 109.
15. Hutchinson, *High Speed Gas,* p. 177.
16. Denis Rooke, the future chairman of British Gas, was the lead figure in the first shipment of liquid methane from Louisiana. See Hutchinson, *High Speed Gas,* pp. 183–186.
17. Ibid., p. 184.
18. NAM was created in 1947 as a fifty-fifty joint venture between Standard Oil of New Jersey (Esso, as Exxon was called then) and Bataafse Petroleum Maatschappij (BPM), a Dutch subsidiary of Royal Dutch Shell, to explore for oil and gas in the Netherlands.
19. On the US experience, see Christopher J. Castaneda and Clarence M. Smith, *Gas Pipelines and the Emergence of America's Regulatory State* (Cambridge: Cambridge University Press, 1996); and Jeff D. Makholm, *The Political Economy of Pipelines: A Century of Comparative Institutional Development* (Chicago: University of Chicago Press, 2012).
20. Tony Judt, *Postwar: A History of Europe since 1945* (New York: Penguin, 2005), p. 67.
21. Daniel Yergin and Joseph Stanislaw, *The Commanding Heights: The Battle between Government and the Marketplace That Is Remaking the Modern World* (New York: Simon and Schuster, 1998).
22. Ken Gladdish writes that the Dutch system of proportional representation tended to lead to the "sometimes painful delivery of unexpected governments via the midwifery of party leaders acting in the wake of elections" (*Governing from the Center: Politics and Policy-Making in the Netherlands* [De Kalb: Northern Illinois University Press, 1991], p. 47. See also K. R. Gladdish, "The Primacy of the Particular," in Larry Diamond and Mark F. Plattner, eds., *Electoral Systems and Democracy* (Baltimore: Johns Hopkins University Press, 2006), pp. 105–120; and K. R. Gladdish, "Governing the Dutch," *Acta Politica* 25 (October 1990): 389–402.
23. Henk Kamp, "The Bright Past and Challenging Future of Natural Gas," speech at the symposium "The Bright Past and Challenging Future of Natural Gas," October 3, 2013, http://www.government.nl/documents-and-publications/speeches/2013/10/03/the-bright-past-and-challenging-future-of-natural-gas.html.

24. Instead, de Pous spent the next twenty years as chairman of the Dutch Council on Economics and Society, where his nickname, "Jan Compromise," sounded increasingly derogatory as the consensus-minded 1960s turned into the more contentious 1970s.
25. Gladdish, *Governing from the Center,* p. 47.
26. Gladdish, *Governing from the Center,* p. 51.
27. Kielich, *Subterranean Commonwealth,* p. 20.
28. The story is told in a memoir by Douglass Stewart and Elaine Madsen, *The Texan and Dutch Gas: Kicking Off the European Gas Revolution* (Victoria, BC: Trafford Publishing, 2006).
29. Kielich, *Subterranean Commonwealth,* p. 20.
30. A useful introductory overview of natural gas pricing is Anthony J. Melling, *Natural Gas Pricing and Its Future: Europe as the Battleground* (Washington, DC: Carnegie Endowment for International Peace, 2010).
31. Peter Neuhaus, RWE, "Herausforderung an die Erdgasvermarktung: Vertrieb in Spannungsfeld liberalisierter Energiemärkte," 2007, http://www.rwe.com/web/cms/contentblob/157880/data/12255/Down4.pdf.
32. Kielich, *Subterranean Commonwealth,* p. 52.
33. Ibid., p. 121.
34. Simon Blakey, personal communication, August 2014.
35. Kielich, *Subterranean Commonwealth,* pp. 72 and 161.
36. Iulii M. Bokserman, *Razvitie gazovoi promyshlennosti SSSR* (Moscow: "Gostoptekhizdat," 1958), p. 47.
37. Ibid., p. 46; see also pp. 47–48.
38. Ibid., p. 46.
39. Per Högselius, *Red Gas: Russia and the Origins of European Energy Dependence* (New York: Palgrave Macmillan, 2013), p. **14.**
40. At the time of Stalin's death, coal still accounted for two-thirds of the Soviet energy supply. In 1955 coal accounted for 64.8 percent, compared to 2.2 percent for natural gas. See ibid., p. 23.
41. The classic work on Soviet-era oil and gas is Robert W. Campbell, *The Economics of Soviet Oil and Gas* (Baltimore: Published for Resources for the Future by the Johns Hopkins Press, 1968).
42. For an overview of the Soviet coal sector, see Leslie Dienes and Theodore Shabad, *The Soviet Energy System: Resource Use and Policies* (New York: John Wiley and Sons, 1979), chapter 4.
43. The share of Ukrainian gas in total Soviet gas output through the mid-1970s was never more than about one-third, but through that time Ukraine was the largest single producing region, and Shebelinka was the largest single field in the European USSR. These features, added to its proximity to Europe and the western Soviet Union, made Ukrainian gas centrally important through the mid-1970s. For detailed production numbers, see ibid., table 18, pp. 70–71.
44. Merle Fainsod, *Smolensk under Soviet Rule* (Cambridge, MA: Harvard University Press, 1958), passim.

45. Kortunov's life and career are described in detail in Högselius, *Red Gas*. A Russian biography of Kortunov is Viktor Andrianov, *Kortunov* (Moscow: "Molodaia Gvardiia," 2007).
46. Neizvestnyi died at a ripe old age in 2016. See his obituary in the *Economist,* August 20, 2016, https://www.economist.com/obituary/2016/08/20/obituary-ernst-neizvestny-died-on-august-9th.
47. A Soviet-era expression meaning "storming to meet the plan targets."
48. Dienes and Shabad, *The Soviet Energy System*.
49. *Gazovaia promyshlennost',* no. 1 (January 1962), p. 2, cited in Högselius, *Red Gas,* pp. 18–19.
50. Mariia Vladimirovna Slavkina, *Triumf i tragediia: Razvitie neftegazovogo kompleksa SSSR v 1960–80-e gody* (Moscow: "Nauka," 2002), pp. 80–82 and 88ff.
51. William Taubman, *Khrushchev: The Man and His Era* (New York: Norton, 2003).
52. The source for this story, as well as the quotes in the two previous paragraphs, is Nikolai Baibakov, "V cherede velikikh svershenii," in O. I. Lobov, ed., *Neftegazostroiteli Zapadnoi Sibiri,* vol. 1 (Moscow: "Rossiiskii Soiuz Neftegazostroitelei," 2004), pp. 10–16.
53. Mariia V. Slavkina, *Baibakov* (Moscow: "Molodaia Gvardiia," 2010), pp. 113ff. The story is also revealing in two other respects, which show how little Khrushchev actually understood about the industry. First, by this time the discovered gas reserves of the Soviet Union were on the verge of becoming the largest in the world. Second, much of the feedstock for the chemical industry came from associated gas as a by-product of the oil industry, and consequently natural gas was only of secondary significance for Khrushchev's chemicals campaign.
54. Bokserman, *Razvitie gazovoi promyshlennosti SSSR*, p. 7.
55. On the dominance of the military-industrial complex in the competition for investment and R&D resources from the 1960s through the 1980s, see Robert Campbell, "Resource Stringency and Civil-Military Resource Allocation," pp. 126–163; Julian Cooper, "The Defense Industry and Civil-Military Relations," pp. 164–191; and Thane Gustafson, "The Response to Technological Challenge," pp. 192–238, all in Timothy J. Colton and Thane Gustafson, eds., *Soldiers and the Soviet State: Civil-Military Relations from Brezhnev to Gorbachev* (Princeton, NJ: Princeton University Press, 1990).
56. Bokserman, *Razvitie gazovoi promyshlennosti SSSR.*

2. The Beginnings of the Gas Bridge

1. The last great "pea-souper" caused by coal occurred in Great Britain in 1962, affecting the entire country over a period of four days. Natural gas was found soon after, and conversion from town gas to natural gas started in 1967.
2. In this book I have generally adhered to the Library of Congress rules for transliterating Slavic names. The exception will be names that are frequently referred to in the Western media under a different spelling—in this case, for example, Tyumen instead of Tiumen'.

3. The account that follows is drawn mainly from M. V. Komgort and G. Iu. Koleva, "Problema povysheniia urovnia industrial'nogo razvitiia Zapadnoi Sibiri i proekt stroitel'stva Nizhneobskoi GES," *Vestnik Tomskogo gosudarstvennogo universiteta,* no. 308 (March 2008): 85–90, http://cyberleninka.ru/article/n/problema-povysheniya-urovnya-industrialnogo-razvitiya-zapadnoy-sibiri-i-proekt-stroitelstva-nizhneobskoy-ges. Earlier Soviet literature made veiled references to the dispute over the proposed hydropower project, but only in post-Soviet times has the full story been told.
4. V. N. Tiurin, "Iamal'skii potentsial," *Oktiabr',* no. 4 (1976): 138.
5. Laurens van der Post, *Journey into Russia* (New York: Vintage, 1964), passim.
6. Viktor Andrianov, *Kortunov* (Moscow: "Molodaia Gvardiia," 2007), pp. 376–377.
7. Shcherbina's rejection of the Nizhneobskaya GES and his conversion to the cause of West Siberian oil and gas marked a key turning point in one of the most remarkable careers in the Soviet era. In 1973, upon the death of Aleksey Kortunov, Shcherbina succeeded him as minister of oil and gas construction. In that position he played a key role in Brezhnev's gas campaign starting in 1980, and in 1983 he was named deputy prime minister responsible for energy. For details and sources, see Thane Gustafson, *Crisis amid Plenty: The Politics of Soviet Energy under Brezhnev and Gorbachev* (Princeton, NJ: Princeton University Press, 1989), pp. 83, 106, and 145. In that position he played a key role in the fight to prevent a nuclear meltdown at Chernobyl, which would have had catastrophic consequences for all of the western Soviet Union and Eastern Europe. Shcherbina died in August 1990 at seventy years of age from the aftereffects of radiation to which he had been exposed at Chernobyl, not having witnessed the end of the Soviet Union. He was one of the many Ukrainians in the upper ranks of the Soviet system. A native of the Khar'kiv region, he rose through the Ukrainian Komsomol and was briefly secretary of the city committee of Khar'kiv before being transferred to Irkutsk—where. as second secretary of the Irkutsk *obkom* of the Party, he oversaw all of the major hydropower projects in the province.
8. On the vigorous battles over natural resources in the Soviet system, see Thane Gustafson, *Reform in Soviet Politics: Lessons of Recent Policies on Land and Water* (Cambridge: Cambridge University Press, 1981).
9. For more on Bogomiakov, see Thane Gustafson, *Crisis amid Plenty* and *Wheel of Fortune: The Battle for Oil and Power in Russia* (Cambridge, MA: Belknap Press of Harvard University Press, 2012).
10. Gennadii Bogomiakov, "Lomaia led nedoveriia," in O. I. Lobov, ed., *Neftegazostroiteli Zapadnoi Sibiri,* vol. 2 (Moscow: "Rossiiskii Soiuz Neftegazostroitelei," 2004), passim.
11. The origins of Kortunov and of Glavgaz were introduced in Chapter 1. In Western literature, the indispensable source is Per Högselius, *Red Gas: Russia and the Origins of European Energy Dependence* (New York: Palgrave Macmillan, 2013).
12. Andrianov, *Kortunov,* pp. 267–281 and 331–333.
13. Medvezh'e, although developed earlier than Urengoy, was discovered later, in 1967.

14. In 1965 Kortunov created a special design bureau to plan gas production and transportation in Tyumen, and in 1966 he launched a dedicated business unit called Tiumengazprom to oversee the operation.
15. For example, Iulii M. Bokserman has an entire chapter in *Razvitie gazovoi promyshlennosti SSSR* (Moscow: "Gostoptekhizdat," 1958) on the state of the Soviet compressor industry as of the late 1950s, before the great West Siberian gas fields were discovered. He was naturally discreet about the reasons for the backwardness of the sector.
16. Derezhov's official biography on the Gazprom website has him starting at Mingazprom in 1966. "Stepan Romanovich Derezhov," http://www.gazprom.ru/press/news/reports/2012/orudzhev/memories/derezhov/ (accessed May 8, 2019). Unfortunately, this source is no longer available on the website.
17. Quoted in Andrianov, *Kortunov,* p. 333.
18. Quoted in Andrianov, *Kortunov,* p. 333.
19. Interview with Iurii Baranovskii and Stepan Derezhov, 2003.
20. Andrianov, *Kortunov,* p. 334.
21. Bruna Bagnato, *Prove di ostpolitik* (Firenze: Olschki, 2003).
22. *Oil and Gas Journal,* June 1962, cited Högselius, *Red Gas,* 28.
23. Kortunov was an interesting exception to the pattern: unusually, he had worked in a wide variety of industrial branches (Högselius, *Red Gas,* pp. 16–17). But in every other respect he was a classic technocrat.
24. Valentin A. Runov and Aleksandr D. Sedykh, *Orudzhev* (Moscow: "Molodaia Gvardiia," 2012), pp. 254–255. The source in the biography is an interview with Derezhov.
25. Wolf Kielich, *Subterranean Commonwealth: 25 Years of Gasunie and Natural Gas* (Amsterdam: Uniepers B.V., 1988), 72.
26. Runov and Sedykh, *Orudzhev,* p. 201.
27. See ibid., pp. 203–208, for detailed descriptions of Orudzhev as a negotiator. Most of these descriptions are hagiographic, but they convey the essential point that Orudzhev was definitely hands on.
28. The Swedish historian Per Högselius explains this in a brilliant study based on exhaustive archival research (*Red Gas,* chapter 4).
29. Ibid., p. 53.
30. Patolichev, a Belorussian by birth, was minister of foreign trade from 1958 through 1985, spanning virtually the entire period from Khrushchev through Brezhnev and up to the beginning of the Gorbachev years.
31. Quoted in Andrianov, *Kortunov,* p. 334.
32. Osipov was well known to US negotiators in the mid-1970s, having led talks on possible US sales of equipment for the development of Far Eastern oil and gas resources, as well as exports of LNG and methanol to the US from north Tyumen. However, none of these deals came to pass.
33. Sorokin, whose ties were to Kortunov, followed the latter when he was removed from the gas ministry and placed in charge of a new ministry of oil and gas construction. This marked a painful split in the gas industry. In the CIA's 1981 directory of national officials, an Aleksey Ivanovich Sorokin appears as deputy

minister of oil and gas construction, appointed in October 1972. He was born on October 28, 1909, so by this time he was already seventy-two. His functions at this time are not indicated. In the 1983 edition of the directory, he's still there. By the 1986 edition, however, after Vladimir Chirskov was named minister of oil and gas construction in place of Batalin in 1984, Sorokin is no longer deputy minister. See the 1981, 1983, and 1986 editions of the *Directory of Soviet Officials: National Organizations* (Washington, DC: Central Intelligence Agency).

34. Stanislav Volchkov was the Soviet trade representative in Cologne and a specialist in machinery.
35. In 1985, Patolichev, in failing health, stepped down; he died four years later. Yet the team he had formed, and particularly Osipov, carried on through 1987.
36. However, the contact almost certainly contained a review or reopener clause, which would mean the price could be adjusted every three years.
37. Högselius, *Red Gas,* pp. 60–62, 92.
38. The conversation between Khrushchev and Ulbricht took place on November 30, 1960. Quoted in Hope M. Harrison, *Driving the Soviets up the Wall: Soviet-East German Relations, 1953–1961* (Princeton, NJ: Princeton University Press, 2005), p. 152.
39. Quoted in ibid., p. 153.
40. Quoted in ibid., p. 164.
41. Quoted in Rainer Karlsch, *Vom Licht zur Wärme: Geschichte der ostdeutschen Gaswirtschaft 1855–2008* (Leipzig: Verbundnetz Gas AG, 2008), p. 133.
42. Quoted in Harrison, *Driving the Soviets up the Wall,* p. 154.
43. Quoted in ibid., p. 154.
44. Quoted in ibid., p. 178.
45. Quoted in ibid., p. 22.
46. Protocol of conversation between Brezhnev and Honecker, July 28, 1970. Quoted in ibid., p. 232.
47. Karlsch, *Vom Licht zur Wärme,* pp. 139ff.
48. Ibid., p. 140.
49. Ibid., pp. 140–141.
50. Karlsch does not mention whether the Poles were supposed to get some of the gas. Presumably they were.
51. Karlsch, *Vom Licht zur Wärme,* p. 142.
52. Ibid., p. 143.
53. This portrait of Brandt is drawn from his memoirs: Willy Brandt, *Erinnerungen* (Frankfurt am Main: Fischer Taschenbuch, 1997)
54. Hans Peter Schwarz, *Axel Springer: Die Biografie* (Berlin: Propyläen Verlag, 2008), p. 379.
55. Brandt, *Erinnerungen,* p.17.
56. Under the Third Transition Law of 1952, the federal government had "financial responsibility" for West Berlin (Brandt, *Erinnerungen,* p. 17).
57. I am indebted to Josephine Moore, a native Berliner, for this observation.

58. Angela Stent, *From Embargo to Ostpolitik: The Political Economy of West German–Soviet Relations, 1955–1980* (Cambridge: Cambridge University Press, 1981), p. 156.
59. The Dutch offered flexibility because they could increase or decrease Groningen production and were close to their markets, which was very valuable in Europe because of the variation between winter and summer. The Soviets had far less opportunity to vary production or delivery levels.
60. The details of the Soviet–West German negotiations are told in vivid detail in Högselius, *Red Gas,* chapter 7.
61. For a brief biography of Schelberger, see Dietmar Bleidick, "Schelberger, Herbert," *Neue Deutsche Biographie* 22 (2005), https://www.deutsche-biographie.de/pnd140465448.html#ndbcontent.
62. Högselius, *Red Gas,* pp. 117–118.
63. Stent, *From Embargo to Ostpolitik,* p. 169.
64. It was to be the last hurrah for the old Kaiserhof, which was torn down four years later. Its former location is now the site of the Linden Center, a banking complex.
65. The ceremony at the Kaiserhof was also the last major appearance of Sorokin in international gas negotiations. In 1972 he followed his friend Kortunov into "exile," as the core of the Kortunov team left the gas ministry for the newly created Ministry of Oil and Gas Construction.
66. Quoted in Högselius, *Red Gas,* p. 129. The account that follows is based primarily on chapter 7 of that work and Stent, *From Embargo to Ostpolitik,* chapter 7.
67. Högselius, *Red Gas,* p. 122.
68. Stent, *From Embargo to Ostpolitik,* p. 169.
69. Högselius, *Red Gas,* pp. 111–112.
70. "Mehr Licht," in *Capital* 10 (1970), cited in Sophie Gerber, *Küche, Kühlschrank, Kilowatt: Zur Geschichte des privaten Energiekonsums in Deutschland, 1945–1990* (Bielefeld: Transcript Verlag, 2015), p. 166.
71. *Rheinische Post,* October 11, 1967, cited in ibid., p. 166.
72. Ibid.
73. Högselius, *Red Gas,* p. 124.
74. Helmut Schmidt, *Menschen und Mächte* (Berlin: Siedler, 1987), p. 61.
75. It is symptomatic that Ulf Lantzke, who had been trained as a jurist and was an expert in competition issues when he was seconded to the Coal and Steel Community, became the first head of the International Energy Agency in 1974, the main mission of which was energy security.
76. Karlsch, *Vom Licht zur Wärme,* pp. 170–173. See also Hilmar Bärthel, *Die Geschichte der Gasversorgung in Berlin: Eine Chronik* (Berlin: Nicolai Publishers on behalf of GASAG Berliner Gaswerke Aktiengesellschaft, 1997).
77. Bärthel, *Die Geschichte der Gasversorgung in Berlin,* pp. 155ff.
78. Ibid., pp. 162–163. The Berlin government then sold minority shares in the unified GASAG to West German companies (Ruhrgas, RWE, and VEBA) while reserving a majority share to itself. A small share was distributed to the company's employees.
79. Gustafson, *Reform in Soviet Politics,* chapter 4.

3. From Optimism to Anxiety

1. Tony Judt, *Postwar: A History of Europe since 1945* (New York: Penguin, 2005), p. 453.
2. Mark Mazower, *Dark Continent: Europe's Twentieth Century* (New York: Vintage, 2000), chapter 10, pp. 327ff.
3. Daniel Yergin, *The Prize: The Epic Quest for Oil, Money, and Power* (New York: Simon and Schuster, 1991), 674–699.
4. Ibid., p. 653.
5. On the renaissance of economics in the Soviet Union in the 1960s, see Thane Gustafson, *Reform in Soviet Politics: Lessons of Recent Policies on Land and Water* (Cambridge: Cambridge University Press, 1981), chapter 4.
6. Mariia V. Slavkina, *Triumf i tragediia: Razvitie neftegazovogo kompleksa SSSR v 1960–80-e gody* (Moscow: "Nauka," 2002), pp. 113ff.
7. See Gustafson, *Reform in Soviet Politics,* especially chapters 2 and 9.
8. This made sense in 1971–1973 because the ratio of gold prices to oil prices soared from a historic level of 20 to 1 through 1970 (that is, an ounce of gold could buy twenty barrels of oil) to over 30 to 1 in 1973. When oil prices quadrupled in 1974, the gold-to-oil ratio sank to about 14 to 1 and remained at an average of about 16 to 1 for the rest of the decade. In short, for the Soviet Union it was rational to export gold in preference to oil through 1973, but less so in 1974 and after.
9. Yegor Gaidar, *Collapse of an Empire: Lessons for Modern Russia* (Washington, DC: Brookings Institution Press, 2007).
10. Inflation officially did not exist in the Soviet Union—it was supposedly a disease of capitalist economies—but it crept into the Soviet economy via the underground economy, as people reacted to shortages by turning to what they called "the economy on the left." Western and Russian economists after the fall of the Soviet Union concluded that there had been a steady hidden level of inflation of over 5 percent per year throughout the period from 1970 to 1985, following which inflation increased sharply as the Soviet government lost control of credit and the money supply.
11. Jonathan P. Stern, *European Gas Markets: Challenge and Opportunity in the 1990s* (London: Royal Institute of International Affairs, 1990), pp. 1–2.
12. The famous phrase "too cheap to meter" was coined by Lewis Strauss, then chairman of the United States Atomic Energy Commission, in a 1954 speech to the National Association of Science Writers. See https://en.wikipedia.org/wiki/Too_cheap_to_meter.
13. I am grateful to Theo Krausz of the Institut des Sciences Politiques Strasbourg for his research paper "Un noyau devenu instable: Que reste-t-il du 'Tout Nucléaire Français?,'" written for my seminar at Georgetown University in the fall of 2016.
14. One authority on commodities drily calls this "the foolishness of bankers" *(la bêtise des banquiers)* (in Philippe Chalmin, ed., *Des ressources et des hommes: Matières premières 1986–2016; Trois décennies de mondialisation et au-delà* (Paris: Editions François Bourin, 2016).

15. This paragraph follows the thinking of Joseph Schumpeter on price cycles. See his essay "Capitalism" in the *Encyclopaedia Britannica,* vol. 4 (Chicago: Encyclopaedia Britannica, 1946), pp. 801–807.
16. See Karl Ditt, "Die Anfänge des Umweltpolitik in der Bundesrepublik Deutschland während des 1960er und frühen 1970er Jahre," in Matthias Frese, Julia Paulus, and Karl Teppe, eds., *Demokratisierung und gesellschaftlicher Aufbruch: Die sechziger Jahre als Wendezeit der Bundesrepublik* (Paderborn, Germany: Ferdinand Schöningh, 2005), pp. 305–347.
17. See Yergin, *The Prize,* p. 544.
18. Robert W. Campbell, *The Economics of Soviet Oil and Gas* (Baltimore, MD: Published for Resources for the Future by the Johns Hopkins Press, 1968).
19. Dieter Helm, *Energy, the State, and the Market: British Energy Policy since 1979* (Oxford: Oxford University Press, 2003), p. 15.
20. A landmark event in the iconography of the environmental movement was the first popular election to the European Parliament. Some historians see the "Cassis" decision of the European Court of Justice as the more significant turning point, in that it marks the emergence of the Single European Market in European legal doctrine. We return to these themes in Chapters 6 and 7.
21. This was largely true of the Eastern European satellites as well. Despite the creation of the Council for Mutual Economic Assistance (Comecon), the nominal aim of which was to foster greater integration between the various East European countries and the Soviet Union, the actual pattern of policymaking in Eastern Europe resembled a spoke and wheel, in which each East European country dealt with the Soviet Union on a bilateral basis and there were few ties linking the East Europeans with one another.
22. Urengoy was discovered in 1966, Medvezh'e in 1967, and Yamburg in 1969. The Yamal region, which covers much of the northern third of Tiumen' Province, is often confused in the West with the Yamal Peninsula, which lies to its north. The first three giants were discovered in the Yamal region; the discoveries on the Peninsula came later, in the 1980s. For a brief recent recap of the history of the development of oil and gas exploration in West Siberia, see Anatolii Brekhuntsov, "Istoriia otkrytiia i osvoeniia mestorozhdenii uglevodorodov v Zapadnoi Sibiri," *Neftegazovaia vertikal',* no. 6 (2016), pp. 17–20.
23. Since this book is primarily about the East-West gas trade, it does not discuss Algeria, which raises different issues.
24. As Angela Stent points in her history of Soviet-German relations, the end of the 1960s and the decade of the 1970s were marked by two seemingly contrary trends: a stabilization of East-West relations (especially Russian–West German relations) as a result of *Ostpolitik* and a destabilization of economic and energy affairs as a result of the tightening of energy supplies and the impact of the oil shock. Both of these trends favored natural gas and the East-West gas bridge—indeed, gas bridges of all sorts. Gas projects flow when the politics, economics, and technologies are aligned. See Angela Stent, *From Embargo to Ostpolitik: The Political Economy of West German–Soviet Relations, 1955–1980* (Cambridge: Cambridge University Press, 1981).

25. The story of the penetration of Dutch gas into Germany is recounted in a memoir by Stewart Douglass and Elain Madsen, *The Texan and Dutch Gas: Kicking Off the European Gas Revolution* (Victoria, BC: Trafford Publishing, 2006).
26. On this point see Vaclav Smil, *Energy and Civilization: A History* (Cambridge, MA: MIT Press, 2017), pp. 387–397.
27. Stern, *European Gas Markets,* p. 29. The relevant document from the European Council is Council Directive 75/405/EEC (*Official Journal,* L 178/24, July 9, 1975). The Netherlands was the first country to reverse this policy and increase its use of gas in the power sector. By the late 1980s, the declining use of gas for power generation had been reversed throughout Europe.
28. The primary driver was this: as a commercial policy, companies like Ruhrgas prioritized sales to customers with large base loads (including power stations) in the first months and years of a new gas supply contract with the clear intention of reducing these sales as the steady, connection-by-connection development of the (higher-value) household and industrial markets proceeded. That is why the numbers show a peak at 32.5 billion cubic meters in 1975, declining steadily thereafter. Then the technology of the combined cycle arrived, which transformed the economics of gas for power.
29. Here and elsewhere, the source is International Energy Agency, *Energy Balances of OECD Countries* (Paris: OECD Publishing, various years).
30. Gustafson, *Crisis amid Plenty.*
31. Jens Hohensee and Michael Salewski, eds., *Energie—Politik—Geschichte: Nationale und internationale Energiepolitik seit 1945* (Stuttgart: Franz Steiner, 1993).
32. Quoted in Yergin, *The Prize,* pp. 598–599.
33. Hohensee and Salewski, *Energie—Politik—Geschichte.*
34. Nadja Daniela Klag, *Die Liberalisierung des Gasmarktes in Deutschland* (Marburg: Tectum Verlag, 2003), p. 124.
35. Ibid.
36. Ibid.
37. Sophie Gerber, *Küche, Kühlschrank, Kilowatt: Zur Geschichte des privaten Energiekonsums in Deutschland, 1945–1990* (Bielefeld: Transcript Verlag, 2015), p. 163.
38. Ibid., p. 169
39. Ibid., pp. 166–168.
40. Klag, *Die Liberalisierung des Gasmarktes in Deutschland,* p. 124, citing Ruhrgas, *Erdgaswirtschaft—Eine Branche im Überblick* (Essen: "Ruhrgas," 2000).
41. See Paul H. Suding, "Policies Affecting Energy Consumption in the Federal Republic of Germany," *Annual Review of Energy* 14 (1989): 223.
42. In 2000 oil continued to hold a strong position, having a one-third share in the market. The remainder was accounted for by coal, mainly in the former East Germany. See Klag, *Die Liberalisierung des Gasmarktes in Deutschland,* p. 125.
43. The numbers that follow are based solely on the Federal Republic of Germany. For statistics on gas consumption in West Germany (as well as combined numbers for Germany as a whole), see Arbeitsgemeinschaft Energiebilanzen e.V., http://www.ag-energiebilanzen.de/12-0-Zeitreihen-bis-1989.html (accessed April 1, 2019).

44. It is currently almost 10 percent in the reunited Germany.
45. Of course, energy policy planners and companies actually looked at this the other way round: Because one cannot use nuclear power and coal directly in the end-use markets, one must use them to make electricity and so one should keep other fuels for the sectors where coal and nuclear power cannot go—using oil for transportation and gas for heating and industrial processes. In addition, to build the transmission lines a gas company needed a high utilization rate with quick, high-volume sales, selling steady load-factor gas to a few big customers. This implied power stations. But neither the gas companies nor the coal companies wanted to compete with each other forever. The gas companies' strategy was therefore to build out the residential and industrial market steadily on the back of the initial sales, pulling gas back from the power stations as they did so—and getting into the higher-value market where they competed with oil. But this was a commercial strategy rather than a political judgment.
46. The one major exception—and the one point of similarity between the environmental responses of France and Germany—is the systematic hostility today to fracking and shale gas.
47. Helm, *Energy, the State, and the Market.*
48. Paris was supplied with Dutch gas from the north as well as Lacq gas from the Aquitaine region, and the first LNG reached Le Havre from Arzew in Algeria as early as 1964. By 1970, plans were well afoot for the next phase of Algerian gas to come to France—with the completion of the much larger Skikda LNG export terminal. For background on the Algerian gas trade with France, see Mark. H. Hayes, "The Transmed and Maghreb Projects: Gas to Europe from North Africa," in David G. Victor, Amy M. Jaffe, and Mark H. Hayes, eds., *Natural Gas and Geopolitics from 1970 to 2040* (Cambridge: Cambridge University Press, 2006), pp. 49–90.
49. The account that follows is based mainly on Boris Dänzer-Kantof and Félix Torres, *L'Energie de la France: De Zoé aux EPR, l'histoire du programme nucléaire* (Paris: Editions François Bourin, 2013).
50. For the early history of the French military and civilian programs, see the memoir of Bertrand Goldschmidt, *Le complexe atomique: Histoire politique de l'energie nucléaire* (Paris: Fayard, 1980).
51. For a description of the events at Wyhl, see Craig Morris and Arne Jungjohann, *Energy Democracy: Germany's Energiewende to Renewables* (London: Palgrave Macmillan, 2016), pp. 15–36. For a discussion that places the protest at Wyhl in its broader political context, see Carol Hager, "The Grassroots Origins of the German Energy Transition," in Carol Hager and Christoph H. Stefes, eds., *Germany's Energy Transition: A Comparative Perspective* (New York: Palgrave Macmillan, 2016), pp. 1–26.
52. Joachim Radkau and Lothar Hahn, *Aufstieg und Fall der Deutschen Atomwirtschaft* (Munich: "Oekom," 2013), p. 105.
53. Goldschmidt, *Le complexe atomique.*
54. Dänzer-Kantof and Torres, *L'Energie de la France.*

55. Roger Karapin, *Protest Politics in Germany: Movements on the Left and Right since the 1960s* (University Park: Pennsylvania State University Press, 2007), p. 117.
56. The German antinuclear protest movement peaked in March 1979, with a 100,000-strong demonstration in Hanover.
57. Karapin, *Protest Politics in Germany,* p. 121.
58. Dänzer-Kantof and Torres, *L'Energie de la France,* pp. 321–333.
59. Quoted in ibid., p. 332.
60. Quoted in ibid., pp. 431 and 435.
61. A French army expression literally meaning "shit in bed." It is used to describe a breakdown of order and discipline.
62. Creys-Malville was probably the most violent event in the history of the antinuclear movement in France and Germany. As a brochure of the time put it, "In comparison with Malville, Brokdorf was like going for a stroll. The police fired smoke, gas, and percussion grenades as in war." (*Against the Atomic State,* quoted in Joachim Radkau, *The Age of Ecology: A Global History* [Cambridge: Polity Press, 2014], p. 153.)
63. An interesting early study, which focuses on the fundamental differences between the French and German political systems as they were in the 1970s, is Dorothy Nelkin and Michael Pollak, *The Atom Besieged: Antinuclear Movements in France and Germany* (Cambridge, MA: MIT Press, 1982).
64. See the description of the role of the courts in the German antinuclear movement in Radkau, *The Age of Ecology,* pp. 314ff. An important landmark was the so-called Würgassen Judgment by the Federal Administrative Court, which ruled in 1972 that "the protective purpose of the Atomic Energy Law, though listed second . . . , takes precedence over the facilitation purpose." This put an important legal weapon in the hands of the antinuclear movement.
65. Edda Müller, *Innenwelt der Umweltpolitik* (Westdeutscher Verlag, 1995), p. 136, quoted in Radkau, *The Age of Ecology,* p. 316.
66. Radkau, *The Age of Ecology,* p. 151.
67. *Stunde Null* ("Zero Hour") is a common German phrase referring to the moment of surrender at midnight on May 8, 1945.
68. Daniel Yergin, *The Quest: Energy, Security, and the Remaking of the Modern World* (New York: Penguin, 2011), chapter 22.
69. Ditt, "Die Anfänge des Umweltpolitik in der Bundesrepublik Deutschland," p. 313. In 1962 the federal Ministry for Health was created, which had responsibility not only for health but also for water and air quality and noise pollution. As Ditt notes, "In this respect the health ministry was a de facto environment ministry." The same was true in Norway, where Gro Brundtland was rising through the ranks of the Labor Party initially as a specialist in public health, which made her nomination as environment minister a logical appointment for a junior minister. Later on, Angela Merkel rose through a similar path in Germany.
70. On the many-sided relations between conservationism and the Nazi regime, see Radkau, *The Age of Ecology,* pp. 57–61. On East Germany, see ibid. and Merrill E. Jones, "Origins of the East German Environmental Movement," *German Studies*

Review 16, no. 2 (May 1993): 235–264, https://www.jstor.org/stable/1431647?seq=1#page_scan_tab_contents.

71. See Gerber, *Küche, Kühlschrank, Kilowatt,* and Ditt, "Die Anfänge des Umweltpolitik in der Bundesrepublik Deutschland."
72. A general characteristic of the period was that public and regulatory pressures tended to arise first in the United States, and companies there were more motivated to respond because of the greater threat of product liability suits.
73. Ditt, "Die Anfänge des Umweltpolitik in der Bundesrepublik Deutschland," p. 334.
74. Ulrich Beck, *Risk Society: Towards a New Modernity,* trans. Mark Ritter (London: Sage Publications, 1992).
75. In France, the same is happening now, at one generation's remove, although in milder form as yet. The interesting puzzle is why it took so long. See Théo Krausz, "Un noyau devenu instable: Que reste-t-il du 'Tout Nucléaire Français?'" (unpublished seminar paper, Georgetown University, available from the author).
76. Stephen Milder, *Greening Democracy: The Anti-Nuclear Movement and Political Environmentalism in West Germany and Beyond, 1968–1983* (Cambridge: Cambridge University Press, 2017), p. 1.
77. Ibid., p. 3.
78. Ibid.
79. Joachim Radkau, *Aufgang und Krise der deutschen Atomwirtschaft 1945–1975: Verdrängte Alternativen in der Kerntechnik und der Ursprung der nuklearen Kontroverse* (Reinbek bei Hamburg: Rowohlt, 1983).
80. Ibid.
81. Andrei S. Markovits and Joseph Klaver, "Alive and Well into the Fourth Decade of Their Bundestag Presence: A Tally of the Greens' Impact on the FRG's Political Life and Public Culture," in Stephen Milder and Konrad H. Jarausch, eds., "Green Politics in Germany," special issue of *German Politics and Society,* vol. 33, no. 4 (Winter 2015): 113.
82. Ibid., p. 131.
83. The antinuclear movement was directed not only against nuclear power plants, but also against nuclear weapons. Nevertheless, most of the literature on the period appears to agree that the protests against civilian power plants were the single most important formative experience in the evolution of the movement from the grass roots to a position in the political establishment of the country.
84. Milder, *Greening Democracy,* p. 7.
85. See Gustafson, *Reform in Soviet Politics.* I experienced this environmental sentiment when I was a *stazher* (an exchange student) at the Geography Institute of the USSR Academy of Sciences in 1972–1973. I became aware of a deep current of environmental radicalism among the graduate students and junior research staff of the institute. Much of their effort was directed against a plan to build a hydropower dam across the Pechora River in the far northwest. In this case they were more successful—probably because the military was not involved—and the project was defeated.

86. As Gorbachev said twenty years later in a 2006 interview, "It might be said that the fall of the Soviet Union began at Chernobyl" (quoted in the 2019 HBO series *Chernobyl*).
87. Nemtsov, a particularly brilliant student, was an undergraduate at the Lobachevskii State University in Gorky from 1976 to 1981 and a graduate student at the State University of Gorky between 1981 and 1985, where he received his *kandidat* degree (the Russian equivalent of a PhD) at the age of twenty-five. He was the very model of the "Gorbachev constituency" that was quietly growing inside elite institutions, even in closed cities. Nemtsov had a successful political career under Boris Yeltsin, serving as deputy prime minister, and he became a prominent member of the opposition to Vladimir Putin. He was murdered in broad daylight in 2015.

4. Norway and the Rise of the North Sea

1. I am grateful to Olav Henke, a talented student in the School of Foreign Service at Georgetown University, for his contribution to this chapter, particularly in the translation of Norwegian-language sources.
2. To be sure, the Norwegians had no choice: Norway was a small country with no experience of fossil fuel development, let alone complex offshore development.
3. See the collection of essays on Soviet technology edited by Ronald Amann and Julian Cooper, *Industrial Innovation in the Soviet Union* (New Haven, CT: Yale University Press, 1982), which contains chapters on individual sectors. For a review of the collection, see Gertrude E. Schroeder, "Conditions of Soviet Technology," *Science* 219 (January 7, 1983): 46–47, https://www.cia.gov/library/readingroom/docs/CIA-RDP85T00153R000100010024-4.pdf. On the chemical industry, see Matthew J. Sagers and Theodore Shabad, *The Chemical Industry in the USSR: An Economic Geography* (Washington, DC: American Chemical Society, 1990).
4. Petter Nore, "The Norwegian State's Relationship to the International Oil Companies over North Sea Oil 1965–75" (PhD diss., Thames Polytechnic, 1979), pp. 88–104. The first Norwegian contracts with international oil companies were modeled on a pathbreaking contract concluded by Iran with the Italian national company ENI in 1957. See the discussion of the role of Evensen later in this chapter.
5. Stephen Howarth and Joost Junker, *Powering the Hydrocarbon Revolution, 1939–1973: A History of Royal Dutch Shell*, vol. 2 (Oxford: Oxford University Press, 2007), p. 199.
6. Ole Gunnar Austvik, *The Norwegian State as Oil and Gas Entrepreneur: The Impact of the EEA Agreement and the EU Gas Market Liberalization* (Saarbrücken, Germany: Verlag Dr. Müller, 2009), pp. 101–102.
7. Ibid. The award was also a demonstration of the close political ties between the Møller group and the Danish political elite.
8. Quoted in James Bamberg, *British Petroleum and Global Oil, 1950–75: The Challenge of Nationalism* (Cambridge: Cambridge University Press, 2000), p. 209.

9. Trevor I. Williams writes, "Analysis of the composition of gas from the Groningen Field suggested that it derived ultimately from coal, and then migrated to the porous Rotliegendes sandstone, and was there trapped by a thick layer of salt." (*A History of the British Gas Industry* [Oxford: Oxford University Press, 1981], pp. 148–149). For a summary of more recent geological work, see Jane Whaley, "The Groningen Gas Field," *GeoExPro* 6, no. 4 (2009), http://www.geoexpro.com/articles/2009/04/the-groningen-gas-field.
10. Howarth and Junker, *Powering the Hydrocarbon Revolution,* p. 465n95.
11. Ibid., p. 43.
12. Ibid., p. 197.
13. Quoted in Daniel Yergin, *The Prize: The Epic Quest for Oil, Money, and Power* (New York: Simon and Schuster, 1991), 669.
14. Hans Veldman and George Lagers, *Fifty Years Offshore* (State College: Pennsylvania State University Press, 1997), quoted in Joseph A. Pratt and William E. Hale, *Exxon: Transforming Energy, 1973–2005* (Austin, TX: Dolph Briscoe Center for American History, 2013), p. 139.
15. The official Exxon history by Pratt and Hale quotes an unnamed engineer: "The North Sea's wave heights . . . are the highest in the world; they're over a hundred feet sometimes, which is amazing. They found [at the Odin field] that the weldings weren't holding and . . . we had to shut down production and spend a bunch of money to redo all the damn welds on the thing" (*Exxon,* p. 151).
16. Kenneth Hutchinson, *High Speed Gas: An Autobiography* (London: Duckworth, 1987), p. 244.
17. Howarth and Junker, *Powering the Hydrocarbon Revolution,* p. 51. See also Pratt and Hale, *Exxon*, which contains interesting observations on the tensions within the Esso-Shell partnership, due largely to the cultural differences between the two companies and their contrasting styles in dealing with foreign governments and managing their internal procedures (pp. 140–141).
18. Howarth and Junker, *Powering the Hydrocarbon Revolution,* p. 212; see also pp. 212–214.
19. Nore, "The Norwegian State's Relationship to the International Oil Companies," p. 7. At the time the Norwegian Foreign Ministry began negotiating the middle-line division of the North Sea with Denmark and Great Britain, the focus on the Norwegian side was on fisheries rather than the prospect of oil and gas, since the presence of the latter had not yet been ascertained. See Dag Harald Claes, "Globalization and State Oil Companies: The Case of Statoil," *Journal of Energy and Development* 29, no. 1 (2003): 47.
20. Quoted in Pratt and Hale, *Exxon,* p. 139. Hagemann, a geologist by training, served as general director of the Norwegian Petroleum Directorate for twenty-five years, from its founding in 1972 to his retirement in 1997 (see https://en.wikipedia.org/wiki/Fredrik_Hagemann). While Hagemann had no direct ties to the Labor Party, Gunnar Berge (who was the second general director, serving from 1997 to 2007) marked a return to the tradition of Labor Party politicians playing leading roles in oil and gas policy (even as the tie to the Labor Party was being weakened at the same time in Statoil) (see https://en.wikipedia.org/wiki/Gunnar_Berge).

21. The full story was considerably more complicated and involved complex negotiations between Denmark and Germany at the same time over the border between the two countries' territorial waters. The episode still causes some bitterness in Denmark and occasional charges that the Norwegians took advantage of the Danes. See Berit Ruud Retzer, *Jens Evensen: Makten, myten og mennesket* (Oslo: BBG Forlag, 1999), pp. 102–107. The biography is also available in a Russian translation: see *Ens Evensen: vlast', mif i chelovek* (Moscow: "Impeto," 2004), pp. 106–110.
22. Quoted in Stig Kvendseth, *Giant Discovery: A History of Ekofisk through the First 20 Years* (Oslo: Phillips Petroleum Company Norway, 1988), p. 13.
23. Curiously, Endacott was not the only future player to have noticed the structures at Slochteren while on vacation. Jens Evensen, the future head of the legal department of the Norwegian Ministry of Foreign Affairs and the chief architect of the delimitation agreement with the British and the Danes, was likewise intrigued by what he witnessed during numerous drives through the Dutch countryside while he was posted in the Netherlands. See Retzer, *Jens Evensen,* pp. 95–96 (or *Ens Evensen,* p. 98).
24. Kvendseth, *Giant Discovery.*
25. See https://en.wikipedia.org/wiki/Jens_Evensen.
26. Berit Ruud Retzer, *Jens Evensen—Mannen som gjorde Norge større* (Oslo: Gyldendal Norsk Forlag, 2017). Just how much Evensen's negotiating achievements changed the size of Norway can be seen from a map available on the website of the Norwegian Petroleum Directorate (Norwegian Petroleum Directorate, FactMaps, http://gis.npd.no/factmaps/html_21/ [accessed April 3, 2019]). See also Retzer, *Jens Evensen* (or *Ens Evensen*).
27. It is worth recalling that at the time of the final investment decision on Ekofisk, oil prices were around $2.50 a barrel.
28. Quoted in Kvendseth, *Giant Discovery,* pp. 25 and 26.
29. Howarth and Junker, *Powering the Hydrocarbon Revolution,* p. 39.
30. Yergin, *The Prize,* p. 669.
31. On the history of the British gas sector, see Hutchinson, *High Speed Gas;* Dieter Helm, *Energy, the State, and the Market: British Energy Policy since 1979* (Oxford: Oxford University Press, 2003); and Williams, *A History of the British Gas Industry.*
32. One will get arguments from Norwegians over just how advanced the Norwegian economy was on the eve of the great oil and gas discoveries. Nore describes Norway in the early 1970s as a "semi-peripheral" player in the world economy, in the sense that "the Norwegian industrial sector consisted of small and generally weak firms, and most Norwegian exports were primary or semi-processed goods" ("The Norwegian State's Relationship to the International Oil Companies," p. 23). Others call the glass half full. As one knowledgable Norwegian (personal communication, n.d.) told me, "Don't exaggerate the 'fish and farming' picture of the pre-oil economy. Norway had a diversified industrial economy, as evidenced especially by the fact that its merchant fleet was active worldwide, using mostly homemade ships, and this of course presupposes a strong metals industry, engineering

skills, etc." This split in perceptions undoubtedly helps explain the strong emotions surrounding Norwegian referendums over joining the European Union.

33. For a description of the UK gas industry on the eve of the era of natural gas, see Hutchinson, *High Speed Gas*, and Williams, *A History of the British Gas Industry.*
34. In 1950 imported oil accounted for only 17 percent of Norway's domestic energy consumption. But by the mid-1960s the low price of oil had caused its share to increase to nearly 35 percent—slightly higher than the average for Western Europe as a whole. The timing of the discovery of oil at Ekofisk could not have been better. Within a short time Norway's own oil had displaced imported oil (see Noreng, *The Oil Industry and Government Strategy in the North Sea,* pp. 37–38).
35. The share of the Norwegian state as an equity owner of Norwegian companies has increased sharply since then. As of the mid-2010s, the Norwegian state owned around one-third of the total value of the companies listed on the Oslo Stock Exchange. See Einar Lie, "Context and Contingency: Explaining State Ownership in Norway," *Enterprise and Society* 17, no. 4 (December 2016): 904–930. However, much of the state's stake consists of passive portfolio positions that it does not actively manage.
36. Nore, "The Norwegian State's Relationship to the International Oil Companies," pp. 21–22.
37. This became a significant issue in the subsequent campaign to break up and denationalize British Gas (the successor to the Gas Council). As one participant recalls, "The original pressure in the early 1980s came from the grass roots of small businessmen in the suburban strongholds of the Conservative party—men whose High Street businesses sold televisions, washing machines, refrigerators, etc. . . . but who could not by law sell gas stoves. It's a fascinating extra element of background to the Lawson-Rooke personality clash, and to the Thatcherite ideology emerging from the think tanks and universities." I am indebted to Simon Blakey for this point.
38. According to Noreng, *The Oil Industry and Government Strategy in the North Sea,* this was typical of Norwegian relations with the international companies, in comparison with the UK mode of administration.
39. Pratt and Hale, *Exxon,* pp. 145–148.
40. Austvik, *The Norwegian State,* pp. 102–103. The actual memorandum was drafted and published in late 1971 and passed by the Storting in June 1972. It was drafted, typically, by a Labor Party politician, the head of the Storting's standing committee on industry.
41. For an example of the reverence with which the Ten Commandments are still treated by the Norwegian government today, see an essay posted by the Norwegian Petroleum Directorate (NPD), "Ten Commanding Achievements," November 19, 2016, https://s3.amazonaws.com/rgi-documents/e3cbbfde7c90c6075 3b477e84627ee06dd50ae25.pdf.
42. According to Nore, "the Norwegians never expected to find gas in the first place, and always considered oil to be the main object of the search on its Continental Shelf" ("The Norwegian State's Relationship to the International Oil Companies," p. 167).

43. Austvik, *The Norwegian State.*
44. There was a technical reason why the landing commandment was difficult to fulfill. Between the main offshore fields and the Norwegian coast lies the so-called Norwegian Trench, a 300-meter-deep canyon that could not be crossed with pipelines until a later generation of technology became available in the early 1980s with the construction of Statpipe, which lands north of Stavanger.
45. Thanks to Norway's abundant hydropower, its small population can be heated by electricity, which is impossible anywhere else in the world. There is no domestic consumption of Norwegian natural gas.
46. See Clive Archer, *Norway outside the European Union: Norway and European Integration from 1994 to 2004* (London: Routledge, 2004), especially pp. 41–64.
47. Arve Johnsen, *Norges evige rikdom: Oljen, gassen og petrokronene* (Oslo: Aschehaug, 2008), p. 36.
48. Fisheries and agriculture were much more prominent than oil or gas in the debate over European Community (EC) membership in 1972. To be sure, the Ten Commandments could be considered a preemptive statement in the context of Norway's EC application, but in fact the only time that energy became an explicit part of the campaign was on the very eve of the referendum in September 1972, when—with singularly bad timing—the EC energy commissioner, Ferdinand Spaak, "announced ideas for a common energy policy that implied that Norway's burgeoning offshore oil and gas discoveries could become a 'Community resource.' The outcry in Norway showed that the feeling was that the EC was about to grab Norway's oil as it would do its fish" (Archer, *Norway outside the European Union,* p. 47). Norway's two referenda were both close, and things could have been quite different, especially if one compares the outcome in Norway with that in Sweden in 1994. In Sweden the margin of victory for the vote was equally narrow, but the "yes" side had the advantage—and thus history turned on a few thousand votes (actually, more like 300,000 in Sweden) in both countries.
49. Hauge served as chairman of the board of Statoil from 1972 to 1975. See Douglas Martin, "Jens Christian Hauge, Guide of Modern Norway, Dies at 91," *New York Times,* November 4, 2006, https://www.nytimes.com/2006/11/04/world/europe/04hauge.html.
50. As Johnsen writes in his memoirs, developing the Labor Party's oil policy was "the single most important case that dominated our work in the late 1960s and in 1970–1971 (*Norges Evige Rikdom,* p. 36). Note that all this precedes both the first oil shock and Norway's discovery of Ekofisk in 1969.
51. Austvik, *The Norwegian State,* p. 104.
52. Kjetil Malkenes Hovland, "Så viktig er Statoil for norsk økonomi," February 6, 2017, http://e24.no/boers-og-finans/statoil/saa-viktig-er-statoil-for-norsk-oekonomi/23917273.
53. Ibid.
54. Interview with Johnsen by Tormund Haugstad, "Former Statoil Chief Arve Johnsen—Changing the Statoil Name Is a Bad Decision," September 19, 2008, https://www.tu.no/artikler/tidligere-statoil-sjef-arve-johnsen-en-feilvurdering-a-fjerne-statoil-navnet/322057.

55. Author's interview with Kaare Willoch, Oslo, June 8, 2017.
56. Ibid.
57. Author's interview with Peter Mellbye, Oslo, June 8, 2017.
58. Austvik, *The Norwegian State,* p. 115. Another important development at about this time was the creation of Petoro as the holder of the state's oil and gas assets.
59. Quoted in Javier Estrada, Arild Moe, and Kåre Dahl Martinsen, *The Development of European Gas Markets: Environmental, Economic, and Political Perspectives* (New York: Wiley and Sons, 1995), pp. 226–227.
60. Austvik, *The Norwegian State,* p. 131.
61. Ibid., p. 110. The old connection with the Labor Party was symbolically severed with Norvik's resignation in 1999. His successor, Olav Fjeld, had no political background, and when he was replaced in 2003 in the wake of a political scandal involving Iran, the next chairman was a Conservative politician, Helge Lund—but the more significant part of his biography, from the standpoint of those who chose him, was that he had been a partner with the McKinsey consultancy and was consequently experienced in how to run a private corporation.
62. Author's interview with Kaare Willoch. The Norwegian government still owns a 67 percent share in Statoil.
63. Finance was available to the companies at less than the going market rates because of the willingness of export finance institutions in several Western European countries to subsidize the purchase the purchase of offshore equipment (Wood Mackenzie, cited in Nore, "The Norwegian State's Relationship to the International Oil Companies," pp. 178–179). However, as much as 70 percent of the capital cost of the first fields was financed by the companies (ibid., p. 179). In the case of the Ekofisk field, Norway had to raise only 5 percent of the cost of the project and the associated pipeline (ibid., p. 209).
64. Kvendseth, *Giant Discovery.* Another equally important development at this time was the joint Norwegian-British development of the Frigg field, with twin pipelines to St. Fergus. This project, and the contract with British Gas that supported it, were as significant as the Ekofisk gas project as a source of experience for the Norwegians.
65. Douglass Stewart and Elaine Madsen, *The Texan and Dutch Gas: Kicking Off the European Energy Revolution* (Victoria, BC: Trafford Publishing, 2006), especially pp. 147–151.
66. As Per Högselius relates, there was initially some resistance to Soviet exports to Germany, particularly from coal interests in the north but also from German gas importers who feared that Soviet gas would compete with Dutch gas (see *Red Gas: Russia and the Origins of European Energy Dependence* [New York: Palgrave Macmillan, 2013], pp. 122–125).
67. Kvendseth, *Giant Discovery,* p. 104.
68. Ibid., p. 102.
69. For details on the construction of Norpipe, see ibid., pp. 92–107. The Danish government, which laid down strict conditions for the burial of the pipeline in the 48 kilometer stretch of Danish territorial waters, delayed matters for two years.

During that time the Emden terminal, which had been completed in 1975, stood empty (ibid., p. 102). The pipeline today is owned by Gassled and is operated by Gassco. The technical service provider is ConocoPhillips (see https://www.norskolje.museum.no/en/norpipe-oil-transport-system/).

70. For example, on the Ekofisk field, "Phillips had to file information about the exact cost of investment in the North Sea with Department of Industry. By the time Phase II of Ekofisk was finished in 1974, which included the installation of five fixed platforms and one storage tank, the Directorate's knowledge of the ongoing investment conditions on the Norwegian shelf was extensive. The Ekofisk costs became of great importance to the Norwegian state as a basis for assessing the likely capital cost of installations in the North Sea" (Nore, "The Norwegian State's Relationship to the International Oil Companies," pp. 213–214). The same was presumably true of the export contracts.
71. Sleipner is named after the eight-legged horse Sleipnir, ridden by the god Odin in Norse mythology.
72. "30 Years of Gullfaks Oil," December 23, 2016, https://www.statoil.com/en/news/30-years-gullfaks-oil.html.
73. Johnsen, *Norges Evige Rikdom,* p. 215.
74. Quoted in "Statpipe Gas Celebrates 20 Years in Operation," *Rigzone,* October 17, 2005, http://www.rigzone.com/news/article.asp?a_id=26068.
75. Quoted in Austvik, *The Norwegian State,* p. 105.
76. Quoted in ibid., p. 119.
77. Author's interview with Peter Mellbye.
78. Jonathan P. Stern, *European Gas Markets: Challenge and Opportunity in the 1990s* (London: Royal Institute of International Affairs, 1990).
79. For background on Sleipner, which consists of separate eastern and western producing fields, see https://www.equinor.com/en/what-we-do/norwegian-continental-shelf-platforms/sleipner.html.
80. On the Reagan embargo, see David S. Painter, "From Linkage to Economic Warfare: Energy, Soviet-American Relations, and the End of the Cold War," in Jeronim Perović, ed., *Cold War Energy: A Transnational History of Soviet Oil and Gas* (London: Macmillan, 2017), pp. 283–318.
81. This at least is the interpretation offered by Torleif Haugland, Helge Ole Bergesen, and Kjell Roland in *Energy Structures and Environmental Futures* (Oxford: Oxford University Press, 1998), pp. 119–120.
82. Quoted in "Ministers Scathing in Letters over Wasted Offshore Resources," *Press and Journal (Highlands and Islands),* http://www.pressreader.com/uk/the-press-and-journal-highlands-islands/20140103/281732677323550.I am indebted to Simon Blakey for this insight.
83. Author's interview with Peter Mellbye.
84. Quoted in Austvik, *The Norwegian State,* p. 118.
85. Ibid., pp. 180–182.
86. James Ball, "The Troll Revolution: An Assessment of the Troll Sales Agreements," *Petroleum Review,* May 1987, pp. 23–26. The gas from Troll was sold on the basis

of netback market prices, which guarantee importers a price equivalent to that of competing fuels in the markets into which they are selling the gas.

87. Stern, *European Gas Markets,* pp. 16–17.
88. Quoted in Pratt and Hale, *Exxon,* p. 148.
89. Ibid., p. 147.
90. In the first half of the 1960s, Norwegian private-sector companies received little encouragement from the state to invest in the petroleum sector. On the contrary, "the government openly discouraged Norwegian shipping and industrial interests that sought to enter the oil industry at this time, due to uncertainty and high risk to invested capital." See Claes, "Globalization and State Oil Companies," p. 47, citing *Report to Parliament,* no. 11 (1968–1969), pp. 6–7. Instead, the government's policy was to invite the international oil companies to assume the risks of early oil and gas exploration.
91. Pratt and Hale, *Exxon,* p. 155.
92. Noreng, *The Oil Industry and Government Strategy in the North Sea.*
93. Ibid., p. 55.
94. Andrew Niel, "Too Big a Maybe," *Economist,* July 26, 1975. Throughout the second half of the 1970s, Neil and his colleagues at the *Economist* wrote a number of articles harshly criticizing British industry for its failure to take advantage of the opportunities in the North Sea oil and gas sector. It was only at the beginning of the 1980s that the British began to catch up. See also Daniel Yergin, "Britain Drills—and Prays," *New York Times Magazine,* November 2 1975.
95. Knut Heidar, *Norway: Elites on Trial* (Boulder, CO: Westview Press, 2001), p. 102.
96. Ibid.
97. Joseph A. Pratt and William E. Hale, *Exxon: Transforming Energy, 1973–2005* (Austin, TX: Dolph Briscoe Center for American History, 2013), p. 153.
98. Ibid., p. 155.
99. Lie, "Context and Contingency." See also Ole Andreas H. Engen, "The Development of the Norwegian Petroleum Innovation System: An Overview" (working paper, Innovation Studies, Centre for Technology, Innovation Culture, University of Oslo, 2007). A published version of this paper essay appears in Jan Fagerberg, David C. Mowery, and Bart Verspagen, eds., *Innovation, Path Dependency, and Policy: The Norwegian Case* (Oxford: Oxford University Press, 2009).
100. Pratt and Hale, *Exxon,* p. 146.
101. Ibid.
102. Hagemann's official successor in 1990 was Gunnar Berge, but since Berge was a minister in the government of the day, Hagemann stayed on as interim director until 1997
103. More recently, Rosneft has encountered similar difficulties in developing offshore equipment, owing to the backwardness of the local military shipyards.
104. Norwegian Petroleum, "Exports of Oil and Gas," http://www.norskpetroleum.no/en/production-and-exports/exports-of-oil-and-gas/ (accessed May 12, 2019).
105. Ibid.

5. Soviet Gas

1. The best account of the economic decline and the onset of crisis is by the former acting prime minister and leader of the market reforms of the early 1990s. See Yegor Gaidar, *Collapse of an Empire: Lessons for Modern Russia* (Washington, DC: Brookings Institution Press, 2007), especially chapter 4.
2. Ibid., p. 88, citing numbers from Leonid Brezhnev.
3. For the story of Brezhnev's agricultural and environmental policies, especially the reclamation campaign, see Thane Gustafson, *Reform in Soviet Politics: Lessons of Recent Policies on Land and Water* (Cambridge: Cambridge University Press, 1981).
4. Quoted in Mariia V. Slavkina, *Baibakov* (Moscow: "Molodaia gvardiia," 2010).
5. Gaidar, *Collapse of an Empire,* p. 95.
6. Ibid.
7. Only Gold, http://onlygold.com/m/Prices/Prices200Years.asp (accessed May 14, 2019). Prices are given in the dollars of the day. The windfall in gold prices continued through much of the 1970s. By 1980 the price of gold had reached $594.90 an ounce. Thereafter it followed the downward slope of commodity prices, not reversing until 2000. The nominal 1980 price was not reached again until 2006.
8. Gaidar, *Collapse of an Empire,* p. 102.
9. Slavkina, *Baibakov,* p. 142.
10. Ibid., p. 141.
11. Valentin A. Runov and Aleksandr D. Sedykh, *Orudzhev* (Moscow: "Molodaia Gvardiia," 2012), pp. 259–262. The authors hint that Kosygin might have been poisoned, but I am unaware of any evidence for that. Kosygin had been in failing health ever since a serious boating accident in 1976.
12. Gaidar, *Collapse of an Empire*. pp. 102–103.
13. This paragraph and the next are a condensation of the discussion in Thane Gustafson, *Crisis amid Plenty: The Politics of Soviet Energy under Brezhnev and Gorbachev* (Princeton, NJ: Princeton University Press, 1989), chapter 2.
14. For details see ibid., especially pp. 23–24.
15. Ibid., table 2.1, drawing on State Statistical Committee of the USSR (Goskomstat SSSR), *Narodnoe khoziaistvo SSSR* (Moscow: Finansy i statistika, various years).
16. The paragraphs that follow are a summary of Gustafson, *Crisis amid Plenty,* chapters 2–4.
17. Per Högselius, *Red Gas: Russia and the Origins of European Energy Dependence* (New York: Palgrave Macmillan, 2013), figure 8.5, p. 145, drawing on official Ukrainian statistics.
18. Gustafson, *Crisis amid Plenty,* p. 27.
19. Ibid., p. 28.
20. U.S. Central Intelligence Agency, *The International Energy Situation: Outlook to 1985,* report number ER77-10240 (Washington, DC: GPO, April 1977); U.S. Central Intelligence Agency, *Prospects for Soviet Oil Production,* report number ER77-10270 (Washington, DC: GPO, April 1977); U.S. Central Intelligence

Agency, *Prospects for Soviet Oil Production: A Supplemental Analysis*, report number ER 77-10425 (Washington, DC: GPO, July 1977).

21. Gustafson, *Crisis amid Plenty,* p. 29.
22. Brezhnev's first major public call for a big increase in gas investment was at the November 1979 plenum of the Central Committee of the Communist Party of the Soviet Union. See *Pravda,* November 28, 1979. See also Gustafson, *Crisis amid Plenty,* p. 30.
23. *Pravda,* February 24, 1981. See also Gustafson, *Crisis amid Plenty,* p. 34.
24. Gustafson, *Crisis amid Plenty,* p. 32.
25. Quoted in *Pravda,* November 18, 1981.
26. Högselius, *Red Gas,* pp. 137–138.
27. Ibid.
28. Ibid., p. 140.
29. Slavkina, *Baibakov,* p. 14.
30. Högselius, *Red Gas,* p. 142.
31. Ibid., pp. 146–147.
32. Quoted in Viktor Andrianov, *Kortunov* (Moscow: "Molodaia Gvardiia," 2007), pp. 474–475.
33. Runov and Sedykh, *Orudzhev,* pp. 259–262. Sedykh was one of those who stayed. He subsequently became deputy minister.
34. Andrianov, *Kortunov,* pp. 475–488. Andrianov's material is especially valuable because he reproduces in full several of Kortunov's memoranda to senior officials such as Veniamin Dymshits, then deputy prime minister and head of Gossnab, and Kosygin.
35. The mocking nickname "Orenburg mafia" was applied above all to Victor Chernomyrdin and his protégé Rem Viakhirev, the ultimate founders of Gazprom as a separate corporation. Chernomyrdin at this time was in the apparatus of the Central Committee. He was by career partly a Party official, having worked in the Party apparatus in Orsk, in Orenburg Province, from 1967 to 1973. In 1973 he was appointed head of the Orenburg gas-processing plant. In 1978 he rose to Moscow to work in the heavy industry department of the Central Committee, where he stayed until 1982. In 1982 he was named deputy gas minister under Orudzhev's successor. He became head of Glavtiumengazprom in 1983, and in 1985 he was named minister of gas. Presumably in his Central Committee job he was one of the people charged with implementing the Brezhnev gas campaign, working initially with Orudzhev's team, until he moved over to a leading position in the gas ministry.
36. *Vneshniaia torgovlia SSSR*, http://istmat.info/node/9321 (accessed May 14, 2019).
37. Högselius, *Red Gas,* p. 159.
38. Central State Archive, Kiev (TsDAVO), quoted in Högselius, *Red Gas,* p. 161.
39. Prime Minister Oleksandr Lyashko, Central State Archive, Kiev (TsDAVO), quoted in Högselius, *Red Gas,* p. 162.
40. Ibid.
41. Högselius, *Red Gas,* p. 164.
42. Ibid., pp. 164–166.

43. The Council for Mutual Economic Assistance (1949–1991) was an economic organization under the leadership of the Soviet Union that included the countries of the Eastern Bloc along with a number of communist states elsewhere in the world. See https://en.wikipedia.org/wiki/Comecon.
44. Runov and Sedykh, *Orudzhev,* pp. 259–262.
45. Rainer Karlsch, *Vom Licht zur Wärme: Geshchichte der ostdeutschen Gaswirtschaft 1855–2008* (Leipzig: Verbundnetz Gas AG, 2008), p. 157.
46. Karlsch observes that in addition to a substantial literature on the Druzhba project, the best sources are Internet websites maintained by the former *trassniki,* who looked back on their time in Soviet Ukraine as one of the most significant experiences of their lives (*Vom Licht zur Wärme,* p. 163).
47. Karlsch, *Vom Licht zur Wärme,* p. 163.
48. Ibid.
49. David S. Painter, "From Linkage to Economic Warfare: Energy, Soviet-American Relations, and the End of the Cold War," in Jeronim Perović, ed., *Cold War Energy: A Transnational History of Soviet Oil and Gas* (London: Macmillan, 2017), pp. 283–318.

6. Crossing the Channel

1. For an account of the situation in Europe at the end of the 1970s, see Daniel Yergin and Joseph Stanislaw, *The Commanding Heights: The Battle between Government and the Marketplace That Is Remaking the Modern World* (New York: Simon and Schuster, 1998).
2. Richard Cockett, *Thinking the Unthinkable: Think Tanks and the Economic Counter-Revolution, 1931–1983* (London: HarperCollins, 1995).
3. Yergin and Stanislaw, *The Commanding Heights,* p. 34.
4. For a useful summary of Röpke's life and influence, see https://mises.org/library/biography-wilhelm-r%C3%B6pke-1899-1966-humane-economist (accessed May 15, 2019).
5. Quoted in Yergin and Stanislaw, *The Commanding Heights,* pp. 34–35.
6. One should perhaps add the word "Celtic" to allow for the contribution of the occasional Irish commissioner, such as Peter Sutherland—who went on to a varied political and business career, as discussed below.
7. According to Yergin and Stanislaw, the Russians were perfectly aware of Hayek and *The Road to Serfdom* and regarded the book as highly subversive. Indeed, after World War II, in occupied Germany, the four-power authority banned the book "at the behest of the Soviet Union" (*The Commanding Heights,* p. 143).
8. For valuable background on the Russian response to neoliberalism, see Peter Rutland, "Neoliberalism in Russia," *Review of International Political Economy* 20, no. 2 (April 2013): 332–362.
9. For background on the evolution of hub-based trading in Europe, see Catherine Robinson and Soufien Taamallah, *Slow and Steady: The Development of Gas Hubs in Europe* (IHS Markit Decision Brief, November 2009).

10. See Dieter Helm, *Energy, the State, and the Market: British Energy Policy since 1979* (Oxford: Oxford University Press, 2003). See also http://www.wikiberal.org/wiki/Stephen_Littlechild.
11. A word about usage: Arthur Cockfield (pronounced "Co-field") was already a peer (that is, a member of the House of Lords) at the time he joined Thatcher's cabinet. In British usage a peer is "Lord Lastname"; the style "Lord Firstname Lastname" is reserved for the sons of dukes and marquesses. In this chapter, however, I refer to him both as "Cockfield" and "Lord Cockfield."
12. Lord Cockfield, *The European Union: Creating the Single Market* (London: Wiley Chancery Law, 1994), p. 59.
13. Ibid., p. 69.
14. Cockfield, *The European Union,* p. 21.
15. Alessandro Olivi and Bino Giacone, *L'Europe Difficile: Histoire politique de la construction européenne* (Paris: Gallimard, 2007), pp. 195–197.
16. Cockfield, *The European Union,* p. 21.
17. In 1979, the European Parliamentary Assembly renamed itself the European Parliament and conducted the first Europe-wide popular election of euro deputies.
18. See Yergin and Stanislaw, *The Commanding Heights,* chapter 11.
19. See Alain Rollat, *Delors* (Paris: Flammarion, 1993), pp. 255ff.; Olivi and Giacone, *L'Europe Difficile,* pp. 195ff..
20. European Court of Justice, 120 / 78 Case Cassis de Dijon, https://eur-lex.europa.eu/legal-content/EN/TXT/?uri=CELEX%3A61978CJ0120.
21. R. Daniel Kelemen, *Eurolegalism: The Transformation of Law and Regulation in the European Union* (Cambridge, MA: Harvard University Press, 2011), p. 19.
22. Cockfield, *The European Union,* p. 24.
23. Ibid.
24. Charles Grant, *Delors—Inside the House that Jacques Built* (London: Nicholas Brearley, 1994), p. 276.
25. Quoted in ibid., p. 273.
26. Ibid., p. 155.
27. Ibid., p. 154.
28. Historical Archives of the European Union, Oral History Collection (1998), Voices of Europe Collection, transcript of interview with Arthur Cockfield, p. 8, archives.eui.eu/en/files/transcript/15166.pdf (available following free registration at https://archives.eui.eu/) (accessed May 15, 2019). Originally Delors regarded the single market as a subsidiary project that would support his great ambition of economic and monetary union and the Citizen's Europe. But with the difficulties that became associated with monetary union, particularly with the onset of the recession of the early 1990s, he realized that he had better have his name attached to something that was a success instead of to economic and monetary union, which everybody at the time thought was going to be a failure.
29. David Allen Green (@davidallengreen), "The other was Lord Cockfield, perhaps the second most significant UK Tory politician of the 1980s," Twitter, August 21, 2017, 5:16 a.m., https://twitter.com/davidallengreen/status/903229985183473664.

30. The full quote is, "Not for nothing was Lord Cockfield once referred to as the only man who spoke like a White Paper" (Anand Menon, *Europe: The State of the Union* [London: Atlantic Books, 2008], p. 52).
31. Sir Roy Denman, "Lord Cockfield," *Guardian,* January 11, 2007, https://www.theguardian.com/news/2007/jan/11/guardianobituaries.obituaries.
32. Ibid.
33. Cockfield, *The European Union,* p. 26.
34. Ibid., p. 36.
35. Ibid., p. 30.
36. Denman, "Lord Cockfield."
37. Margaret Thatcher, *The Autobiography* (London: Harper, 2013), p. 551.
38. Cockfield, *The European Union,* p. 19. By "Treaties" Cockfield has in mind the Treaty of Rome.
39. Cockfield, *The European Union,* p. 21.
40. Ibid.
41. Ibid., p. 26.
42. Ibid., pp. 39–40.
43. Ibid., p. 41.
44. Quoted in Grant, *Delors,* p. 68.
45. Cockfield, *The European Union,* pp. 84–85.
46. Ibid., pp. 54–55.
47. Quoted in ibid., pp. 50–51.
48. Cited in ibid., p. 50.
49. Thatcher, *The Autobiography,* p. 551.
50. Cited in Cockfield, *The European Union,* p. 63.
51. Delors, *Mémoires,* quoted in Stephen Wall, *A Stranger in Europe: Britain and the EU from Thatcher to Blair* (Oxford: Oxford University Press, 2008), p. 56.
52. Cockfield, *The European Union,* p. 63.
53. Wall, *A Stranger in Europe,* p. 56.
54. Olivi and Giacone, *L'Europe Difficile,* pp. 207–209.
55. Thatcher, *The Autobiography,* p. 553.
56. Charles Moore, *Margaret Thatcher: The Authorized Biography,* vol. 2: *Everything She Wants* (London: Penguin, 2016), p. 401.
57. Wall, *A Stranger in Europe,* p. 57.
58. Moore, *Margaret Thatcher,* vol. 2.
59. Quoted in Wall, *A Stranger in Europe,* p. 41.
60. Cockfield, *The European Union,* p. 38.
61. It is one of the ironies of the situation that the head of a current task force on Brexit is none other than Sir Jonathan Faull. For a brief biography of Faull, see https://en.wikipedia.org/wiki/Jonathan_Faull.
62. Sir Jonathan Faull, personal communication, n.d.
63. Author's interview with Philip Lowe, n.d.
64. John Shelley, "PROFILE: Adrian Fortescue 'Official and a Gentleman,'" *Politico,* October 10, 2001, https://www.politico.eu/article/profile-adrian-fortescue/.

65. Cockfield, *The European Union,* pp. 47–48. After Cockfield's departure from Brussels in 1989, Fortescue went on to develop a new expertise in a branch that one might not think of at first as part of the Single European Market—the increasing spread throughout the increasingly wide-open community of organized crime and terrorism. There was at that time no specialized directorate for justice and law enforcement, but Fortescue virtually single-handedly created one. From a tiny unit with just three members, he gradually expanded his team until in 1999 "justice and home affairs" became a directorate-general in its own right. *Politico* once described Fortescue as "the backroom boy who more than anyone else has been responsible for putting in place the cross-border cooperation which is the cornerstone of the Union's fight against terrorism and organised crime" (Shelley, "PROFILE: Adrian Fortescue"). Fortescue went on to head the new directorate-general until 2003, a total of fourteen years.
66. See the so-called Sutherland Report, presented to the Commission in 1992 by the High Level Group on the Functioning of the Internal Market, presided over by Peter Sutherland ("The Internal Market after 1992: Meeting the Challenge," October 31, 1992, http://aei.pitt.edu/1025/1/Market_post_1992_Sutherland_1.pdf).
67. Grant, *Delors,* pp. 160, 162, and 165.
68. Brittan appears to have been one of the few Treasury officials to migrate to Brussels. By the early 1980s the Treasury had become converted to neoliberal thinking, but few young Treasury officials appear to have made the move.
69. Through such "Howe alumni" as Brittan, Treasury Secretary Geoffrey Howe and migrants from the Department of the Treasury had at least as much influence over EU Commission policy as Whitehall, and perhaps more.
70. Dave Keating and Tim King, "Leon Brittan, Former European Commissioner, Dies," *Politico,* January 22, 2015, http://www.politico.eu/article/leon-brittan-former-european-commissioner-dies-aged-75/.
71. Grant, *Delors,* p. 159.
72. Ibid., p. 162.
73. Speech by Leon Brittan to the Institution of Civil Engineers, April 18, 1991, European Commission Press Release Database, February 19, 2018, http://europa.eu/rapid/press-release_SPEECH-91-40_en.htm.
74. Grant, *Delors,* pp. 106–107.
75. Ibid., pp. 159–160.
76. Ibid., p. 161.
77. Ibid.
78. Ibid., p. 162.
79. Ibid., p. 139.

7. Brussels

1. See Jonathan P. Stern, *Competition and Liberalization in European Gas Markets: A Diversity of Models* (London: Royal Institute of International Affairs, 1998), especially chapter 2 (on European gas markets from 1960 to 1990). On the liberal-

ization of the UK gas market, see Dieter Helm, *Energy, the State, and the Market: British Energy Policy since 1979* (Oxford: Oxford University Press, 2003), chapters 6 and 13.

2. Michael Stoppard, *A New Order for Gas in Europe,* OIES Papers on Natural Gas, no. NG2 (Oxford: Oxford Institute for Energy Studies, 1996), p. xi and chapters 1 and 2.
3. Ibid., p. 1.
4. Stern, *Competition and Liberalization,* p. 91.
5. Jonathan P. Stern, *Third-Party Access in European Gas Industries: Regulation-Driven or Market-Led?* (London: Royal Institute of International Affairs, 1992), p. ix.
6. "Commission Decision of 13 February 1985 on the Preferential Tariff Charged to Glasshouse Growers for Natural Gas in the Netherlands," http://eur-lex.europa.eu/legal-content/EN/TXT/PDF/?uri=CELEX:31985D0215&qid=1508767536648&from=EN (accessed May 16, 2019).
7. On this point see R. Daniel Kelemen, *Eurolegalism: The Transformation of Law and Regulation in the European Union* (Cambridge, MA: Harvard University Press, 2011).
8. As Stoppard wrote in 1996, "The ideological drive behind reform has come largely from outside the gas industry" (*A New Order for Gas in Europe,* p. 44).
9. Reforms in the electricity industry served in many respects as a template for the reforms in the gas industry, which followed soon after. See Francis McGowan, *The Struggle for Power in Europe: Competition and Regulation in the EC Electricity Industry* (London: Royal Institute of International Affairs, 1993).
10. Stoppard, *A New Order for Gas in Europe,* p. 46.
11. Stephen Padgett, "The Single European Energy Market: The Politics of Realization," *Journal of Common Market Studies* 30, no. 1 (March 1992): 57.
12. Stephen Padgett, "Between Synthesis and Emulation: EU Policy Transfer in the Power Sector," *Journal of European Public Policy* 10, no. 2 (April 2003): 231.
13. Ibid. Cardoso's alliance with commissioners Leon Brittan (competition) and Martin Bangemann (industry) was to be crucial in getting the first draft energy directive through the College of Commissioners.
14. Speech by Leon Brittan to the Institution of Civil Engineers, April 18, 1991, European Commission Press Release Database, February 19, 2018, http://europa.eu/rapid/press-release_SPEECH-91-40_en.htm.
15. Padgett, "The Single European Energy Market," p. 55.
16. Author's interview with Jonathan Faull, London, October 17, 2017.
17. Simon Blakey, *Third-Party Access for Gas—The Evolving Options* (IHS Markit Decision Brief, November 1992).
18. Daniel Yergin, "Overview," in *When Gas Markets Diverge* (IHS Markit European Gas Watch, April 2000).
19. Kelemen, *Eurolegalism,* p. 154.
20. Author's interview with Jonathan Faull, October 17, 2017.
21. The single most valuable source of information and data on the competition directorate is its annual reports. For a complete listing, together with downloadable PDF files for each year, currently from 2017 back to 1971, see European

Commission, "Report on Competition Policy," http://ec.europa.eu/competition/publications/annual_report/#rep_1980.

22. See Tim Büthe, "The Politics of Competition and Institutional Change in the European Union: The First Fifty Years," in Sophie Meunier and Kathleen McNamara, eds., *Making History: European Integration and Institutional Change at Fifty* (Oxford: Oxford University Press, 2007), pp. 175–193; and Francis McGowan, "Competition Policy," in Helen Wallace and William Wallace, eds., *Policy-Making in the European Union,* 4th ed. (Oxford: Oxford University Press, 2000), pp. 116–146. See also Kelemen, *Eurolegalism.*
23. Kelemen, *Eurolegalism.* p. 143.
24. The coverage of Vestager and her directorate in major media such as the *Financial Times* is well-nigh adulatory. See, for example, Alex Barker, "Lunch with the FT: Margrethe Vestager: 'I Tell You, the Fights We Had,'" *Financial Times,* September 23–24, 2017, available by subscription **at** https://www.ft.com/content/aa9e1468-9f20-11e7-8cd4-932067fbf946.
25. Büthe, "The Politics of Competition and Institutional Change in the European Union," pp. 176–177.
26. Quoted in ibid., p. 181.
27. In the light of later events, it is interesting to note that the German government was divided in the *Grundig* case, and a key role behind the scenes was played by "some ordo-liberal officials . . . who sought at the European level the strong competition and antitrust rules they had been unable to get through at the domestic level" (ibid.).
28. Philip Lowe, personal communication.
29. Ibid.
30. The source for these numbers is European Commission, "Cartel Statistics," http://ec.europa.eu/competition/cartels/statistics/statistics.pdf (accessed April 9, 2019).
31. As one example, every cartel decision generates an average of four to five appeals to the courts. Kelemen, *Eurolegalism,* p. 175.
32. "Completing the Internal Market," White Paper from the European Commission to the European Council (June 1985), p. 41, quoted in ibid., p. 158, http://europa.eu/documents/comm/white_papers/pdf/com1985_0310_f_en.pdf.
33. Ibid., p. 158.
34. Ibid., pp. 158–159.
35. Lowe, personal communication.
36. Author's interview with Jonathan Faull.
37. Residential consumers paid public tariffs, often agreed on or imposed by national authorities. These were perfectly well-known and accessible. Only large industrial prices, in some countries, were not known.
38. This paragraph is drawn mainly from Stern, *Third-Party Access in European Gas Industries* and *Competition and Liberalization;* Stoppard, *A New Order for Gas in Europe;* and Simon Blakey, *The European Gas Transit Directive* (IHS Markit Decision Brief, November 1990).
39. Blakey, *The European Gas Transit Directive..*
40. Ibid.

41. Burckhard Bergmann, "Natural Gas in the United Germany," paper presented at the Sixth European Gas Conference, Oslo, May 15, 1991, quoted in Stern, *Third-Party Access in European Gas Industries,* p. 71.
42. See McGowan, "Competition Policy," pp. 121 and 143–145.
43. That is not to say that all was quiet by any means. The 1998 report on competition shows a case brought against Électricité de France, a gas distribution case in Denmark (both of these cases involving state aid); a case against GDF's dominant position; and a merger case involving Neste's proposed acquisition of Gasum. A long-standing exemption for energy-sector cases ended in that year, leading directly to the First Gas Directive in that year. During his years as competition commissioner, van Miert spent much of his time on sectors other than energy, such as air transportation.
44. In the late 1980s, similar case-based cracks had played a similar role in Great Britain, notably the Sheffield Forgemasters case against British Gas. Cases such as these marked the rise of the Ofgas regulator and the gradual humbling of British Gas that culminated in the breakup of that company. See Helm, *Energy the State and the Market,* pp. 244–245.
45. The onslaught against destination clauses (which became such a prominent feature of the DG-COMP cases) began shortly after this.
46. Mario Monti, "Foreword by Mario Monti," in European Commission, *XXXIst Report on Competition Policy 2001* (Luxembourg: Office for Official Publications of the European Communities, 2002), http://ec.europa.eu/competition/publications/annual_report/2001/en.pdf. Of course, in the case of network industries this view was strongly disputed by the established companies such as GDF, Eni, and Ruhrgas.
47. *Financial Times,* February 11, 1998.
48. Kelemen, *Eurolegalism,* p. 160.
49. As Lowe (personal communication) notes, one should not confuse the reorganization of DG-COMP with the process of modernization of EU antitrust law: "The move to modernise EU antitrust law and procedures started under Karel van Miert and was pursued actively to its fruition under Mario Monti. It involved essentially three things: first, moving to a US-inspired 'self-assessment' approach where it was up to firms in the first instance to ensure that their agreements did not infringe the law. This freed DG-COMP from a mountainous backlog of investigations, largely created by firms who used notification to protect their agreements against attack by any competition authority in the EU, pending the Commission's decision on them; secondly, sharing with national competition authorities in the EU the power to take decisions on antitrust cases under EU law, depending on whether the center of gravity of the transaction was national or genuinely cross-border. This gave national competition authorities far more freedom to act and extended considerably the enforcement of EU law beyond what DG-COMP could ever do by itself; thirdly, priority application of EU over national competition law in the majority of situations. This modernization was enshrined in Regulation 1/2003 and entered into force in January 2004, the same date as the enlargement of the EU to include ten countries, mainly from Eastern Europe. Under Mario Monti, I undertook a major reorganiza-

tion of DG-COMP in 2003. This was in the wake of some major Court defeats for the Commission on mergers, but also paved the way for the implementation of the new Modernization Regulation on antitrust work. Modernization of EU antitrust law and reorganization of DG-COMP are linked but were not the same thing."

50. Kelemen, *Eurolegalism,* p. 167.
51. For a bar chart giving the total cartel fines levied by DG-COMP by the five-year period corresponding to each commissioner, see European Commission, "Cartel Statistics," March 5, 2019, http://ec.europa.eu/competition/cartels/statistics/statistics.pdf.
52. Quoted in Nikki Tait, "Woman in the News: Neelie Kroes," *Financial Times,* November 6, 2009.
53. Neelie Kroes, "Tackling Cartels—A Never-Ending Task," European Commission, October 8, 2009, http://europa.eu/rapid/press-release_SPEECH-09-454_en.htm?locale=en. For a complete list of the speeches from 1999 to 2014 given by Kroes as commissioner, together with those of her predecessors and successors, see European Commission, "Speeches and Articles by the Commissioner," http://ec.europa.eu/competition/speeches/index_speeches_by_the_commissioner.html (accessed April 10, 2019).
54. Neelie Kroes, "What's Wrong with Europe's Energy Markets?," March 1, 2006, European Commission, http://europa.eu/rapid/press-release_SPEECH-06-137_en.htm?locale=en.
55. And it was also a none-too-gentle way of telling the Commission that the previous twenty years had been wasted.
56. Neelie Kroes, "More Competitive Energy Markets: Building on the Findings of the Sector Inquiry to Shape the Right Policy Solutions," European Commission, September 19, 2007, http://europa.eu/rapid/press-release_SPEECH-07-547_en.htm?locale=en.
57. Ibid. In fairness to the gas industry, Kroes's comments about cross-border trade applied to the electricity sector but not to gas, where after all cross-border gas flows represented about 20 percent of the whole market—compared to about 4 percent for the power sector. The gas industry had brought gas not just across borders, but across whole continents, from deep in the Sahara to Milan and Madrid, and from Siberia and the Norwegian North Sea to Frankfurt, Prague, and Paris. The origin of the industry lay in the gas trade between the Netherlands, Belgium, Luxembourg, France, Germany, and Italy—the six founding members of the European Union—within five or six years of the signing of the Treaty of Rome. Needless to say, their feelings toward Kroes were not exactly warm when they heard remarks like this.
58. Ibid.
59. European Commission, "DG Competition Report on Energy Sector Inquiry," January 10, 2007, http://ec.europa.eu/competition/sectors/energy/2005_inquiry/full_report_part1.pdf.
60. For more on the Madrid forum, see European Commission, "Madrid Forum," https://ec.europa.eu/energy/en/events/madrid-forum (accessed April 10, 2019).
61. Author's interview with Philip Lowe, London, September 14, 2017.

62. For a slightly longer version of this quote, see Thane Gustafson, *Capitalism Russian-Style* (Cambridge: Cambridge University Press, 1999), pp. 19–20.
63. Anatolii Chubais, ed., *Privatizatsiia po-rossiiski* (Moscow: "Vagrius," 1999), pp. 19–20. In retrospect it is revealing that Chubais did not mention competition.
64. Evgeniia Pis'mennaia, *Sistema Kudrina: Istoriia kliuchevogo ekonomista putinskoi Rossii* (Moscow: Mann, Ivanov, Ferber, 2013), pp. 17–18.
65. Vitalii Naishul', *Drugaia zhizn'* (Moscow: Elektronnaia Biblioteka RoyalLib.com, 1987), p. 1, https://royallib.com/read/nayshul_vitaliy/drugaya_gizn.html#0.
66. Vitali Naishul, interview by Philip Nichols, "Vitali Naishul and the Snake Hill Gang: An Insider's Look at Russia Today," YouTube, November 22, 2011, https://www.youtube.com/watch?v=tx8SYWySfHo (accessed May 16, 2019).
67. At the time the Chilean model was of interest to the Russian reformers because it was a model of how radical economic change needed concentrated state power to carry it out—a popular theme among the Russian reformers in the early 1990s.
68. Gregory Grossman, "Capital Intensity: A Problem in Soviet Planning" (PhD diss., Harvard University, 1952), p. 174.
69. In 1992 I participated in a World Bank mission to Russia, the principal aim of which was to acquaint the Russians with Western experience in the liberalization of the gas industry and gas markets. These missions brought to Russia teams of expert Western consultants—in this case, a past chairman of British Gas—whose responsibility was to tutor promising Russians recruited from strategic institutions such as banks, research institutes, legislative committees of the Duma, and ministries in the basics of market liberalization. (The privatization of British Gas, as I recall, was offered as a model worthy of particular emulation.) Many of the Russian "alumni" of these missions went on to occupy important policymaking positions and played key roles in various market reform initiatives.
70. We have seen above in this chapter the role of the Madrid forum as a vehicle for consensus building in the European gas industry. From the annual participants' lists of the biannual forum meetings one can see that from the mid-2000s on, these were occasionally attended by officials from GazExport, who were there as observers but were exposed to detailed discussions of the main regulatory issues of the moment. Some of these officials were fairly senior, such as Mikhail Mal'gin, who attended the forum in 2005. At the time he was responsible for all of Gazprom's gas exports to Germany. Since Russian officials were not normally invited to the forum, one suspects that the occasional Russian observer attended as special invitees of the companies—in Mal'gin's case, Gazprom's partner Wintershall or perhaps Ruhrgas.
71. The speaking and writing of Andrei Konopolyanik, who served as the deputy secretary general of the Energy Charter secretariat in Brussels in the 2000s, is emblematic of such channels of knowledge transfer. See "Andrey A. Konoplyanik, Dr. of Science, Professor," http://www.konoplyanik.ru/en/ (accessed April 10, 2019). Konoplyanik, who is one of Russia's leading energy economists and a particular expert on price formation in energy, continues to play a prominent role in

Russian-European energy relations, notably as a member of the Gas Advisory Council. See, for example, Andrey Konoplyanik, *Tsena energii: Mezhdunarodnye mekhanizmy formirovaniia tsen na neft i gaz* (Brussels: Energy Charter Secretariat, 2007).

8. The Battle for Germany

1. For a picture of the German gas industry at the turn of the century, see Simon Blakey et al., *The Hub of Competition: The Future of the German Gas Market* (joint multiclient study by IHS Markit, April 2003).
2. By contrast, the liberalization of the telecommunications, postal, and power sectors took place more rapidly and smoothly.
3. Note that the nuclear phaseout is intended as a response not to climate change, but to opposition to nuclear power chiefly because of issues of safety and waste disposal.
4. For an overview of the German *Energiewende,* see Stephen Padgett, "Energy and Climate Protection Policy," in Stephen Padgett, William E. Paterson, and Reimut Zohlnhöfer, eds., *Developments in German Politics 4* (Basingstoke: Palgrave Macmillan, 2014), pp. 241–261.
5. See Josephine Moore and Thane Gustafson, "Where to Now? Germany Rethinks Its Energy Transition," *German Society and Politics* 36, no. 3 (Autumn 2018): 1–22.
6. This description of the traditional structure of the German gas industry is drawn from Heiko Lohmann, *The German Path to Natural Gas Liberalization: Is It a Special Case?* (Oxford: Oxford University Press, 2006), p. 8.
7. Ibid., p. 20.
8. Only the largest regional company, EWE Oldenburg, had large enough storage capacity to be able to provide load balancing by using storage. Only a handful of the others had any significant storage capacity. See ibid., pp. 10–12 and table 15.
9. For background, see "Das A und O: BASF gegen Ruhrgas," *Der Spiegel,* November 27, 1989, http://www.spiegel.de/spiegel/print/d-13498852.html.
10. Wintershall's initial plan was to build a 560-kilometer pipeline from Emden, the landing point for Norwegian gas in Germany, through Wintershall's base in Kassel, to BASF's headquarters in Ludwigshafen.
11. "Gazprom i BASF: Dvoe na odnogo," *Kommersant,* October 15, 1990, https://www.kommersant.ru/doc/266775.
12. See "Das A und O."
13. The Russians had long complained, even back in Soviet times, that Ruhrgas was extracting more than its fair share of the Russian gas's value. Sabit Orudzhev, the Soviet gas minister, once teased Ruhrgas's CEO, Klaus Liesen, about their luxurious new corporate headquarters in Essen, saying to him, "One of your new chandeliers costs more than your entire old office. See how profitable our partnership is!" (Valentin A. Runov and Aleksandr D. Sedykh, *Orudzhev* [Moscow: "Molodaia Gvardiia," 2012], p. 205.)
14. See "Zu viel gemauschelt," *Der Spiegel,* October 22, 1990, p. 153, https://www.spiegel.de/spiegel/print/d-13502669.html.

15. Author's interview with Gert Maichel, n.d.
16. See "Mit Gold gefüllt," *Der Spiegel,* December 17, 1990, https://www.spiegel.de/spiegel/print/d-13502624.html; and "Wintershall-Chef scheiterte an Ruhrgas," *Der Spiegel,* August 12, 1991, https://www.spiegel.de/spiegel/print/d-13488293.html. See also Jonathan P. Stern, *Competition and Liberalization in European Gas Markets: A Diversity of Models* (London: Royal Institute of International Affairs, 1998), pp. 139–142.
17. Author's interview.
18. Quoted in "Raushalten, fertigmachen," *Der Spiegel,* January 24, 1994, pp. 84–85, https://www.spiegel.de/spiegel/print/d-13683672.html.
19. Quoted in ibid.
20. Not surprisingly, Detharding was always a strong supporter of Gazprom in the German media—for example, in an interview in *Die Welt* during the economic crisis of 1998, when the International Monetary Fund made its support for the Russian government contingent on a reform of Gazprom ("Eine der wertvollsten Firmen der Welt," *Die Welt,* June 30, 1998, https://www.welt.de/print-welt/article622058/Eine-der-wertvollsten-Firmen-der-Welt.html.
21. Author's interview with Rainer Seele, n.d.
22. Lohmann, *The German Path to Natural Gas Liberalization,* p. 23.
23. Quoted in ibid., p. 1.
24. Cited in Andreas Busch, "Globalization and National Varieties of Capitalism: The Contested Viability of the German Model," in Kenneth Dyson and Stephen Padgett, eds., *The Politics of Economic Reform in Germany: Global, Rhineland, or Hybrid Capitalism?* (London: Routledge, 2006), pp. 25–47.
25. Lohmann, *The German Path to Natural Gas Liberalization,* figure 2, p. 9.
26. On the Schröder reforms, see Eric Langenbacher, and David Conradt, eds., *The German Polity,* 11th ed. (Lanham, MD: Rowman and Littlefield, 2017), especially pp. 153ff.
27. Lohmann, *The German Path to Natural Gas Liberalization,* pp. 5–6.
28. For a fascinating testimonial to the attitudes of the major German utilities on the eve of liberalization, see "Wir schaffen es schneller," *Der Spiegel,* September 9, 1996, https://m.spiegel.de/spiegel/print/d-9089520.html.The article reports a debate between Dietmar Kuhnt, CEO of the utility RWE, and his counterpart Ron Summer of Deutsche Telekom. Kuhnt perceives his own industry as highly competitive, while he portrays Deutsche Telekom as a monopoly that he intends to compete with. Yet ironically, at this time the beginnings of competition in telecommunications were already leading to lower prices, while the price of electricity was rising—for reasons that Kuhnt insists have nothing to do with lack of competition.
29. The feed-in tariff is a preferential rate at which power utilities are obligated to buy back power generated by solar installations. For an account of the politics of the new Energy Law between 1996 and 1998, see Nadja Daniela Klag, *Die Liberalisierung des Gasmarktes in Deutschland* (Marburg: Tectum Verlag, 2003), pp. 249–253. The Greens, while still in opposition, had objected to the fact that the feed-in tariff was initially limited to 10% (p. 249).

30. Heiko Lohmann, *The German Gas Market Post 2005: Development of Real Competition,* Report NG-33 (Oxford: Oxford Institute for Energy Studies, 2009), p. 4, https://www.oxfordenergy.org/wpcms/wp-content/uploads/2010/11/NG33-ThegermangasMarketPost2005DveleopmentofRealCompetition-HeikoLohmann-2009.pdf.
31. The phrase comes from ibid., p. 27.
32. The technical details and the blow-by-blow progress of the talks are discussed at length in Lohmann, *The German Path to Natural Gas Liberalization,* chapters 3 and 4. As Lohmann points out, it was not only the content of the regulatory regime that was at stake, but there was also an ideological component. The German associations representing industry and industrial consumers were opposed in principle to government intervention. Hence they clung to the negotiating table long after it was apparent that compromise could not be achieved. In contrast, local distributors and the municipalities, which had seen regulation work in the power sector, had no such objections in principle. See ibid., pp. 31 and 33.
33. Lohmann observes, however, that the lack of formal agreements did not prevent third-party access from happening in practice (ibid., pp. 34–38). But it happened "not because the system was transparent with clear obvious rules, but rather because a small number of people and companies gained competence and managed to find solutions to their individual transportation problems." (ibid., p. 38). Nevertheless, he argues, this was far from an overall success story. See his conclusions in ibid., pp. 47–48.
34. Ibid., p. 32. The Ministry has changed names repeatedly throughout its history. For the sake of simplicity, I refer to it in this book as the Federal Economics Ministry.
35. Ibid., p. 47.
36. The technical measure of liquidity is the so-called churn ratio—that is, the ratio of financial transactions in the market to the underlying physical transactions. A market with a churn ratio of about fifteen is considered reliably liquid. For background on this point, see Simon Blakey, "The Appeal for Liquidity in German Gas Markets" (IHS Markit White Paper, July 2004), esp. pp. 11–12.
37. Long-term contracts between producer-exporters and importing utilities tended to be 20–25 years, but contracts between the utilities and their domestic customers had shorter and more variable durations.
38. As explained in Chapter 1, the principle of *Anlegbarkeit* meant that that the price of gas must not exceed that of competing fuels, which in practice meant a basket of fuels consisting mainly of heating oil.
39. Contract duration was in most cases reduced to two years or less. However, import contracts between Gazprom and European buyers remain long-term. Existing contracts between Gazprom and Germany will remain a stabilizing factor until the last ones expire in 2035.
40. Lohmann, *The German Path to Natural Gas Liberalization,* p. 127.
41. The plants were Brokdorf, Grohnde, Unterweser, and Isar 1 and 2. Unterweser and Isar 1 have since been decommissioned.

42. The details of the takeover are discussed in Lohmann, *The German Path to Natural Gas Liberalization,* chapter 7.
43. The key step, E.ON's purchase of BP's stake in Ruhrgas, was an interesting example of how a buyer and seller can come together despite opposite outlooks. For BP, Ruhrgas was a company without a future, and when approached by E.ON, BP was eager to sell in exchange for E.ON's oil assets. E.ON calculated that there were still ten years of monopoly rents left in Ruhrgas, which made it a good buy. Even so, E.ON's move took German observers by surprise, since only a short time before E.ON's management had hailed its oil assets as the key to a successful diversification strategy. For an overview of E.ON's strategy as perceived—and criticized—by the German business community at the time, see Frank Dohmen, "Lieber schnell was Neues," *Der Spiegel,* July 23, 2001, https://m.spiegel.de/spiegel/print/d-19699010.html.
44. Lohmann, *The German Path to Natural Gas Liberalization,* p. 112. The out-of-court settlement was closely watched by the German media. See, for example, "Alle neune!," *Manager Magazin,* January 30, 2003, http://www.manager-magazin.de/unternehmen/artikel/a-232974.html; and "Der Ruhrgas-Deal ist besiegelt," *Manager Magazin,* January 31, 2003, http://www.manager-magazin.de/unternehmen/artikel/a-233079.html. It was rumored at the time that Schröder intervened personally at the last minute with a phone call to the Finnish government, following which one of the main complainants, Fortum, withdrew its suit. See "Der Kanzler half," *Der Spiegel,* February 10, 2003, http://www.spiegel.de/spiegel/print/d-26329180.html.
45. These are described in Lohmann, *The German Path to Natural Gas Liberalization,* chapter 7.
46. See Frank Dohmen and Christian Reiermann, "Ganz, ganz schwierige Kiste," *Der Spiegel,* July 1, 2002, https://www.spiegel.de/spiegel/print/d-23011339.html. It was said at the time that the favorable decision of the Ministry for Economics was the result of a political favor by a loyal ally of Schröder, who approved the merger through a so-called minister decision that bypassed normal procedures. Articles in the German media also hinted strongly that Schröder's decision to support the merger was strongly influenced by his past connections to the energy industry. See Cerstin Gammelin, "Die Erdgas-Connection," *Die Zeit,* December 15, 2005, https://www.zeit.de/2005/51/Erdgas.
47. Lohmann, *The German Path to Natural Gas Liberalization,* p. 109.
48. One of the takeovers was that of the East German company VNG. By a series of steps too complex to describe here, by 2008 VNG had become the joint property of two West German utilities, EWE and EnBW. It was not a happy marriage, and by that year there were rumors that VNG might be a takeover target by Gazprom. See Peter Dinkloh, "German VNG May Be Gazprom's First EU Takeover," Reuters, June 20, 2008, https://www.reuters.com/article/us-gazprom-germany/german-vng-may-be-gazproms-first-eu-takeover-idUSL2059012120080620. But Gazprom showed no interest. After more adventures EWE sold its interest to EnBW, leaving EnBW with a 75 percent share of VNG—as remains the case today.

49. Eva Ruffing, "How to Become an Independent Agency: The Creation of the German Federal Network Agency," *German Politics* 23, nos. 1–2 (2014): 43–58.
50. Ruffing, "How to Become an Independent Agency," p. 43.
51. On the contrast between liberalization in telecommunications and electricity in Germany, see Peter Humphreys and Stephen Padgett, "Globalization, the European Union, and Domestic Governance in Telecoms and Electricity," *Governance* 19, no. 3 (July 2006): 383–406.
52. Lohmann, *The German Gas Market Post 2005,* p. 6.
53. See, for example, Karsten Langer, "Die Davids müssen kämpfen," *Der Spiegel,* February 19, 2006, http://www.spiegel.de/netzwelt/web/bundesnetzagentur-chef-kurth-die-davids-muessen-kaempfen-a-401553.html.
54. Lohmann, *The German Gas Market Post 2005.*
55. Ibid.
56. Stefan Lennardt, *Strategische Kommunikatsion in regulierten Märkten* (Münster, Germany: Litverlag Dr. W. Hopf, 2009), pp. 77–78. Schmidt was nearing retirement age when he undertook the network access case as his last major project. He had experienced the long opposition of the industry associations at first hand and complained of the personal attacks they had mounted against the regulators in the media.
57. Lohmann, *The German Gas Market Post 2005.*
58. Ibid., p. 3n3. By 2007 this had become a stock theme in Bergmann's public addresses. For example, "'We see currently a real revolution in our business and we want to promote competition in the German gas market,' said Burckhard Bergmann, the CEO of E.ON Ruhrgas, at the company's annual press conference yesterday" (James Ball, "E.ON Ruhrgas Predicts Revolution in Business Model," *Gas Matters,* May 2007). For background on Burckhard Bergmann, see James Ball, "The Bergmann Era Draws to a Close," *Gas Matters,* October 2007.
59. Personal communication from Simon Blakey, who witnessed the conversation.
60. This is the major theme of an important article by Kirsten Westphal of the German Institute for International and Security Affairs. Her key point is that institutional and regulatory policies in Europe and in Germany, as well as the changes associated with the *Energiewende,* have predominated over security concerns. The result, in Westphal's view, is a loss of balance in German gas policy, which could become source of vulnerability in the future. This book agrees with all but the last point, which is explored in Chapter 12, on Russian-German relations. See Kirsten Westphal, "Institutional Change in European Natural Gas Markets and Implications for Energy Security: Lessons from the German Case," *Energy Policy* 74 (2014): 35–43.

9. Gazprom Survives and Gets Away

1. Clifford Gaddy and Barry Ickes, "Russia's Virtual Economy," *Foreign Affairs* 77 (1998): 53–67.
2. Thane Gustafson, Vadim Eskin, and Aleksandr Rudkevich, *Gazprom's Dilemma: Too Much Gas or Too Little?* (IHS Markit Private Report, June 1993).

3. To be more precise, while natural gas plays only a minor role in the sparsely settled eastern two-thirds of Russia, where Gazprom's pipeline system does not reach, the western third of the country, where most of the country's population and wealth are concentrated, depends on natural gas.
4. Thane Gustafson, *Wheel of Fortune: The Battle for Oil and Power in Russia* (Cambridge MA: Harvard University Press, 2012).
5. Quoted in Valerii Paniushkin, Mikhail Zygar', and Irina Reznik, *Gazprom: Novoe russkoe oruzhie* (Moscow: "Zakharov," 2008), p. 21.
6. Quoted in ibid., p. 11.
7. Quoted in ibid., p. 17.
8. Quoted in ibid.
9. Paniushkin, Zygar', and Reznik give a detailed account of the way Chernomyrdin overcame the resistance of Prime minister Nikolai Ryzhkov (ibid., pp. 17–19). Rem Viakhirev confirms the episode: "He didn't understand what we were after—you have a ministry, he said, and you want to make some kind of collective farm. In the end he gave up, saying, 'Go to the devil. Do what you like'" (quoted in Irina Malkova and Valerii Igumenov, "Poslednee interv'iu Rema Viakhireva: 'Putin kogda uslyshal, chto ia ukhozhu, tak obradovalsia,'" *Forbes Russia,* September 11, 2012, http://www.forbes.ru/sobytiya/lyudi/116511-eksklyuzivnoe-intervyu-rema-vyahireva-putin-kogda-uslyshal-chto-ya-uhozhu-tak-.
10. See Michael McFaul, *Russia's Unfinished Revolution: Political Change from Gorbachev to Putin* (Ithaca, NY: Cornell University Press, 2001).
11. After a modest twenty-year career in the Ministry of Instrument Making (one of the nine ministries of the military-industrial complex), by the late 1980s Chernogorodskii had gravitated to the staff of the Council of Ministers. But in October 1990 he moved over to the fledgling Russian government, which under Yeltsin was beginning to challenge the Soviet government under Gorbachev. Chernogorodskii took over the newly created Anti-monopoly Committee. After the attempted coup of August 1991 and the subsequent reorganization of the government, Chernogorodskii turned the Anti-monopoly Committee into a radical voice for market reform. Once Chernomyrdin became prime minister at the end of June 1992, however, Chernogorodskii had clearly fallen out of step. He was dismissed from the committee in July 1992. From that point on he worked in the private sector, notably as head of Optimum-Finans. He died at age sixty-two in October 2002.
12. "Anti-Monopoly Committee Slams the Door on Gazprom," *Russian Petroleum Investor,* April 1992, pp. 18ff.
13. In the spring of 1992, while on a World Bank mission to Russia, I had occasion to travel with a former director of British Gas who had played a lead role in the initial liberalization of the British gas industry. His faith in the doctrines of liberalization and his conviction that Gazprom would shortly follow the same path were utterly unshakable, even after the sobering experience of meeting with Gazprom officials.
14. Elena Chernova and Leonid Skoptsov, "Moroz po kozhe ot reformy v gazovoi otrasli," *Moskovskie Novosti,* no. 6, 1992, p. 4.

15. Lobov was a longtime Yeltsin associate from the Sverdlovsk party apparatus who was first deputy prime minister from April through November 1991. Though linked to Yeltsin, he had little sympathy for the radical reform program and was out of office during Gaidar's tenure. He returned to the government under Prime Minister Chernomyrdin, subsequently as chairman of the Security Council (September 1993–June 1996). He left the government again after the reformers returned to office in the spring of 1997. Subsequently he began a successful new career as a real estate developer in Moscow.
16. Sergei Emel'ianov, "Eksport rossiiskogo gaza: Istoriia, sostoianie, perspektivy," *Neftegazovaia Vertikal',* no. 6 (2003). Gorbachev had begun loosening the Ministry of Foreign Trade's monopoly control of foreign trade as early as 1988, and it is likely that Soiuzgazeksport had begun drifting under Gazprom's influence well before the actual incorporation occurred in 1991.
17. Chernova and Skoptsov, "Moroz po kozhe ot reformy v gazovoi otrasli." Technically speaking, the Gazprom spokesman who undoubtedly supplied this scenario had a point: gas men agree that one of most challenging tasks in running a marketized gas system is devising various balancing mechanisms so that supply and demand will end up equal, not only over one month or year but also day by day and hour by hour. In that sense a gas system is not different in essence from a power grid, except that part of the gas can be stored underground and part in the pipeline.
18. Pavel Gorbenko, "Eshche odna rossiiskaia birzhevaia struktura reshila nazhat' na gaz," *Kommersant,* no. 8 (February 17–24, 1992), p. 14.
19. Malkova and Igumenov, "Poslednee interv'iu Rema Viakhireva."
20. Suleimanov and the other gas generals were not entirely without independent resources. In the early 1990s they built a profitable sideline exporting gas condensate to Finland by rail. Condensate was traditionally treated as a nearly worthless by-product by the gas industry and effectively dumped into the nearest crude oil pipeline, thus wasting its value as a petrochemical feedstock. But in the early 1990s the gas generals built up a stock of private railcars to export their condensate on their own account. However, this had no significant impact on the overall balance of power between the gas generals and Gazprom's corporate center.
21. "Problemy i perspektivy TEK glazami vedushchikh rossiiskikh politikov," *Neft' i kapital,* no. 11 (November 1995), p. 11.
22. Quoted in ibid.
23. I may be showing my American colors here. A European would consider that natural mineral resources are national assets and find bizarre the American notion that what is under the subsoil is yours just like your backyard is yours. In this sense, Gaidar's "property of the entire people" is perhaps not so different from the way an Englishman might say something belongs to the Crown or a Norwegian would speak of inalienable state rights. For Europeans, private rights to produce, own, and sell oil and / or gas (or diamonds or coal) are concessions to individuals from the collective, where natural ownership belongs.
24. Georgii Smirnov, "Dve pravdy ob effektivnosti eksporta gaza iz Rossii," *Neft' i kapital,* no. 1 (January 1996), p. 59. The actual arrangement was a little more so-

phisticated than Smirnov describes in this sentence. What Gazprom actually did was to claim high transportation costs, which it assigned to its wholly owned transportation subsidiaries. As a result, Gazprom systematically understated its true profits, and at least through 1995, the government "looked through its fingers" (as the Russian saying goes) at this arrangement. Even today, Gazprom takes a large bite out of the export value chain through its control of the transportation system and its assignment of transportation tariffs as costs to its pipeline subsidiaries.

25. Quoted in ibid.
26. Gustafson, Eskin, and Rudkevich, *Gazprom's Dilemma.*
27. For the details of how this happened, see Thane Gustafson, *Wheel of Fortune: The Battle for Oil and Power in Russia* (Cambridge, MA: Belknap Press of Harvard University Press, 2012), pp. 75–78.
28. Responding to a presidential decreee dated October 27, 1992, declaring Gazprom a joint-stock corporation (see Dmitrii Chernov-Andreev, "Gazprom izmenil svoi status," *Kommersant-Daily,* October 28, 1992, p. 2), on November 28 the Antimonopoly Committee protested that the decree violated the terms of the 1992 privatization program, according to which firms (*kontserny*) containing state enterprises were not eligible for conversion into joint-stock corporations (see "Antimonopol'nyi komitet sporit s Gazpromom," *Kommersant-Daily,* December 12, 1992, p. 4, https://www.kommersant.ru/doc/33209). However, the protest was ignored by both the government and the parliament.
29. Presidential Decree No. 1333, November 5, 1992. The presidential decree was then made official by a government decree (No. 138, February 17, 1993). The latter date marked the official founding date of Gazprom as a joint-stock corporation. This decree, by the way, was the first of many to affirm the principle of third-party access"—open access to the gas pipeline system for all users—discussed in Chapter 8).
30. Quoted in Malkova and Igumenov, "Poslednee interv'iu Rema Viakhireva."
31. Among many books on the subject, see Joseph R. Blasi, Maya Kroumova, and Douglas Kruse, *Kremlin Capitalism: Privatizing the Russian Economy* (Ithaca, NY: Cornell University Press, 1997). See also Thane Gustafson, *Capitalism Russian-Style* (Cambridge: Cambridge University Press, 1999).
32. There are some parallels here to the first phase of the reform of the British gas industry. In 1986 British Gas was privatized and sold to mass ownership as a monopoly. At that time there was no liberalization, third-party access was a dead letter, and the beginnings of a spot market in natural gas were a good three years in the future.
33. The results of the auctions varied widely by region, depending on local demand but also on the supply of shares made available by Gazprom. Thus in Tiumen' Province one voucher fetched 2,100 Gazprom shares, each with ten-ruble face value. In Altai Province the exchange rate went as high as 6,000 shares, whereas in Perm' one voucher was worth only 16 shares. See Sergei Savushkin, "Vy mozhete kupit' sebe mnogo aktsii 'Gazproma,' no vladet' kontsernom vse ravno budet gosudarstvo," *Neft' i kapital,* no. 1 (January 1995), pp. 37–40.

34. However, it is an interesting question whether the Viakhirev management team took their shares with them (and if so, how many shares), when they left the company after Viakhirev's dismissal in 2001.
35. Background file on Gazprom maintained on the website of *Ekspert* magazine.
36. The "shares for loans" episode in the oil industry is described in Gustafson, *Wheel of Fortune*, Chapter 3.
37. Presidential decrees Nos. 399 of 26 March 1996 and 599 of 22 April 1996. Source: Gazprom website, op. cit.
38. The story of Itera is discussed in Chapter 11.
39. For an excellent if critical portrait of Viakhirev, see Elizaveta Osetinskaia, Iuliia Bushueva, and Tat'iana Lisova, "Kto v strane glavnyi," *Vedomosti,* June 1, 2001, https://www.vedomosti.ru/newspaper/articles/2001/06/01/kto-v-strane-glavnyj.
40. The episode is discussed in Viakhirev's last interview, ten years after his removal. See Malkova and Igumenov, "Poslednee interv'iu Rema Viakhireva."
41. Putin press conference, Sochi, March 27, 2004, http://kremlin.ru/events/president/transcripts/22400.
42. Jonathan P. Stern, *The Russian Natural Gas Bubble: Consequences for European Gas Markets* (London: Royal Institute of International Affairs, 1995).
43. Quoted in Malkova and Igumenov, "Poslednee interv'iu Rema Viakhireva."
44. A further source of hemorrhage of value was the government's pricing policy for nonindustrial users—that is, municipal and residential consumers. Gas prices for nonindustrial customers were not indexed to inflation but were revised at irregular intervals. Thus, in March 1995, even after the nonindustrial price was raised tenfold at one bound, it was still 5.5 times lower than the industrial price. This became an increasingly serious source of losses as time went by, because in the second half of the 1990s residential and municipal consumption became the fastest-growing sectors of gas demand.
45. On the epidemic of nonpayments and its impact on Gazprom's cash flow, see Jonathan P. Stern, *The Future of Russian Gas and Gazprom* (Oxford: Oxford University Press, 2005), pp. 49–50. See also Thane Gustafson and Vadim Eskin, *Russia's Gazprom Turns Inward: Can It Bring the Domestic Market under Control?* (IHS Markit Private Report, March 1997).
46. See Stern, *The Future of Russian Gas and Gazprom,* pp. 38–39.
47. One should not imagine, by the way, that ordinary Russian consumers and businesses simply scoffed at their gas bills and enjoyed free gas and heat. Standing between Gazprom and the final burner-tip consumers was a tier of middlemen, the so-called municipal gas *(gorgaz)* and province gas *(oblgaz)* companies. These were local distributors that took gas from the Gazprom high-pressure trunk lines and brought it to the final consumer. The potential of these players (which had been insignificant local agencies in Soviet times) to extract rent was not lost on local influentials, who soon took them over and used their local power and their position as middlemen to extract payment from local consumers—but not necessarily to pass the proceeds along to Gazprom. Since the distributors were well connected to local politicians, whose good will Gaz-

prom needed, there was initially little Gazprom could do about this. But as the 1990s went on Gazprom fought to bypass the *oblgazy* and *gorgazy* or take them over.

48. Vadim Eskin and Thane Gustafson, *Gazprom Goes for the Gold* (IHS Markit Private Report, February 1996).
49. Interview with Vladimir Rezunenko, then a key member of Gazprom's management committee, quoted in Ol'ga Bolmatova, "'Gazprom' sozdaet sebe usloviia dlia spokoinoi raboty," *Neft' i Kapital,* no. 1 (January 1995), p. 63.
50. Ibid.
51. Vadim Eskin and Thane Gustafson, *Russian Gas: From Feast to Famine* (IHS Markit Private Report, February 1996); Vadim Eskin, Thane Gustafson, and Matthew J. Sagers, *Outlook for Gas Production from Gazprom's "Big Three" Fields: What Will Be the Rate of Decline?* (IHS Markit Decision Brief, December 2002).
52. In the summer of 1994 there were persistent rumors that Viakhirev was about to be removed, and even Russia's premier news anchor, Evgenii Kiselev, announced his imminent departure on Russia's prime-time news program, *Itogi.* In November 1994 the State Property Committee, a longtime enemy of Gazprom, circulated a draft presidential decree ordering the breakup of Gazprom. See Aleksandr Ol'gin, "Igry vzroslykh liudei," *Neft' i kapital,* no. 0 (1994), pp. 8–9. But such reports were so common that they were practically background noise.
53. Even as prime minister Chernomyrdin routinely accompanied Gazprom delegations on trips to Russia's major gas partners. Thus, in the first half of 1994 he traveled with Gazprom representatives to Finland and Italy.
54. Marina Latysheva, "Novaia eksportnaia poshlina oboidetsia 'Gazpromu' v 340 millionov dollarov," *Neft' i Kapital,* no. 3 (March 1995), pp. 64–66. The increase came in the form of a presidential decree, but there is no suggestion that it originated anywhere but in the prime minister's office. The decree contained some exemptions. For example, gas delivered under state-to-state agreements, in payment for sovereign credits, was not subject to the export tax. But where it did apply, the export tax came straight out of Gazprom's bottom line, since gas prices in long-term supply contracts were pegged to oil-based reference baskets and could not be raised to recover the tax.
55. Sergei Kolchin, "Chei dom—Gazprom?," *Neft' i kapital,* no. 12 (December 1995), p. 35.
56. Ibid. Gazprom's purchasing agents were the numerous open and hidden daughter companies set up by Gazprom in the late 1980s and the first half of the 1990s. When Aleksey Miller replaced Viakhirev in the spring of 2001, it was discovered that there were thousands of such companies.
57. Ibid.
58. Gustafson and Eskin, *Russia's Gazprom Turns Inward.*
59. Derezhov made this remark at a meeting at Gazprom headquarters in 1994 that I attended.

60. WATAN (Turkmen Television) press release, December 17, 1999. I am assured by people who saw the exchange on television that this transcript, which was provided to me by a colleague in Ashgabat, is genuine.
61. Niyazov could almost be said to be a member of the Saint Petersburg elite himself, having been educated as a hydropower engineer in Leningrad.
62. On the rise and fall of the trading houses, see Jonathan P. Stern, "The Impact of European Regulation and Policy on Russian Gas Exports and Pipelines," in James Henderson and Simon Pirani, eds., *The Russian Gas Matrix: How Markets are Driving Change* (Oxford: Oxford University Press for the Oxford Institute for Energy Studies, 2014), pp. 88–90.
63. Ibid., pp. 82–83.

10. Gazprom under Pressure

1. One major and welcome exception is the work of the Oxford Institute of Energy Studies, and in particular the studies of Jonathan Stern, James Henderson, Simon Pirani, Katja Yafimava, and their team over the past three decades. I have drawn inspiration from their work here and throughout this book. Much of it is available on the website of the Institute. See, in particular, James Henderson and Simon Pirani, eds., *The Russian Gas Matrix: How Markets Are Driving Change* (Oxford: Oxford University Press, 2014). Another noteworthy exception is the brilliant history by Per Högselius, *Red Gas: Russia and the Origins of European Energy Dependence* (New York: Palgrave Macmillan, 2013), on which I have relied heavily throughout this book.
2. Tatiana Mitrova and Tim Boersma, *The Impact of US LNG on Russian Natural Gas Export Policy* (New York: Columbia University Center on Global Energy Policy, 2018), p. 27.
3. The full story is a bit more complicated. Medvedev was initially posted to Vienna to work for Donaubank, one of the three "Soviet foreign banks" *(Sovzagranbanki).* Donaubank served as a channel to fund various USSR-friendly organizations. After the breakup of the Soviet Union, many Donaubank staff became actual investment bankers, repackaging funds from the Communist Party into international investments.
4. In February 2019, Medvedev was abruptly dismissed, along with his domestic counterpart, Valeriy Golubev. Medvedev has been replaced as the senior head of international relations and deputy chairman of the management committee by Burmistrova, who is said to have won out over a rival, Pavel Oderov, the current head of the International Business Department and director of GM&T. Both can be considered part of the fourth generation.
5. On the subsequent evolution of Russia's Central Asian gas imports, see Stanislav Yazynin, Matthew Sagers, Julia Nanay, and Anna Galtsova, *Russia's Need for Central Asian Gas Diminishing, but Has It Disappeared?* (IHS Markit Insight, November 2014).
6. Valerii Paniushkin, Mikhail Zygar', and Irina Reznik, *Gazprom: Novoe russkoe oruzhie* (Moscow: "Zakharov," 2008), p. 104.

7. For a portrait of Gref and his role in the early 2000s, see Thane Gustafson, *Wheel of Fortune: The Battle for Oil and Power in Russia* (Cambridge, MA: Belknap Press of Harvard University Press, 2012), pp. 254–255. Since 2007 Gref has been the CEO of Sberbank.
8. From the beginning Putin appears to have been more interested in gas than in electricity. The answer may lie in the fact that from the first he saw gas as an instrument of geopolitical influence (and resource rents), which electricity was not. He was conscious from the first of the difficult dependence of the gas industry on Ukraine and Central Asia, which was not the case to the same degree with electricity. Gas earned export revenue for the state, while electricity did not. Putin is rumored to have said as early as 1999 that he wanted to be the head of Gazprom, not prime minister or president. For a presentation of valuable background on Putin's policy on electricity, which takes a different view of Putin's priorities, see Peter Rutland, "Power Struggle: Reforming the Electricity Industry," in Robert Orttung and Peter Reddaway, eds., *The Dynamics of Russian Politics: Putin's Reform of Federal-Regional Relations,* vol. 2 (Lanham, MD: Rowman and Littlefield, 2005), pp. 267–294.
9. The term "Family" (capitalized) refers to a group of state officials and family members close to Yeltsin who were instrumental in recruiting the then largely unknown Putin as a successor to Yeltsin. For a lively account based on interviews with the participants, see Mikhail Zygar', *All the Kremlin's Men: Inside the Court of Vladimir Putin* (New York: PublicAffairs, 2016).
10. Paniushkin, Zygar', and Reznik, *Gazprom,* p. 103.
11. No relation to Aleksandr Medvedev.
12. Quoted in ibid., pp. 104–105.
13. Quoted in ibid., p. 105. A month before, the government had greased the skids under Viakhirev by pushing through an amendment to Gazprom's charter. Up to that time the CEO could be replaced only by a unanimous vote of the board, which in effect guaranteed Viakhirev's place in perpetuity. But under the new rule a simple majority was enough (see ibid., pp. 105–106).
14. Quoted in ibid., pp. 111–112. For a similar description, see Boris Vishnevskii, "Ten', znaiushchaia svoe mesto," *Moskovskie novosti,* September 24, 2004, http://www.compromat.ru/page_11389.htm. Vishnevskii in particular dismisses accounts that Miller played any significant role in attracting Western investors to Saint Petersburg. And although he was eventually named deputy chairman of the committee, it was in the wake of a scandal that caused the dismissal of his direct boss, Alexander Anikin (see Vishnevskii, "Ten', znaiushchaia svoe mesto"). A lengthy profile of Miller and his management of Gazprom appears in Aleksey Grivach and Andrei Denisov, "Upravliaiushchii natsional'nym gazovym dostoianiem," *Vremia novostei,* May 25, 2006, p. 8, http://www.vremya.ru/2006/89/8/152813.html.
15. This account is based largely on Paniushkin, Zygar', and Reznik, *Gazprom,* but it meshes with identical accounts from numerous other sources. For a description of the Committee on External Relations under Putin, see Gustafson, *Wheel of Fortune,* chapter 6.

16. "Aleksey Miller Elected as Gazprom Management Committee Chairman for Another 5-Year Term," February 16, 2016, http://www.gazprom.com/press/news/2016/february/article266936/.
17. This trend is not confined to Gazprom. The dominance of the "Saint Petersburgers" in top personnel appointments, which had been pronounced in the 2000s, began to fade in the 2010s all across the government.
18. Quoted in Paniushkin, Zygar', and Reznik, *Gazprom,* 115ff.
19. For details see, Grivach and Denisov, "Upravliaiushchii natsional'nym gazovym dostoianiem."
20. Paniushkin, Zygar', and Reznik, *Gazprom,* pp. 119–120.
21. See Konstantin Smirnov and Alena Miklashevskaia, "Forbes opiat' soschital vsekh bogatyx," *Kommersant-Daily,* June 23, 2001, p. 3; and Fedor Kotrelev, "Bogatye stali bednee—a russkie bogache," *Kommersant-Daily,* March 2, 2002.
22. Maria Braslavkaia, "Spisok Millera," *Rusenergy,* May 20–26, 2002. *Rusenergy,* a newsletter of news and commentary, was a valuable source of analysis about Gazprom and other energy issues during the 2000s and early 2010s. Its founder and chief editor, Mikhail Krutikhin, remains an important independent voice in energy journalism today.
23. Paniushkin, Zygar', and Reznik, *Gazprom,* pp. 120–121.
24. One of the few exceptional reporters in the general Western press was David Hoffman of the *Washington Post,* who was shortly to publish *The Oligarchs,* his pathbreaking book that devoted ample space to Gazprom under Viakhirev. See for example, David Hoffman, "Foreign Investors Criticize Deals by Russian Gas Giant," *Washington Post,* December 24, 2000, https://www.washingtonpost.com/archive/politics/2000/12/24/foreign-investors-criticize-deals-by-russian-gas-giant//e98cf51e-efd1-45f5-b4e8-84d2edec8f77/?utm_term=.3b90aa0aba6c. Concerning shareholdings by Viakhirev and Chernomyrdin's relatives and allies, Hoffman wrote in that article: "These include Viakhirev, whose daughter, Tatiana Dedikova, owns 6.4 percent of Stroitransgaz; Chernomyrdin's sons Vitaly and Andrei, each of whom owns 5.96 percent; and three children of Arngolt Bekker, who is president of Stroitransgaz and a board member of Gazprom. Bekker owns 20 percent outright, and his children own between 2.6 percent and 6.9 percent each. A relative of yet another Gazprom board member and deputy chief executive officer, Vyacheslav Sheremet, owns another 6.4 percent of Stroitransgaz." Hoffman's primary source appears to have been the late Boris Fyodorov, former finance minister and a prominent reformer, who at the time had just been elected as the Kremlin's representative on the board and was a prominent voice in opposition to Viakhirev's regime.
25. "Leader: Fire Up Gazprom," *Financial Times,* May 29, 2001.
26. For a description of the reform program, see Rudiger Ahrend and William Tompson, "Unnatural Monopoly: The Endless Wait for Gas Sector Reform in Russia," *Europe-Asia Studies* 57, no. 6 (September 2005):. 801–821.
27. It was only during Putin's second term that he began to allow other interests—notably, those connected with Gennadii Timchenko and some of the leading

shareholders of Bank Rossiia—to take growing roles in the gas sector, and Gazprom's control over its assets began once more to weaken.

28. This section is based on my interactions with the reformers in the early 2000s, including conversations with Gref, then economics minister, and his principal deputy in charge of the gas restructuring program at that time, Andrei Sharonov.
29. At this point, on the eve of 2005 and for a year or two afterward, Gazprom appeared to have reached the height of its power. For two well-informed analyses that sum up the situation at the time, see "Kto ne s Gazpromom, tot protiv nego," *Neftegazovaia vertikal'*, No. 25 (December 16, 2005); and Viktor Somov and Nikolai Sudarev, "Kak podelit' 'Gazprom': Minekonomrazvitiia vpervye predlozhilo real'nyi plan sozdaniia gazovogo rynka," *Rusenergy Praim-Onlain,* March 27, 2003.
30. The 1999 federal law, "On Gas Supply in the Russian Federation" had set this limit. However, two presidential decrees of the late 1990s changed the numbers slightly. Presidential Decree No. 529 of May 28, 1997, limited foreign ownership to 9 percent of Gazprom. Presidential Decree No. 943 of August 10, 1998, loosened the limits somewhat by allowing the sale to foreigners of another 5 percent.
31. In March 2003 Boris Fyodorov, a Gazprom board member, said in an interview with Interfax that a full liberalization of Gazprom's shares would make it possible to "at least double" the company's market capitalization within a year. Quoted in "Gazprom Makes First Step towards Liberalizing Share Market," *Interfax Petroleum Report* 12, no. 10 (March 14–20, 2003), p. 16. Since in the event full liberalization took two years longer than anticipated, Fyodorov's prediction was not realized until 2005.
32. Ibid.
33. See Varvara Aglamish'ian and Aleksey Tikhonov, "El'tsin poterial 'Gazprom,' a Putin ego vernul," *Izvestiia,* September 15, 2004, https://iz.ru/news/294144.
34. The details of the abortive takeover are recounted in Gustafson, *Wheel of Fortune,* pp. 341–351, and need not concern us here.
35. For the story of the repeated postponement of development of the Yamal Peninsula, see Thane Gustafson, *Crisis amid Plenty: The Politics of Soviet Energy under Brezhnev and Gorbachev* (Princeton, NJ: Princeton University Press, 1989), chapter 5.
36. A. V. Epishev, "The Impact of Geographical Conditions and the Need for Environmental Protection in the Development of the Natural Gas Industry in the Northern USSR," *Izvestiia AN SSSR* (seriia geograficheskaia), No. 4 (1979), pp. 52–63; translated in *Soviet Geography* 22, no. 2 (February 1981): 67–80. See also "Uroki Bovanenkovo," *Rusenergy,* November 28, 2012.
37. E. G. Altunin, "Strategiiu vybrat' segodnia," *Ekonomika i organizatsiia promyshlennogo proizvodstva*, No. 2 (1979), p. 18.
38. For background on the Obskaya-Bovanenkovo railroad and the Naks, see Liudmila Iudina, "Severnyi reportazh: Doroga zhizni," *Trud*, June 2, 2000. Vladimir Nak was one of the legendary builders of the far north. When he died in 2010, both Miller and Putin paid tribute to his accomplishments. By that time the railroad had finally been completed. See Tat'iana Netreba, "Vladimir Nak: Posles-

lovie k ispovedi," *Rossiiskaia gazeta*, February 24, 2010. But the project was dogged by scandal right up to the end, including accusations that the Naks had been displaced by interests associated with the Miller management and that the costs of the railroad had run far above budget. See Roman Shleinov, "Kak obschitali 'Gazprom,'" *Vedomosti*, August 26, 2013, p. 20, https://www.vedomosti.ru/newspaper/articles/2013/08/26/kak-obschitali-gazprom; and "Rodstvenniki-podriadchiki," *RBK Daily*, February 27, 2015, p. 13.

39. Thane Gustafson, Matthew J. Sagers, and Sergej Mahnovski, *Conquering Yamal: Gazprom's Strategy for Developing the Next Generation of Russian Gas Supply* (IHS Markit Private Report, September 2007).
40. The first pipe for the pipeline from Bovanenkovo to Ukhta was delivered in time for the spring construction season of 2008. This was a point of considerable pride, because unlike the earlier generations of Soviet and Russian pipelines, this one was built entirely with Russian-made large-diameter pipe. For details, see Sergei Smirnov, "'Gazprom' poluchil pervye truby dlia 'Yamala'," *Vremia novostei*, April 22, 2008, http://www.vremya.ru/2008/69/4/202360.html.
41. James Henderson, "Non-Gazprom Russian Producers: Finally Becoming Truly Independent?," in James Henderson and Simon Pirani, eds., *The Russian Gas Matrix: How Markets Are Driving Change* (Oxford: Oxford University Press, 2014), p. 314.
42. With respect to the smaller fields, see Matthew J. Sagers and Thane Gustafson, *Small Is Big: The Last of Russia's Cheap Natural Gas* (IHS Markit Private Report, July 2005). The story of Gazprom's turn to the independents is told in more detail in Henderson, "Non-Gazprom Russian Producers," pp. 314–346. The story is brought up to date through 2016 in James Henderson and Arild Moe, "Russia's Gas 'Triopoly': Implications of a Changing Gas Sector Structure," *Eurasian Geography and Economics* 58, no. 4 (2017): 442–468.
43. In this regard Gazprom's behavior was no different from that of its Western European counterparts in Great Britain, Germany, and Italy, as these faced the first legislative demands to open access to their pipeline systems.
44. For background on the evolution of the domestic Russian gas market during the 2000s, see Philip Vorobyov, *Waking Giant: The Russian Domestic Gas Market* (IHS Markit Decision Brief, August 2004); Vitaly Yermakov, *Russia's "New Deal" in the Domestic Gas Market: The End of Cheap Gas* (IHS Markit Decision Brief, January 2008); and Vitaly Yermakov, Matthew J. Sagers, and Tatiana A. Mitrova, *Near-Term and Long-Term Outlooks for Russian Gas Consumption: Did 2006 Launch a New Trend?* (IHS Markit Private Report, June 2007).
45. See Aleksey Grivach, "Nastoiashchaia nezavisimost'," *Vremia novostei*, December 7, 2005, http://www.vremya.ru/2005/228/8/140695.html. For the rapidly evolving picture two years later, see Mikhail Krutikhin, "Davidy i Goliaf: Razmer uchastiia nezavisimykh kompanii v balanse gaza v Rossii stanovitsia kriticheskim," *Rusenergy*, October 15, 2008. And for the scene two years after that, see "Zavisimost' ot nezavisimykh: 'Gazpromu' prorochat sokrashchenie doli v dobyche gaza," *Rusenergy*, October 19, 2010.

46. Krutikhin, "Davidy i Goliaf." The reason for this is that the domestic market is not uniform. The independents were making money by supplying baseload volumes to large, reliable customers such as steel mills, fertilizer plants, and power stations. Gazprom was losing money by selling gas at subsidized prices to the population, serving regions with weak payment discipline and having to accommodate the seasonal swing that was becoming higher and higher with independents taking away steady customers.
47. "Zavisimost' ot nezavisimykh."
48. Quoted in Irina Reznik, "INTERV'IU: Leonid Mikhelson, gendirektor i sovladelets kompanii 'NOVATEK,'" *Vedomosti*, November 2, 2005, https://www.vedomosti.ru/newspaper/articles/2005/11/02/intervyu-leonid-mihelson-gendirektor-i-sovladelec-kompanii-novatjek.
49. This was followed by several years of rising domestic prices, as Russia adopted the goal of bringing domestic prices up to the level of export netbacks. This created a favorable price environment for the independents.
50. Troika Dialog Research, *Novatek: Gas-Fired Growth,* December 2004.
51. Reznik, "INTERV'IU: Leonid Mikhelson."
52. For example, when Novatek opened a new treatment plant for condensate, it welcomed gas from Gazprom. See Oleg Smirnov, "Partnery: 'Gazprom' + 'Novatek,'" *Rossiia*, July 14, 2005.
53. Ivan Igor-Tismenko, "Ne upuskaia moment: Rossiiskii kondensat vospolnit spros na vysokokachestvennuiu neft' v Evrope," *Rusenergy,* June 30, 2005.
54. "Russia: Total Enters into a Strategic Partnership with the Independent Gas Company Novatek," March 2, 2011, https://www.total.com/en/media/news/press-releases/russie-total-sengage-dans-un-partenariat-strategique-avec-lindependant-gazier-novatek. Novatek had been looking for an international partner for some time, without success. With Total there had been a long courtship: as noted in the text, the French company had come close to acquiring a stake in Novatek as early as 2004. However, Total withdrew its bid to acquire a stake in Novatek after the Anti-Monopoly Service failed to approve the deal (See https://sputniknews.com/analysis/2005080541096318/) and Novatek conducted an initial public offering in London instead. The focus of the Total deal at that time was condensate rather than LNG. In contrast, the South Tambey LNG project (discussed below in the text) was still widely believed to be uneconomic. See "Ledovaia obstanovka: Dlia iamal'skogo proekta SPG razrabatyvaiut kontseptsiiu," *Rusenergy*, May 13, 2010, and "Bystro ne poluchitsia: Proektu 'Yamal SPG' ne khvataet gazovykh resursov i inostrannykh partnerov," *Rusenergy*, September 29, 2011.
55. Dena Sholk, Mikhail Kuznetsov, and Matthew J. Sagers, *NOVATEK's Yamal LNG: Russian "Mega" Project Remains on Track Despite Challenges* (IHS Markit Strategic Report, January 2017).
56. See Simon Blakey and Thane Gustafson, *Securing the Future: Making Russian-European Gas Interdependence Work* (IHS Markit Special Report, September 2007).
57. Aleksey Grivach, "Yuriy Komarov: My davno gotovimsia k liberalizatsii rynka v Evrope," *Vremia novostei,* January 30, 2004, p. 6. At this point, Komarov was still

the head of Gazeksport and a deputy chairman, although he was soon to be replaced in both positions by Medvedev. As the former cohead of Wingas, he was at that time the senior Gazprom official with the longest sustained experience of the liberalization of the European gas market. As he states in his interview with Grivach, one of the objectives of Gazprom's expanded presence in Great Britain downstream was to prepare the way for the "North European Gas Pipeline," which Gazprom at the time envisioned as terminating in Great Britain. (In truncated form, this pipeline eventually became Nord-Stream-1 and terminated in Germany instead.)

58. I remember visiting GM&T in the mid-2000s, when its headquarters was on the Thames, and being told that there were only three Russians in the on-site management.

59. Aleksey Grivach, "'Gazprom' priobrel 600 klientov v Velikobritanii," *Vremia novostei,* June 23, 2006, p. 8, http://www.vremya.ru/2006/108/8/154910.html. Gazprom's new client base was acquired from Pennine Natural Gas. See James Wilson, "Pennine Puts Backbone into Gazprom," *Financial Times,* June 23, 2006.

60. In early 2018 Vasilev left Gazprom, in the wake of a restructuring of its export operations that reportedly involved a plan to move GM&T headquarters to Saint Petersburg.

61. "Gazprom Pursues UK but Brown Issues Warning," *Financial Times,* June 7, 2006.

62. Christopher Adams, "Russian Energy Chief Seeks to Melt MPs' Hearts on Gas Supplies," *Financial Times,* July 19, 2006.

63. For a sober summary of the successes and failures of Gazprom's international strategy as of late 2007–early 2008, see Aleksey Grivach and Andrei Denisov, "Uspekhi i neudachi 'energeticheskoi sverkhderzhavy,'" *Rossiia v global'noi politike* 6, no. 2 (2008): 101–112.

64. For background on the competition between LNG and Russian pipeline gas in European markets, see Shankari Srinivasan and Michael Stoppard, *Testing the Boundaries for Gas to Europe: Russian Pipeline versus LNG* (IHS Markit Private Report, July 2007).

65. Joseph T. Kosnik, *Natural Gas Imports from the Soviet Union: Financing the North Star Joint Venture Project* (New York: Praeger, 1975). On the US gas bubble of the late 1970s, see Daniel Yergin, *The Quest: Energy, Security, and the Remaking of the Modern World* (New York: Penguin, 2011), chapter 16.

66. Tat'iana Golubovich and Il'ia Khrennikov, "'Gazprom' otkroet Ameriku," *Vedomosti,* March 11, 2003. See also Aleksey Grivach, "Otchet o komandirovke: Poezdka v Tokio tolknuka 'Gazprom' na mirovoi rynok szhizhennogo gaza," *Vremia novostei,* June 16, 2003, p. 8. During this period Miller was conducting exploratory talks with Conoco Phillips's CEO, James Mulva, about a possible LNG project for export to the United States. (Lou Noto of Mobil had also discussed it with Gazprom in the 1990s.)

67. Mikhail Krutikhin, "V ozhidanii gazovoi ataki: Desheveiiushchii szhizhennyi gaz ugrozhaet truboprododnym ambitsiiam 'Gazproma,'" *Rusenergy,* June 25, 2003.

68. The saga of production-sharing agreements in general and the fortunes of the one in Sakhalin in particular are recounted in Gustafson, *Wheel of Fortune,* chapters 4 and 11.
69. This episode is recounted in Gustafson, *Wheel of Fortune*, pp. 397–399.
70. Krutikhin, "V ozhidanii gazovoi ataki." The company was also divided internally over the priority to be given to dry versus wet gas. Ananenkov and the old guard favored dry Cenomanian gas, whereas Ryazanov, being responsible for liquids, was a strong backer of developing liquids from Achimov gas.
71. On the long and (so far) failed odyssey of Shtokman, see Thane Gustafson, *Off Again, On Again . . . The Ever-Surprising Saga of Shtokman* (IHS Markit Decision Brief, November 2007).
72. For background on the US shale gale, its origins, and its significance for the United States and the world, see Yergin, *The Quest,* chapter 16. The first news of the rise of shale gas in the United States began to reach the US gas community in 2007–2008. The early impact of the "shale gale" is analyzed in Robert Ineson et al., *Fueling North America's Energy Future: The Unconventional Natural Gas Revolution and the Carbon Agenda* (IHS Markit Special Report, January 2010), https://assets.publishing.service.gov.uk/government/uploads/system/uploads/attachment_data/file/43227/1296-ihs-cera-special-report.pdf.
73. The name of Gazeksport was changed to GazpromEksport in 2006.
74. Aleksey Grivach, "Shtokman zatiagivaetsia," *Vremia novostei*, October 9, 2009, p. 8, http://www.vremya.ru/2009/186/8/239283.html.
75. Alexander Vertiachikh, "Indeks nedeli," *Sankt-Peterburgskie vedomosti*, February 24, 2010, p. 3.
76. For examples of the Russian media coverage of the shale gale and its implications for Russian LNG policy and exports to Europe, see Ol'ga Khvostunova, "Slantsevyi klondaik," *The New Times,* March 11, 2013, pp. 40–43, https://newtimes.ru/articles/detail/63966. The earliest reference I have been able to find to the prospects for competition between US LNG from shale gas and Russian pipeline gas for the European market is in Oleg Nikiforov, "Bitva za Evropu," *Nezavisimaia gazeta,* May 31, 2010, p. 9. In the more specialized media, such as *Interfax Petroleum Report,* references to shale gas begin to appear in late 2009, but the frequency increases slowly at first, peaking in 2012 and declining thereafter—as though shale gas was no longer news.
77. My notes from the Saint Petersburg conference, June 2010.
78. The full name of the commission is Commission for Strategic Development of the Fuel and Energy Sector and Environmental Security.
79. See the transcript of the meeting of the commission, available at the Presidential website: "Zasedanie Komissii po voprosam strategii razvitiia TEK i ekologicheskoi bezopasnosti," October 23, 2012, http://kremlin.ru/events/president/news/16702. At that meeting Sechin was sharply critical of Gazprom's export strategy.
80. Samuel J. Andrus, *Shale Gas Reloaded: The Evolving View of North American Natural Gas Resources and Costs* (IHS Markit Strategic Report, February 2016). For a recent update, see Daniel Yergin and Samuel Andrus, *The Shale Gale*

Turns 10: A Powerful Wind at America's Back (IHS Markit Strategic Report, July 2018).

81. Olga Tanas, "Three Things Keeping Gazprom Managers Awake at Night," *Bloomberg News,* February 28, 2019, https://www.bloomberg.com/news/articles/2019-02-28/three-things-keeping-gazprom-managers-awake-at-night.
82. On the history of the "black swan" metaphor, see "Black Swan Theory," https://en.wikipedia.org/wiki/Black_swan_theory.
83. Jonathan Stern and Howard Rogers, "The Transition to Hub-Based Gas Pricing in Continental Europe" (Oxford Institute for Energy Studies, March 2011), https://www.oxfordenergy.org/wpcms/wp-content/uploads/2011/03/NG49.pdf.
84. Mitrova and Boersma, *The Impact of US LNG on Russian Natural Gas Export Policy,* p. 17. See also Tatiana Mitrova, Vyacheslav Kulagin, and Anna Galkina, "The Transformation of Russia's Gas Export Policy in Europe," *Proceedings of the Institution of Civil Engineers—Energy* 168, no. 1 (2015): 30–40.
85. Mitrova and Boersma, *The Impact of US LNG on Russian Natural Gas Export Policy,* p. 16.
86. Production finally began in late 2012, after several postponements. "Uroki Bovanenkovo," There were also major infrastructural challenges to overcome, chiefly the unstable nature of permafrost. Gazprom's engineers call the development of Bovanenkovo as a process of trial and error.
87. See Henderson and Moe, "Russia's Gas 'Triopoly.'"
88. Thane Gustafson, Vitaly Yermakov, and Nicholas Naroditski, *Pivot to the East: Russia's New Emerging LNG Strategy* (IHS Markit Private Report, December 2013), p. 6.
89. The term "independent" is used somewhat loosely in Russia. In practice, it means any gas producer other than Gazprom. Thus Rosneft, even though it is a large state-owned company, is called an "independent."
90. Matthew Sagers, Dena Sholk, Anna Galtsova, and Thane Gustafson, *Russia's New LNG Strategy: Breaking the Ice* (IHS Markit Strategic Report, April 2018).
91. "Perechen' poruchenii po itogam soveshchaniia o razvitii proektov proizvodstva szhizhennogo prirodnogo gaza," December 25, 2017, http://kremlin.ru/acts/assignments/orders/56501.
92. Yurii Barsukov, "'Gazprom' vidit v SNG istochnik poter' biudzheta," *Kommersant,* April 8 2019. Novatek pays value-added tax (VAT), but it is exempt from export tax and Mineral Extraction Tax (MET).
93. Statement by Leonid Mikhel'son, "'Novatek' ne vidit konkurentsii svoego SPG s gazom 'Gazprom,'" reported by RIA Novosti, April 9 2019.
94. "Putin Orders Energy Ministry to Study Idea of Arctic National Project," *Interfax Russia and CIS Oil and Gas Weekly,* no. 9 (February 28 2019), p. 7.
95. Miller included in his statement the prospect of pipeline exports from the Yamal to China, but this is unrealistic unless Gazprom builds the so-called Altay pipeline to western China. Gazprom has promoted this project for years, but the Chinese have shown no interest. Yet if there is no pipeline outlet from Yamal to Asia, Europe alone would not justify dedicating the Yamal Peninsula to pipeline gas alone.

96. Mitrova and Boersma, *The Impact of US LNG on Russian Natural Gas Export Policy*, p. 28.
97. Simon Blakey, Alun Davies, Laurent Ruseckas, Shankari Srinivasan, and Michael Stoppard, *European Natural Gas—The New Configuration* (IHS Markit Strategic Report, April 27, 2018), p. 1.
98. Simon Blakey, Alun Davies, and Shankari Srinivasan, *The Future of Long-Term Contracts and the European Midstream* (IHS Markit Strategic Report, January 2016).

11. Russia and Ukraine

1. I am grateful to Philip Vorobyov for his patient reading of drafts and his wise advice throughout the writing of this chapter, and indeed of the rest of the book. I also wish to thank Simon Pirani of the Oxford Institute of Energy Studies and Laurent Ruseckas of IHS Markit for their insightful comments on drafts of this chapter.
2. See Thane Gustafson and Anna Galtsova, *The Changing Future of the Gas Problem in Russian-Ukrainian Relations* (IHS Markit Private Report, November 2014).
3. In the concluding chapter I will argue that there could be a revival of Russian transit through Ukraine in the second half of the 2020s or early 2030s, if European demand for Russian gas continues to increase and exceeds the capacity of the bypass pipelines. But by that time any Russian transit through Ukraine will hopefully take place under European Union (EU) rules, which will create a quite different situation.
4. This was the case at least in the western third of the country. The eastern two-thirds of the Soviet Union were unconnected to the rest of the gas network and remain so today.
5. The third interruption, which lasted from mid-June through the end of October 2014, was initiated by Ukraine. For background on the gas trade implications of the 2014 crisis, see Laurent Ruseckas, Matthew J. Sagers, Shankari Srinivasan, Michael Stoppard, Thane Gustafson, and Daniel Yergin, *Ukraine Crisis: What It Means for Europe's Gas Supply* (IHS Markit Special Report, March 2014). For data on imports and transit under the 2009 contracts, see Simon Pirani, *After the Gazprom-Naftogaz Arbitration: Commerce Still Entangled in Politics* (Oxford: Oxford Institute of Energy Studies, March 2018), p. 3.
6. The involvement of leading politicians and oligarchs has been a source of fascination for the media in both Russia and Ukraine, and there is a huge literature of *gazovyi kompromat* (i.e., supposedly compromising material related to gas), none of which can be trusted but which in the aggregate makes a not implausible picture. New angles keep surfacing. For a recent example, see Mariia Zholobova and Roman Badanin, "Peterburgskie znakomye Putina okazalis' benefitsiarami ukrainskogo gazovogo transita v 2000x.," *Dozhd'*, April 19, 2018, https://tvrain.ru/articles/peterburgskie_znakomye_putina_okazalis_benefitsiarami_ukrainskogo_gazovogo_tranzita_v_2000_h-462135/?utm_source=facebook&utm_medium=social&utm_campaign=news&utm_term=462135.

7. One could make a good argument that arms and metals have been equally important sources of corrupt rents in Russian-Ukrainian relations, and perhaps coal as well. But in the 1990s these were essentially part of the same complex chain of barter. As an example, in the late 1990s and early 2000s Yulia Tymoshenko was the subject of an Interpol arrest warrant originated by Moscow, because she had failed to supply the uniforms and other materiel to the Russian Ministry of Defense specified in a contract in exchange for gas. (Needless to say, this existing warrant was an awkward element in the conversation when Tymoshenko made her first visit to Moscow as prime minister in 2005.)
8. On the structure of the Ukrainian gas market and the potential for Ukrainian indigenous production, see Philip Vorobyov, "Ukraine at the Crossroads," *Petroleum Economist,* November 2014, pp. 11–15, and "Lilliputians in the Land of Giants," *Petroleum Economist,* June 2013, pp. 4–8.
9. We come back to this point later in the chapter. There is, of course, no way to tag the molecules that are coming into Ukraine from the West. The key point is that once they have been purchased by European buyers, under EU law Russia no longer has any legal means of limiting where they are resold.
10. This does not necessarily mean that Russian gas will no longer transit through Ukraine—that remains to be seen—but that the leverage associated with a high Ukrainian share of transit, together with Russia's initial lack of alternatives, will no longer have the same strategic implications or the same poisonous effect on the overall relationship.
11. For one of the best analyses of Ukrainian politics, though alas a decade out of date, see Paul D'Anieri, *Understanding Ukrainian Politics: Power, Politics, and Institutional Design* (Armonk, NY: M. E. Sharpe, 2007). See also Paul D'Anieri, ed., *Orange Revolution and Aftermath: Mobilization, Apathy, and the State in Ukraine* (Washington, DC: Woodrow Wilson Center Press, 2010). One could argue that this very weakness makes Ukrainian politics more competitive and prevents a Putin-style recentralization of control.
12. For a description of the barter economy in Russia as it was up to the end of the 1990s, see Clifford Gaddy and Barry Ickes, *Russia's Virtual Economy* (Washington, DC: Brookings Institution Press, 2002). In Ukraine the barter system worked in essentially the same way as in Russia, and for the same fundamental reasons—namely, the disruption of established commercial and legal relations and the general lack of credible currencies.
13. There are broadly two schools of thought about Ukrainian politics. One focuses on the weaknesses of Ukrainian political institutions, as seen particularly in Paul d'Anieri (see note 11 above). The other points to a systematic failure of leadership at the presidential level, leading to repetitive patterns of collapse. For excellent essays, see Lucan Way, *Pluralism by Default: Weak Autocrats and the Rise of Competitive Politics* (Baltimore: Johns Hopkins Press, 2015), pp. 43–91; and Serhiy Kudelia and Taras Kuzio, "Nothing Personal: Explaining the Rise and Decline of Political Machines in Ukraine," *Post-Soviet Affairs* 31, no. 3 (2015): 250–278. The answer is clearly that both are to blame.

14. Dnepropetrovsk ("Peter's city on the Dnieper River") is an industrial town in east central Ukraine, and historically it has been the fief of some of the most important players in the Ukrainian elite, both in Soviet times and since—including Brezhnev. See Orysia Kulick, "When Ukraine Ruled Russia: Regionalism and Nomenklatura Politics after Stalin, 1944–1990" (PhD diss., Stanford University, 2017). In 2016 Dnepropetrovsk was renamed Dnipro, but I have kept the earlier name in this book.
15. Jonathan Stern, *The Russian Natural Gas "Bubble": Consequences for European Gas Markets* (London: Royal Institute of International Affairs, 1995).
16. An important source for information about the 1990s and the 2000s is the work of Margarita Balmaceda, who has thoroughly researched the energy trade of the eastern countries of the Former Soviet Union. For a reconstruction of the Russian-Ukrainian gas trade in this period, see her *Energy Dependency, Politics and Corruption in the Former Soviet Union: Russia's Power, Oligarchs' Profits, and Ukraine's Missing Energy Policy, 1995–2006* (London: Routledge, 2008), and *The Politics of Energy Dependency: Ukraine, Belarus, and Lithuania between Domestic Oligarchs and Russian Pressure* (Toronto: University of Toronto Press, 2013), chapter 4.
17. Jonathan Stern, *The Future of Russian Gas and Gazprom* (Oxford: Oxford University Press, 2005), p. 88.
18. For a helpful overview of the triangular period of the Former Soviet Union gas trade, see ibid., pp. 66–108.
19. As of the spring of 2019, Turkmenistan has resumed exporting gas to Russia on a limited basis.
20. This was the second Ukrainian bypass pipeline to Europe. The first, Yamal-Europe (or Europol), began operation in 1997, as sections of it were constructed back from the German-Polish border (in contrast to previous Soviet-era export pipelines, which were built forward from the producing fields). Planning for the Yamal-Europe pipeline started in 1992. Intergovernmental agreements between Russia, Belarus, and Poland were signed in 1993. In 1994, Wingas, the joint venture of Gazprom and Wintershall, a subsidiary of BASF, started building the German section of the pipeline. The first gas was delivered to Germany through the Belarus-Polish corridor in 1997. The Belarus and Polish sections were completed in September 1999, and the pipeline reached its rated annual capacity of about 33 billion cubic meters of natural gas in 2005, after completion of all compressor stations. See https://en.wikipedia.org/wiki/Yamal%E2%80%93Europe_pipeline.
21. For more on Lazarenko's career, see the biography by Sergei Leshchenko, *Amerikans'ka saga Pavla Lazarenka,* serialized in *Ukrains'ka Pravda* in Ukrainian in 2012, https://www.pravda.com.ua/articles/2012/09/13/6972637/.
22. It is impossible to do justice to the successive stages of Tymoshenko's extraordinary career, with all their colorful twists and turns, in a few brief paragraphs. The biographical material in this section is drawn from Dmitrii Popov and Il'ia Mil'shtein, *Oranzhevaia printsessa: Zagadka Iulii Timoshenko* (Moscow: Izdatel'stvo Ol'gi Morozovoi, 2006).

23. For a lengthy report on Itera and Igor Makarov, see Jeanne Whalen, "Pipe Dream: How Gas Firm Itera Got So Huge, So Fast," *Wall Street Journal,* October 24, 2000.
24. For a Russian account of the rise of the barter trade and the origins of Itera's role, see Valerii Paniushkin and Mikhail Zygar', *Gazprom: Novoe russkoe oruzhie* (Moscow: "Zakharov," 2008), pp. 148–156.
25. Something of the high color of Makarov's beginnings and subsequent rise and fall can be found in an article by Irina Mokrousova, "Miller dlia nego byl nikto," *Vedomosti,* May 20, 2013.
26. There is considerable debate among scholars over whether the Orange "Revolution" was really a revolution, or simply a popular revolt. In retrospect, it is more the latter, but throughout this chapter I have used the term without quotes since that is the established usage.
27. See Thane Gustafson, *Capitalism Russian-Style* (Cambridge: Cambridge University Press, 1999), especially chapter 2.
28. Pirani, *After the Gazprom-Naftogaz Arbitration,* p. 11.
29. International Energy Agency, *Ukraine Energy Policy Review* (Paris: International Energy Agency, 2006), pp. 75–77.
30. Pirani, *After the Gazprom-Naftogaz Arbitration,* p. 13.
31. Stern, *The Future of Russian Gas and Gazprom,* p. 90. Stern provides details of the agreements reached by the two presidents and their prime ministers on the eve of the Orange Revolution.
32. Thane Gustafson and Matthew J. Sagers, *Gas Transit through Ukraine: The Struggle for the Crown Jewels* (IHS Markit Private Report, April 2003).
33. Was the Orange Revolution a revolution as we conventionally understand it, or a forerunner of a new political phenomenon, on the model of internet revolutions in the twenty-first century? For an exposition of the latter view, which takes us far beyond the scope of this book, see Mark R. Beissinger, "The Semblance of Democratic Revolution: Coalitions in Ukraine's Orange Revolution," *American Political Science Review* 107, no. 3 (August 2013): 574–592. For an opposite view, see in particular Anders Åslund and Michael McFaul, *Revolution in Orange: The Origins of Ukraine's Democratic Breakthrough* (Washington, DC: Carnegie Endowment for International Peace, 2006), and Anders Åslund, *How Ukraine Became a Market Economy and Democracy* (Washington, DC: Peterson Institute for International Economics, 2009).
34. The standing joke in Moscow at the time went like this: "Can there be a color revolution in Moscow? No—they've run out of colors."
35. For a discussion of the Munich speech, see Angela Stent, *The Limits of Partnership: U.S.-Russian Relations in the Twenty-First Century* (Princeton, NJ: Princeton University Press, 2014), chapter 6, and *Putin's World: Russia against the West and with the Rest* (New York: Twelve, 2019).
36. On the evolution of the intra-CIS gas trade in the 2000s, see Simon Pirani and Katja Yafimava, "CIS Gas Markets and Transit," in James Henderson and Simon Pirani, eds. *The Russian Gas Matrix: How Markets Are Driving Change* (Oxford: Oxford University Press, 2014), pp. 181–216.
37. See ibid., table 7.2, p. 186.

38. See Simon Pirani, "Russo-Ukrainian Gas Wars and the Call on Transit Governance," in Caroline Kuzemko et al., eds., *Dynamics of Energy Governance in Europe and Russia* (Houndmills, UK: Palgrave Macmillan, 2012), pp. 169–186.
39. Reportedly he died of a heart attack, although there are darker explanations.
40. Jonathan Stern, *The Russian-Ukrainian Gas Crisis of January 2006,* Oxford Institute of Energy Studies, https://www.oxfordenergy.org/wpcms/wp-content/uploads/2011/01/Jan2006-RussiaUkraineGasCrisis-JonathanStern.pdf (especially pp. 3–6).
41. See ibid.
42. These numbers are not as large as they may sound. Loss of a quarter of daily supplies is not very meaningful in midwinter, when gas is being drawn from storage. It probably has hardly more effect than changing the linepack (i.e., the gas already contained in the pipeline). Thus it is difficult to sustain the view that the effect was severe—though it was powerfully symbolic.
43. Stern, *The Russian-Ukrainian Gas Crisis of January 2006,* p. 8. The gas issues as they stood in the aftermath of the 2006 dispute are analyzed in Christine Telyan and Thane Gustafson, *Russia and Ukraine's New Gas Agreement: What Does It Mean and How Long Will It Last?* (IHS Markit Decision Brief, January 2006); and Christine Telyan and Matthew Sagers, *Energy and the Ukrainian Economy: Obstacles and Opportunities Ahead* (IHS Markit Private Report, October 2006).
44. Gazprom has consistently used the word "theft" to condemn such offtakes.
45. For a detailed analysis of the 2009 dispute, see Simon Pirani, Jonathan Stern, and Katja Yafimava, "The Russo-Ukrainian Gas Dispute of January 2009: A Comprehensive Assessment" (Oxford Institute for Energy Studies, NY-27, February 2009), https://www.oxfordenergy.org/wpcms/wp-content/uploads/2010/11/NG27-TheRussoUkrainianGasDisputeofJanuary2009AComprehensiveAssessment-JonathanSternSimonPiraniKatjaYafimava-2009.pdf.
46. Prices were known to be about to decline in the next quarter, as the time lags in their indexation formulas kicked in to reflect the late 2008 collapse in world oil prices. It was in their commercial interest to lift less Russian gas in January, draw down storage rapidly, and refill their storages with cheaper gas in April. I am indebted to Simon Blakey for these important points.
47. Deepening the mystery of Putin's actions is the fact that in the days leading up to the cutoff the CEO and deputy CEO of Ruhrgas had spent a lot of time in Moscow and Kiev, respectively. They knew what to expect from Russia and were trying to persuade Ukraine to pay. On the 2009 cutoff and its aftermath, see Simon Blakey and Thane Gustafson, *Russian-Ukrainian Gas: Why It's Different This Time* (IHS Markit Decision Brief, January 2009), and *Lessons for Europe of the Russian-Ukrainian Gas Crisis* (IHS Markit Decision Brief, February 2009); Thane Gustafson and Simon Blakey, *It's Not Over Till It's Over: The Russian-Ukrainian Gas Crisis in Perspective* (IHS Markit Decision Brief, January 2009).
48. For details on the rumors of Firtash's possible early ties to organized crime, see Glenn R. Simpson, "U.S. Probes Possible Crime Links to Russian Natural-Gas Deals," *Wall Street Journal,* December 22, 2006. According to material that appeared in WikiLeaks from an interview with US ambassador William Taylor,

Firtash acknowledged early contacts with the crime boss Semen Mogilevich but justified them on the ground that "in the 1990s one needed Mogilevich's support to do business" (ibid.). See also "Arest Firtasha: Novyi povorot v gazovoi voine," Forbes.ru, July 7, 2011, https://www.forbes.ru/mneniya-column/konkurentsiya/252012-arest-firtasha-novyi-povorot-v-gazovoi-voine.

49. For an extensive if not necessarily reliable report on Firtash and his relations with Russian interests, see Stephen Grey, Tom Bergin, Sevgil Musaieva, and Roman Anin, "Putin's Allies Channelled Billions to Ukraine Oligarch," Reuters, November 26, 2014, https://www.reuters.com/article/russia-capitalism-gas-special-report-pix/special-report-putins-allies-channelled-billions-to-ukraine-oligarch-idUSL3N0TF4QD20141126.

50. This account of Firtash's early years is drawn largely from Andrei Krasavin, "V nuzhnoe vremia," LentaCom.ru, December 11, 2013, http://www.lentacom.ru/print/news/22043.html.

51. RUE was created in 2004 and obtained a franchise to import gas from Central Asia to Ukraine. It was financed at the beginning by a loan from Gazprombank. See Mariia Rozhkova and Irina Reznik, "Truboukladchik Firtash," *Vedomosti,* May 30, 2006. By 2006 RUE had a franchise to import all Central Asian gas through Gazprom's pipeline system, plus up to 17 billion cubic meters of surplus gas for subsequent sale in Europe. Its capital value at the time was estimated at over $7 billion. It had contracts to export gas to Poland, Slovakia, and Hungary. See Irina Reznik, "Gazoobraznoe sostoianie," *Vedomosti,* May 2, 2006.

52. Opinions differ on the extent to which Firtash transferred his allegiance to Yushchenko. According to Simon Pirani, "The 2006 deal, which strengthened Rosukrenergo's position as an intermediary, was signed under Yushchenko's nose and he couldn't do anything about that. Firtash was always aligned with the Party of Regions which was the largest party in the Ukrainian parliament throughout Yushchenko's presidency. [A good indicator] is the career of Yuri Boiko, Firtash's closest political ally, who held numerous important positions under Yushchenko's presidency, rising to the level of deputy prime minister under Yanukovich" (Simon Pirani, personal communication, June 1, 2018).

53. Strictly speaking, Gazprom provided Firtash with access to Turkmen gas, transit services for it through Russia to Ukraine, and access to the gray scheme under which he sold it in Central European countries—all very valuable, but indirect, forms of support.

54. Elena Mazneva and Vasilii Kashin, "Vtoraia gazovaia," *Vedomosti*, January 11, 2009 (interview conducted by Irina Reznik).

55. By 2008 Central Asia had largely ceased to export gas to Ukraine. The following year, Firtash lost control of Emfesz, which had become essentially worthless. See Irina Reznik and Viktoriia Sunkina, "Firtash proigryvaet," *Vedomosti,* August 5, 2009.

56. An investigation by Reuters claims on the basis of a review of documents in Moscow and Cyprus that the Kremlin arranged for a new line of credit of $11 billion in exchange for Firtash's support for Yanukovych. These claims cannot be

verified, but Firtash's support for Yanukovych is not in doubt. See "Reuters: Firtash kupil svoi aktivy na kredity ot Putina," November 26, 2014, https://glavcom.ua/news/202182-firtash-kupil-svoi-aktivy-na-dengi-putina---reuters.html.

57. On the impact of the undeclared war in the eastern provinces on Firtash's assets, see RBK-Ukraina, October 13, 2016, https://daily.rbc.ua/rus/show/dmitriy-firtash-segodnya-skazat-neobhodim-1476301489.html. This is the second of two interviews, the first of which was RBK-Ukraina, October 12, 2016, https://daily.rbc.ua/rus/show/dmitriy-firtash-menya-printsipialno-dokazat-1476224423.html.
58. David Herszenhorn, "Even if Cease-Fire Holds, Money Woes Will Test Kiev," *New York Times,* February 13, 2015, https://www.nytimes.com/2015/02/14/world/europe/even-with-cease-fire-economy-in-ukraine-is-crumbling.html.
59. Roman Kazmin, "Details of Gazprom's Contracts with NAK Emerge," ICIS Heren, January 23, 2009, https://www.icis.com/resources/news/2009/01/23/9309290/details-of-gazprom-s-contracts-with-nak-emerge/.
60. See Auyezov, "Ukraine-Russia Gas Deal."
61. See, in particular, Simon Pirani, "Adversity and Reform: Ukrainian Gas Market Prospects" (Oxford Institute for Energy Studies, Energy Insight No. 7, March 2017), https://www.oxfordenergy.org/wpcms/wp-content/uploads/2017/03/Adversity-and-reform-Ukrainian-gas-market-prospects-OIES-Energy-Insight.pdf.
62. See Anna Galtsova and Thane Gustafson, *Gas Market Reform in Ukraine: Moving into High Gear or Barely Moving?* (IHS Markit Strategic Report, July 2016.)
63. For an overview of the reform process and its achievements through early 2017, see Pirani, "Adversity and Reform.".
64. Yuriy Vitrenko, "Naftogaz of Ukraine: What Are We Fighting For?," *Politico,* March 14, 2019, https://www.politico.eu/sponsored-content/naftogaz-of-ukraine-what-are-we-fighting-for/.
65. Fabrice Deprez, "Interview: Naftogaz's Fragile Success," BNE Intellinews, March 6, 2018, https://www.intellinews.com/interview-naftogaz-s-fragile-success-137808/.
66. This involved resurrecting a disused pipeline via the interconnection point at Budince, since all the capacity at the other interconnection point, Velke Kapusany, was booked (although not all used) by Gazprom for gas going east to west.
67. It is interesting in retrospect to note that these initial contracts were negotiated under Yanukovych—the 2014 contract was concluded literally as his government was collapsing. Typically, Yanukovych was placing his bets in all directions.
68. European Commission, *Quarterly Report on European Gas Markets,* vol. 9, issue 1 (fourth quarter 2015 and first quarter 2016), available in the reports archive of the following section of the European Commission's website, https://ec.europa.eu/energy/en/data-analysis/market-analysis.
69. European Commission, *Quarterly Report on European Gas Markets,* vol. 10, issue 4 (fourth quarter 2017), https://ec.europa.eu/energy/sites/ener/files/documents/quarterly_report_on_european_gas_markets_q4_2017_final_20180323.pdf.

70. See Alla Eremenko and Leonid Unigovskii, "Tranzit i revers: Skovannye odnoi tsep'iu?," *Zerkalo nedeli,* December 23, 2016, https://zn.ua/energy_market/tranzit-i-revers-skovannye-odnoy-cepyu-.html.
71. Anne-Sophie Corbeau, Shankari Srinivasan, and Simon Blakey, *Legal Unbundling: Disassembling the European Gas Puzzle* (IHS Markit Decision Brief, April 2004).
72. Naftogaz was the sole counterparty in the 2009 transit agreement with Gazprom, and Gazprom reportedly refused to allow any assignment of this contract to a new entity. To strengthen his direct control of Ukrtransgaz, Kobolev fired the manager and purged the leadership of the organization. See Dmitri Riasnoi, "'Naftogaz' smenil rukovodstvo 'Ukrtransgaza,'" *Ekonomicheskaia pravda,* April 16, 2018, https://www.epravda.com.ua/rus/news/2018/04/16/636055/.
73. In 2017 Naftogaz's pretax profits from transit in 2017 outweighed its losses from gas sales by five to one. See Naftogaz financial statements for the year ended December 31, 2017, http://www.naftogaz.com/files/Zvity/2017%20ENG%20Naftogaz%20stand%20alone%20FS.pdf.
74. Laurent Ruseckas, *Future Transit of Russian Gas through Ukraine: Risks May Be Larger Than They Appear* (IHS Markit Strategic Report, November 2018).
75. See Vorobyov, "Lilliputians in the Land of Giants."
76. Anna Galtsova, Thane Gustafson, and Matthew Sagers, *Ukraine's Gas Production on the Rise in 2017: Is the Goal of Energy Independence within Reach?* (IHS Markit Insight, November 2017). See also Vorobyov, "Lilliputians." Ukraine's official gas production program, "Energy Strategy to 2035," projects 27.6 billion cubic meters in 2020, compared to 19.6 billion cubic meters in 2016.
77. Vorobyov, "Ukraine at the Crossroads."
78. See "Stockholm Arbitration Orders Gazprom to Pay Naftogaz $4.63 Bln for Insufficient Transit Shipments," *Interfax Russia and CIS Oil and Gas Weekly* no. 9, March 1–6, 2018, pp. 4–16.
79. See Arbitration Institute of the Stockholm Chamber of Commerce, "About the Arbitration Institute," http://www.sccinstitute.com/about-the-scc/ (accessed April 28, 2019).
80. Quoted in "Gazprom, Ukraine Spin Arbritration Outcome," *Nefte Compass,* 26, no. 51 (December 28, 2017): 10.
81. Quoted in "Gazprom Hoping New Stockholm Litigation Will Correct Mismatch of Interests with Naftogaz," *Interfax Russia and CIS Oil and Gas Weekly* no. 10, March 7–14, 2018, p. 4.
82. As for imports to Ukraine, the tribunal's decision was that Gazprom should sell 4 billion cubic meters per year to Naftogaz for the duration of the contract. After discussions with Gazprom, Naftogaz sent an advance payment for the first month. Gazprom accepted the payment but then, a few minutes before deliveries were to resume, and on one of the coldest days of the year, returned the cash and said that it would not make the deliveries.
83. For an update as of late 2018, see Simon Pirani, "Russian Gas Transit through Ukraine after 2019: The Options," Oxford Institute for Energy Studies, November 2018, https://www.oxfordenergy.org/wpcms/wp-content/uploads/2018/11/Russian-gas-transit-through-Ukraine-after-2019-Insight-41.pdf.

84. See Pirani and Yafimava, "CIS Gas Markets and Transit." On the changing role of Central Asian gas in the Russian and Ukrainian equation following the completion of the Turkmenistan-China pipeline, see Simon Pirani, "Central Asian and Caspian Gas for Russia's Gas Balance," in Henderson and Pirani, *The Russian Gas Matrix,* pp. 347–367. Between 2008 and 2009 Turkmen exports to Russia abruptly dropped by three quarters, from 42.3 to 11.8 billion cubic meters (ibid., table 14.1, p. 348).
85. For a brilliant discussion of Lithuania, see Balmaceda, *The Politics of Energy Dependency*, chapter 6.
86. Pirani, "Russo-Ukrainian Gas Wars and the Call on Transit Governance," p. 169.
87. On earlier phases of the decline in consumption, before the recent price reforms were implemented, see Thane Gustafson, Matthew J. Sagers, and Sergej Mahnovski, *Ukrainian Gas Consumption Collapses in 2009: The Implications for Ukraine's Dependence on Russian Gas and for the EU-Ukrainian Gas Accord* (IHS Markit Decision Brief, September 2009); and Anna Galtsova, Matthew J. Sagers, Thane Gustafson, and Nick Naroditski, *Ukraine's Declining Gas Consumption: Where Does It Go From Here?* (IHS Markit Decision Brief, November 2014).
88. The unknown in the investment story is the immense amount of shadow capital concealed under mattresses in Ukraine, which some economists value at some $80 billion. This may help explain why the Ukrainian economy has kept on growing in recent years, despite the lack of foreign direct investment.

12. Russian-German Gas Relations

1. Since 2015, the German Federal Office for Economic Affairs and Export Control (known by its German acronym, BAFA) no longer publishes a breakdown of gas imports to Germany by country. In 2015, 35 percent of German gas imports came from Russia. See Sören Amelang and Julian Wettengel, "Germany's Dependence on Imported Fossil Fuels," Clean Energy Wire, March 8, 2018, https://www.cleanenergywire.org/factsheets/germanys-dependence-imported-fossil-fuels.
2. Aurélie Bros, Tatiana Mitrova, and Kirsten Westphal, "German-Russian Gas Relations: A Special Relationship in Troubled Waters" (Stiftung Wissenschaft und Politik [SWP], German Institute for International and Security Affairs, Research Paper RP 13, December 2017), https://www.swp-berlin.org/fileadmin/contents/products/research_papers/2017RP13_wep_EtAl.pdf.
3. For a German comment on the implications for the Nord Stream 2 pipeline, which illustrates well the combination of economic and political dilemmas currently faced by Germany in its diplomatic relations with Russia, see Peter Carstens and Konrad Schuller, "Gas und Rosen," *Frankfurter Allgemeine Sontagszeitung,* May 20, 2018.
4. Bros, Mitrova, and Westphal, "German-Russian Gas Relations."
5. This is partly a function of the professional orientation of the researcher-scholar. The work of the Oxford Institute for Energy Studies clearly belongs in the economic community. In contrast, the German SWP, one of whose major functions is to advise the chancellor's office on security affairs, belongs primarily in the

security community. The same might be said of the French Institut Français des Relations Internationales and Chatham House in Great Britain.

6. Stephen F. Szabo, *Germany, Russia, and the Rise of Geo-Economics* (London: Bloomsbury, 2015), p. 62. In his pathbreaking study of the two communities in Germany, Szabo is one of the few analysts to have focused on the divide and to have attempted to bridge it in his own work.
7. According to the Russian Statistical Service (Rosstat), in 2016 Germany accounted for 7.4 percent of Russian exports (behind only China and the Netherlands, which had the most Russian exports mainly owing to oil exports, reflecting the critical role of Dutch North Sea terminal facilities in the reexport of Russian oil) and 10.7 percent of Russian imports (behind only China). See the Rosstat statistical handbook, *Russia in Figures, 2017,* p. 482, http://www.gks.ru/free_doc/doc_2017/rusfig/rus17e.pdf.
8. Gazprom's contracts with Wingas run to 2031, while its four contracts with Uniper (the successor of E.ON, and Ruhrgas before that) run to 2035. Gazprom's twenty-year contract with Shell Europe is the longest, expiring only in 2038. In addition, Gazprom sells a small quantity of gas directly on Germany's gas hubs.
9. Angela Stent, *Putin's World: Russia against the West and with the Rest* (New York: Twelve Books, 2019), p. 99.
10. Kirsten Westphal, "Institutional Change in European Natural Gas Markets and Implications for Energy Security: Lessons from the German Case," *Energy Policy* 74 (2014): 41.
11. Bros, Mitrova, and Westphal, "German-Russian Gas Relations," p. 14.
12. Ibid.
13. Ibid.
14. For an insightful discussion of the ways the classic long-term contract mitigated and managed risks, see Simon Blakey, *The Future of Long-Term Gas Contracts and the European Midstream* (IHS Markit Strategic Report, January 2016).
15. Angela Stent, *The Limits of Partnership: U.S. Russian-Relations in the Twenty-First Century* (Princeton, NJ: Princeton University Press, 2014).
16. See the revealingly titled monograph by Elena Korosteleva, *The European Union and Its Eastern Neighbours: Toward a More Ambitious Partnership?* (New York: Routledge, 2012).
17. Kirsten Westphal, "Germany and the EU-Russia Energy Dialogue," in Pami Aalto, ed., *The EU-Russian Energy Dialogue: Europe's Future Energy Security* (Farnham UK: Ashgate, 2008), p. 98.
18. *Verflechten* means to braid or interweave. *Verflechtung* was meant to describe the active process of weaving broader economic links.
19. Bros, Mitrova, and Westphal, "German-Russian Gas Relations," p. 15.
20. Paraphrased by Szabo, *Germany, Russia, and the Rise of Geo-Economics,* from Angela Stent, *Russia and Germany Reborn: Unification, the Soviet Collapse, and the New Europe* (Princeton, NJ: Princeton University Press, 1999), pp. 148–149.
21. For an account of the causes of the Russian default of 1998, see Thane Gustafson, *Capitalism Russian-Style* (Cambridge: Cambridge University Press, 1999), chapters 9 and 10.

22. For the oil boom of 2000–2014 and its impact on Russia, see Thane Gustafson, *Wheel of Fortune: The Battle for Oil and Power in Russia* (Cambridge, MA: Harvard University Press, 2012).
23. Quoted in Szabo, *Germany, Russia, and the Rise of Geo-Economics,* p. 62.
24. Hans-Joachim Spanger, "Die deutsche Russlandpolitik," in Thomas Jäger, Alexander Höse, and Kai Oppermann, eds., *Deutsche Aussenpolitik* (Springer: Wiesbaden, 2010), p. 662.
25. Stent, *Putin's World,* p. 81.
26. Ibid.
27. The story of the Putin-Schröder meetings in 2005 is told in Mikhail Zygar', *All the Kremlin's Men: Inside the Court of Vladimir Putin* (New York: Public Affairs, 2016), pp. 119–120. Zygar' is well acquainted with the Russian gas industry and the Russian-German gas relationship. A decade earlier, as a correspondent for *Kommersant,* he coauthored a well-regarded analysis of Gazprom at the beginning of the 2000s and the Viakhirev-Miller succession. See Valerii Paniushkin and Mikhail Zygar', *Gazprom: Novoe russkoe oruzhie* (Moscow: Zakharov, 2008).
28. Quoted in "Polish Defense Minister's Pipeline Remark Angers Germany," *Voice of America Online,* May 3, 2006, https://www.voanews.com/a/a-13-polish-defense-minister-pipeline-remark-angers-germany/327455.html.
29. Zygar', *All the Kremlin's Men,* p. 120.
30. Westphal, "Germany and the EU-Russia Energy Dialogue," p. 106.
31. Hannes Adomeit, "Deutsche Rußlandpolitik: Ende des 'Schmusekurses'?," *Russie. Cei.Visions,* No. 6(b), September 2005.
32. Mikhail Krutikhin, "Vershki i koreshki," *Rusenergy Praim-Onlain,* June 21, 2003. An additional cause of delay may have been that Gazprom was negotiating in parallel with potential Russian partners, initially Itera and subsequently Surgutneftegaz.
33. Mikhail Krutikhin, "Gostepriimnyi khoziain: Rossiia obeshchaet pustit' inostrantsev v gazovye proekty," *Rusenergy,* April 12, 2005.
34. In addition, after E.ON acquired Ruhrgas, E.ON became interested in investing in the Russian power sector. From this point on, electricity became a more powerful motive in E.ON's Russian upstream strategy than gas.
35. Interview with Burckhard Bergmann, n.d.
36. "Gazprom setzt Eon under Druck," *Süddeutsche Zeitung,* March 27, 2006.
37. See Westphal, "Germany and the EU-Russia Energy Dialogue," 102–103.
38. Andrew E. Kramer, "Gazprom and E.ON to Swap Assets," *New York Times,* July 13, 2006, https://www.nytimes.com/2006/07/13/business/worldbusiness/13iht-eon.2194072.html.
39. Stent, *The Limits of Partnership,* Chapters 4–6.
40. Bros, Mitrova, and Westphal, "German-Russian Gas Relations."
41. John Lough, "Germany's Russia Challenge" (NATO Defense College, Fellowship Monograph no. 11, February 2018), http://www.ndc.nato.int/news/news.php?icode=1139.
42. Stefan Kornelius, *Angela Merkel: The Chancellor and Her World* (London: Alma Books, 2013), pp. 182–183.

43. Angela Merkel, "Regierungserklärung von Bundeskanzlerin Merkel," March 13, 2014, https://www.bundesregierung.de/breg-de/themen/buerokratieabbau/regierung serklaerung-von-bundeskanzlerin-merkel-443682.
44. See Tuomas Forsberg, "From *Ostpolitik* to 'Frostpolitik'? Merkel, Putin, and German Foreign Policy toward Russia," *International Affairs* 92, no. 1 (2016): 21–42.
45. Angela Merkel, "Rede von Bundeskanzlerin Merkel beim Festakt zum 20-jährigen Bestehen der Bundesnetzagentur am 29. Mai 2018 in Bonn," May 29, 2018, https://www.bundeskanzlerin.de/bkin-de/suche/rede-von-bundeskanzlerin -merkel-beim-festakt-zum-20-jaehrigen-bestehen-der-bundesnetzagentur-am-29 -mai-2018-in-bonn-1141088.
46. Jonathan Stern, "Russian Responses to Commercial Change in European Gas Markets," in James Henderson and Simon Pirani, eds., *The Russian Gas Matrix: How Markets Are Driving Change* (Oxford: Oxford University Press, 2014), pp. 52 and 58–59.
47. Ibid, p. 58.
48. Sergei Komlev, "Pricing the 'Invisible Commodity," Gazprom Export Discussion Paper, January 11, 2013. (The paper is no longer available online, but can be obtained from the author.)
49. Jonathan Stern and Howard Rogers, "The Transition to Hub-Based Pricing in Continental Europe: A Response to Sergei Komlev of Gazprom Export" (Oxford Institute for Energy Studies, February 12, 2013), https://www.oxfordenergy.org /wpcms/wp-content/uploads/2013/02/Hub-based-Pricing-in-Europe-A -Response-to-Sergei-Komlev-of-Gazprom-Export.pdf.
50. Quoted in Stern, "Russian Responses to Commercial Change in European Gas Markets," p. 64.
51. Quoted in Jonathan Stern and Howard Rogers, "The Dynamics of a Liberalised European Gas Market: Key Determinants of Hub Prices, and Roles and Risks of Major Players" (Oxford Institute for Energy Studies, NG94, December 2014), https://www.oxfordenergy.org/wpcms/wp-content/uploads/2014/12/NG-94.pdf.
52. For background, see Simon Blakey, *Diversity and Security in European Energy: The Case of the Nord Stream Pipeline* (IHS Markit Private Report, January 2008); and "Nord Stream," https://en.wikipedia.org/wiki/Nord_Stream. See also https://www .atlanticcouncil.org/images/publications/Nord_Stream_2_interactive.pdf.
53. Nord Stream 1 was included in the "Trans-European Energy Networks" priority projects guidelines and was approved by the European Parliament and the Council in 2006.
54. Technically, at the time Nord Stream AG was still called North Transgas Oy, a reflection of the fact that the original project had been planned as a joint venture between Gazprom and the Finnish company Neste / Fortum. The pipeline and the operating company were renamed Nord Stream AG in October 2006.
55. For a detailed review of the OPAL affair through the end of 2013, see Kash Burchett and Thane Gustafson, *The OPAL Dispute: Resolution in the Pipeline?* (IHS Markit Decision Brief, January 2014); and Thane Gustafson and Kash Burchett, *Russia-EC Gas Relations in the Midst of the Ukrainian Crisis: Is Continued Progress*

Possible? (IHS Markit Private Report, March 2014). The regulatory and legal issues are reviewed in Katja Yafimava, "The OPAL Exemption Decision: Past, Present, and Future" (Oxford Institute for Energy Studies, Paper NG 117, January 2017), https://www.oxfordenergy.org/wpcms/wp-content/uploads/2017/01/The-OPAL-Exemption-Decision-past-present-and-future-NG-117.pdf.

56. All three quotes come from the same speech by Putin on November 28, 2010, available on the prime minister's website at http://archive.premier.gov.ru/events/news/13118/. An excerpt appears in Maksim Tovkailo, Oksana Gavshina, and Evgeniia Pis'mennaia, "Zashchitnik rossiiskogo," *Vedomosti*, November 29, 2010.
57. Evgeniia Pis'mennaia and Margarita Liutova, "Iskliuchenie dlia 'Gazproma,'" *Vedomosti*, February 25, 2011, p. 3.
58. Ibid.
59. Margarita Liutova and Maksim Glikin, "Dogovor dlia 'Gazproma,'" *Vedomosti*, December 24, 2012.
60. Elena Khodiakova, "Evrokomissiia ne otdast OPAL," *Vedomosti,* April 19, 2013.
61. Margarita Liutova and Maksim Tovkailo, "Trubnyi vybor," *Vedomosti*, December 17, 2012.
62. Burchett and Gustafson, *The OPAL Dispute.*
63. For an update and analysis through September 2017, see Yafimava, "The OPAL Exemption Decision." When the unbundling laws were applied in Germany, Wintershall and Gazprom had to divide up their assets so that they each remained on the right side of the new law. They reached a commercial agreement in which Wintershall retained the gas supply and trading business, and Gazprom took ownership of the pipeline assets. Wintershall was compensated for the loss of its assets by being granted greater access to the Achimov formations in the Siberian upstream.
64. In the wake of the 2016 compromise, Jonathan Stern wrote, in words that could stand as a battle cry for the regulatory community, "Objections to Russian gas supplies and pipelines on political and security grounds have a long history and should be argued on their own merits." He then asked "whether it is valid for such objections to be used to distort a natural gas regulatory framework which has been many years in the making." (introduction to Yafimava, "The OPAL Exemption Decision."
65. As we shall see in the Conclusion, however, there could be a second life for Ukrainian transit in the late 2020s, if Russian pipeline exports to Europe grow beyond their present level, and if political relations allow.
66. All the necessary permits have been obtained from transit countries for Nord Stream 2 to pass through their territorial waters, with the exception of Denmark. If Denmark withholds permission, the pipeline will have to be rerouted to avoid Danish waters. That will delay the project, but it will not stop it.
67. For a valuable analysis of the legal issues involved in Nord Stream 2, see Alan Riley, "A Pipeline Too Far? EU Law Obstacles to Nordstream 2," *International Energy Law Review,* March 2018.
68. Reply from Rainer Baake, then State Secretary of the Federal Economics Ministry, addressed to Norbert Lammert, president of the German Bundestag, April 4, 2016, https://www.bmwi.de/Redaktion/DE/Parlamentarische-Anfragen/2016/18-7952.pdf?__blob=publicationFile&v=4.

69. Quoted in Stephanie Bolzen, Christoph B. Schiltz, and Andre Tauber, "Ostseepipeline reisst neue Gräben in der EU auf," *Die Welt,* December 18, 2015, https://www.welt.de/wirtschaft/energie/article150129711/Ostseepipeline-reisst-neue-Graeben-in-der-EU-auf.html.
70. "Meeting with Vice-Chancellor and Minister of Economic Affairs and Energy of Germany Sigmar Gabriel," October 28, 2015, http://en.kremlin.ru/events/president/news/50582.
71. Quoted in Severin Weiland, "Robust Richtung Russland," *Der Spiegel,* April 17, 2018, http://www.spiegel.de/politik/deutschland/heiko-maas-robust-richtung-russland-a-1203189.html.
72. Peter Carstens and Konrad Schuller, "Gas und Rosen," *Frankfurter Allgemeine Sontagszeitung,* May 20, 2018.
73. Tom Weingärtner, "EU-Gremien einigen sich auf Gasrichtlinie," *Energie und Management Daily,* February 14, 2019, p. 6.
74. Quoted in Dana Heide, "Altmaier sucht beim Thema Nord Stream 2 den Dialog mit den USA," *Handelsblatt,* January 22, 2019, https://www.handelsblatt.com/unternehmen/energie/gaspipeline-altmaier-sucht-beim-thema-nord-stream-2-den-dialog-mit-den-usa/23895502.html.
75. "AKK kritisiert USA: 'Nicht der beste Umgang zwischen Freunden,'" *Die Zeit,* February 14, 2019, https://www.welt.de/politik/deutschland/article188763687/Nord-Stream-2-AKK-kritisiert-USA-Nicht-der-beste-Umgang-zwischen-Freunden.html.
76. Alan Posener, "Es gibt Alternativen!," *Internationale Politik 2* (March–April 2019), https://zeitschrift-ip.dgap.org/de/ip-die-zeitschrift/archiv/jahrgang-2019/es-gibt-alternativen. In addition to his work for *Die Welt,* Posener is one of Germany's most influential bloggers.
77. Deutsche Welle Russian-language service, Vadim Shatalin, "Zhiteli Germanii podderzhivaiut stroitel'stvo 'Severnogo potoka—2,'" February 27, 2019, https://p.dw.com/p/3EDss; and "Mehrheit des Deutschen hält Bau der Ostseepipeline Nord Stream 2 für richtig," *Handelsblatt,* January 21, 2019, https://www.handelsblatt.com/politik/deutschland/umfrage-mehrheit-der-deutschen-haelt-bau-der-ostseepipeline-nord-stream-2-fuer-richtig/23891892.html?ticket=ST-46851-H0gfr4yCQdc9YImeTVhr-ap1.
78. "Zapasy Rossii: kogda zakonchatsia neft' i gaz," *Gazeta,* February 26, 2019, https://www.gazeta.ru/interview/nm/s12198073.shtml?refresh.
79. Mathias Brüggmann, "Streit um Nord Stream 2: Russland hintertreibt Merkel-Plan," *Handelsblatt,* February 27, 2019, https://www.handelsblatt.com/politik/international/energieversorgung-streit-um-nord-stream-2-russland-hintertreibt-merkel-plan/24045082.html.
80. For the full text of Medvedev's interview with *Luxemburger Wort* correspondent Stefan Scholl, see "Interv'iu Dmitriia Medvedeva Liuksemburgskomu izdaniu 'Liuksemburger vort,'" March 5, 2019, http://government.ru/news/35917/.
81. "A Contentious Year Awaits the EU in 2019," Stratfor, November 16, 2018, https://worldview.stratfor.com/article/contentious-year-awaits-eu-2019,

82. Thane Gustafson, *Russia Sanctions Year Five: Deepening Conflict among US Decision-Makers over Implementation* (IHS Markit Strategic Report, March 2019).
83. "CDU-Chefin kritisiert USA im Streit über Gaspipeline," *Der Spiegel,* February 14, 2019, https://www.spiegel.de/forum/politik/nord-stream-2-cdu-chefin-kritisiert-usa-im-streit-ueber-gaspipeline-thread-864746-1.html.
84. To add one more layer of complexity, there is an inherent conflict between two goals of reaching a new transit agreement quickly and of completing Ukraine's compliance with EU legislation. The easiest way to reach the first goal would be for Gazprom and Naftogaz to agree on a multiyear contract at an agreed-upon fixed tariff. But under EU rules this would not be allowed. For an excellent analysis, see Laurent Ruseckas, *Future Transit of Russian Gas through Ukraine: Risks May Be Larger Than They Appear* (IHS Markit Strategic Report, December 2018).

13. Battle Joined, War Averted

1. Alex Barker, Christian Oliver, and Jack Farchy, "Keeping the Taps Open," *Financial Times,* March 16, 2015.
2. See Jonathan Stern and Katja Yafimava, "The EU Competition Investigation of Gazprom's Sales in Central and Eastern Europe: A Detailed Analysis of the Commitments and the Way Forward" (Oxford Institute for Energy Studies, Paper NG-121, July 2017), https://www.oxfordenergy.org/wpcms/wp-content/uploads/2017/07/The-EU-Competition-investigation-of-Gazproms-sales-in-central-and-eastern-Europe-a-detailed-analysis-of-the-commitments-and-the-way-forward-NG-121.pdf.
3. Quoted in Barker, Oliver, and Farchy, "Keeping the Taps Open."
4. "EU Raids Gazprom Subsidiaries," *Nefte Compass,* September 29, 2011, p. 4.
5. *Interfax Russia and CIS Oil and Gas Weekly,* September 13–19, 2013, pp. 5–6.
6. Aleksey Grivach, "Rassledovanie s sokhraneniem litsa: Aleksey Grivach o tom, pochemu Evrokomissiia zagovorila s 'Gazpromom' o sniatii antimonopol'nykh pretenzii," *Gazeta.ru,* December 11, 2013, https://www.gazeta.ru/comments/2013/12/11_a_5797577.shtml (accessed May 23, 2019). For a longer version of Grivach's argument, see his *Reformirovanie gazovogo rynka Evropy: Chuzhie tufli mogut nateret' mozoli* (Moscow: National Fund for Energy Security, October 2013).
7. Quoted in Alex Barker, "Brussels on Course to Issue Gazprom Antitrust Charges,"*Financial Times,* December 4, 2013. See also *Interfax Russia and CIS Oil and Gas Weekly,* December 5–11, 2013, pp. 31–32.
8. Quoted in Rochelle Toplensky, "Margrethe Vestager, EU Competition Commissioner," *Financial Times,* December 10, 2016.
9. Its contents have become partially public, owing to a leak that took place three years later.
10. European Commission, "Antitrust: Commission sends Statement of Objections to Gazprom—Factsheet," April 22, 2015," http://europa.eu/rapid/press-release_MEMO-15-4829_en.htm.

11. Quoted in Thane Gustafson and Simon Blakey, "The Settlement of Gazprom's Anti-Competition Case," *IHSMarkit PGCR Regional Insight*, September 2018.
12. Dena Sholk and Matthew J. Sagers, *Lithuania's Gas Market: The Difficulties of Supply Diversification* (IHS Markit Decision Brief, February 2016).
13. For a detailed analysis of the legal procedures involved, see Alan Riley, "Commission vs. Gazprom: The Antitrust Clash of the Decade?" (Centre for European Policy Studies, Policy Brief No. 285, October 31, 2012), https://www.ceps.eu/system/files/PB%20No%20285%20AR%20Commission%20v%20Gazprom_0.pdf.
14. Quoted in Thane Gustafson and Kash Burchett, *Russia-EC Gas Relations in the Midst of the Ukrainian Crisis: Is Continued Progress Possible?* (IHS Markit Private Report, March 2014), p. 9.
15. DG-COMP had already failed to prove unfair pricing in the Lithuanian arbitration. Note that DG-COMP had changed its wording in the Gazprom case, where it referred instead to "competitive pricing."
16. That was made much easier by the time of the negotiations because Gazprom had already accepted hub pricing in most of its contracts with Western European buyers.
17. For an analysis of the preliminary concession agreement, see Laurent Ruseckas, *The European Commission and Gazprom Reach a Deal: What Does It Mean in Practice?* (IHS Markit Insight, March 2017).
18. Rochelle Toplensky, "Russia's Gazprom Dodges Fine in EU Antitrust Settlement," *Financial Times,* May 24, 2018, https://www.ft.com/content/02a15d08-5f3f-11e8-9334-2218e7146b04.

Conclusion

1. This is admittedly a controversial point. Modern coal plants can ramp up and down and follow intermittent load just as well as gas plants. And increasingly sophisticated grid management and demand-side management with the Internet of Things are touted to deliver balance for wind and solar in the future. Nevertheless, the future of coal is in doubt, and demand-side management is still only a bright question mark on the horizon. On the whole, I continue to bet on gas.
2. Simon Blakey, Alun Davies, Laurent Ruseckas, Shankari Srinivasan, and Michael Stoppard, *European Natural Gas—the New Configuration: Introducing the IHS Markit Pan-European Gas Flows and Price Differentials Model* (IHS Markit Strategic Report, April 2018).
3. The efficient gas market may enhance the market power of the seller, who—in the absence of buyer-nomination rights in long-term contracts—will have daily make-or-buy choices at the hubs. The key point here is that sellers concerned to maintain their longer-term market share will not abuse their day-to-day market power.
4. The concept of a multisided "coalition of the willing" is at the heart of Per Högselius's outstanding history of the origins of the gas bridge, *Red Gas: Russia and the Origins of European Energy Dependence* (New York: Palgrave Macmillan, 2013).

5. See Angela Stent, *Putin's World: Russia against the West and with the Rest* (New York: Twelve Books, 2019).
6. James Henderson, foreword, in Jonathan Stern, "The Future of Gas in Decarbonising European Energy Markets: The Need for a New Approach" (Oxford Institute for Energy Studies Paper NG 116, January 2017), https://www.oxfordenergy.org/wpcms/wp-content/uploads/2017/01/The-Future-of-Gas-in-Decarbonising-European-Energy-Markets-the-need-for-a-new-approach-NG-116.pdf. The Oxford Institute has led the way in exploring the implications of the politics of climate change for the European gas industry. See in particular Jonathan Stern, "Narratives for Natural Gas in Decarbonising European Energy Markets" (Oxford Institute for Energy Studies Paper NG 141, February 2019), https://www.oxfordenergy.org/wpcms/wp-content/uploads/2019/02/Narratives-for-Natural-Gas-in-a-Decarbonisinf-European-Energy-Market-NG141.pdf.
7. The rapidly declining costs of renewables offer the promise that wind and solar, as well as biomass, can compete in the future without new subsidies or regulatory privileges.
8. Josephine Moore and Thane Gustafson, "Where to Now? Germany Rethinks Its Energy Transition," *German Politics and Society* 36, no. 3 (Autumn 2018): 1–22.
9. Daniel Wetzel, "Bundesregierung will den Deutschen das Gas abdrehen," *Welt,* October 31, 2016, https://www.welt.de/wirtschaft/article159149712/Bundesregierung-will-den-Deutschen-das-Gas-abdrehen.html.
10. *Grünbuch Energieeffizienz: Discussionspapier des Bundesministeriums für Wirtschaft und Energie,* https://www.bmbf.de/pub_hts/gruenbuch_energieeffizienz.pdf (accessed April 12, 2017).
11. Tessa Coggio and Thane Gustafson, "When the Exit? The Difficult Politics of German Coal," *German Politics and Society* (forthcoming).
12. See the IHS Markit *European Gas Hub Tracker,* a regularly updated publication of the IHS Markit Global Gas service consisting of tables and charts in Excel format documenting European gas hub dynamics. See also Blakey et al., *European Natural Gas.*

Acknowledgments

This book could not have been written without Simon Blakey, who taught me everything I know about European gas. Over the course of many visits and conversations, on both sides of the Atlantic, he introduced me to the core European themes of this book, and he read every draft multiple times. My knowledge still falls far short of his, but I hope I have learned enough to write a coherent story, and any mistakes, needless to say, are mine and not his.

The second major influence on this book, again a virtual coauthor, was Philip Vorobyov, who read and reread all the chapters of this book and offered valuable insights and criticism. His contribution was especially important on Russia, Ukraine, and Russian-Ukrainian relations. His work in both countries has given him an intimate knowledge of how gas and politics actually work there, which he has shared with me. I am immensely grateful for his friendly counsel and support.

My good friends and colleagues, Daniel Yergin at IHS Markit and Angela Stent at Georgetown, have been staunch sources of support and inspiration for *The Bridge,* as they have for all of my previous books. They both read draft chapters and offered valuable insights and criticism. Warm thanks to them both.

John Webb, with his usual obliging care and inexhaustible patience, provided the essential foundation on which this books rests, as he did for my previous book, *Wheel of Fortune.* John, as always, was the perfect teammate. He checked every line of the text and footnotes, maps, and tables, not once but several times, with cheerful endurance. (I estimate that we examined together over 20,000 revisions from the copy editors!) Again, any remaining mistakes are mine and not his.

I am especially grateful to those who read and criticized earlier chapters and helped the book to evolve. Bob Otto read the whole book and offered trenchant criticism, backed up by useful Russian sources. Tim Boersma, Howard Chase, Ed Chow, Sir Jonathan Faull, Sir Philip Lowe, Tatiana Mitrova, Simon Pirani, Peter Reddaway, Jonathan Stern, Sergey Vakulenko, and Julian West all read individual chapters and offered helpful criticism and insights, often over pleasant lunches. I also thank Dominique Arel, Craig Kennedy, Sergei Komlev, John Lough, Daniel Poneman, Peter Rutland, and Jesse Scott for their insights, and particularly David Rennie, who first alerted me to the special role of Lord Cockfield. Special thanks also go to Margarita Balmaceda, who organized and led a wonderful conference on Russian-European gas in Germany, and helped to crystallize the whole project in my mind.

I thank the many friends, businesspeople, officials, and experts whom I was privileged to meet during my research in Europe, and with whom I had the opportunity to discuss the themes of this book. In Norway, I would like to recognize Lars Erik Aamot, Pål Eitrheim, Ole-Anders Lindseth, Peter Mellbye, Willy Olsen, Berit Ruud Retzer, Henrik Seip, Espen Svensen, Dag and Kjore Villoch, and Geir Westgaard. Special thanks are due to Eric Bjelland, Arild Moe, and Petter Nore, who read and commented on the draft chapter on Norway. Per Högselius and his masterpiece, *Red Gas,* were important sources of inspiration. In Germany, I am grateful to Herbert Detharding, Uwe Fip, Wolfgang Knell, Dieter Pfaff, Klaus Schäfer, and Reinier Zwitserloot. Special recognition is due to Gert Maichel and Rainer Seele, who kindly read and criticized the chapters on Germany. I have also benefited enormously from Heiko Lohmann's two great books about the transformations in the German gas industry, and I remember with pleasure meeting with the author together with Simon Blakey.

I am indebted to my colleagues and teammates at IHS Markit, especially the Russian and Caspian Energy team under the able leadership of Matt Sagers. Special kudos are due to Anna Galtsova, who has blossomed in recent years into an internationally recognized expert on Russian gas. I am grateful also to the IHS Markit Global Gas team, led by the ever-capable Shankari Srinivasan, and her colleagues Alun Davies, Deborah Mann,

Catherine Robinson, Michael Stoppard, and especially Laurent Ruseckas, who read and commented on the Ukrainian chapter. I also thank Philip Frangulies for his friendly support as head of the Power, Gas, Coal, and Renewables group, as well as Atul Arya, who heads Energy Insight and directs the Energy Research Council, a highly useful forum. The Moscow office of IHS Markit provided invaluable backup. My warm thanks to Irina Zamarina, and of course to the incomparable Igor.

Georgetown University's Government Department has been my home for thirty-five years, and I am grateful as always for the productive and collegial atmosphere that reigns there. I owe special thanks to the Department's talented chairman, Charles King. My research assistants at Georgetown, Tessa Coggio, Olav Henke, Josephine Moore, and Emily Sandys, provided essential support in exploring the German and Norwegian gas industries, as well as the Russian connections. Finally, my students provide the lifeblood that keeps us all young. I thank them all.

Mike Aronson, Jeff Dean, and Janice Audet, my editors at Harvard University Press, supported this book project from the earliest days, and read and commented encouragingly on every draft as it came in. Emeralde Jensen-Roberts and Megan Posco have been valuable allies in promoting the book. John Donohue and Jeanne Ferris at Westchester Publishing Services did the copyediting, as they did for *Wheel of Fortune,* with extraordinary speed and care. Jerome Cookson did the marvelous maps. Thanks to them for their dedication and skill.

Lastly, I am grateful to my wonderful family, Nil, Peri, Farah, and Kenan. They are the reason, and the reward, for everything I do.

Index